THE
TECHNICAL
MANAGER'S
HANDBOOK
A Survival Guide

THE
TECHNICAL MANAGER'S HANDBOOK

A Survival Guide

Melvin Silverman, Ph.D., P.E.

CHAPMAN & HALL

I(T)P   An International Thomson Publishing Company

New York • Albany • Bonn • Boston • Cincinnati • Detroit • London • Madrid • Melbourne
Mexico City • Pacific Grove • Paris • San Francisco • Singapore • Tokyo • Toronto • Washington

Art direction: Andrea Meyer, emDASH inc.

Copyright © 1996
Chapman & Hall

Printed in the United States of America

For more information, contact:

Chapman & Hall
115 Fifth Avenue
New York, NY 10003

Thomas Nelson Australia
102 Dodds Street
South Melbourne, 3205
Victoria, Australia

Nelson Canada
1120 Birchmount Road
Scarborough, Ontario
Canada, M1K 5G4

International Thomson Editores
Campos Eliseos 385, Piso 7
Col. Polanco
11560 Mexico D.F.
Mexico

Chapman & Hall
2-6 Boundary Row
London SE1 8HN
England

Chapman & Hall GmbH
Postfach 100 263
D-69442 Weinheim
Germany

International Thomson Publishing Asia
221 Henderson Road #05-10
Henderson Building
Singapore 0315

International Thomson Publishing-Japan
Hirakawacho-cho Kyowa Building, 3F
1-2-1 Hirakawacho-cho
Chiyoda-ku, 102 Tokyo
Japan

1 2 3 4 5 6 7 8 9 10 XXX 01 00 99 97 96

Library of Congress Cataloging-in-Publication Data

Silverman, Melvin.
  The technical manager's handbook : a survival guide / by Melvin Silverman
    p.   cm.
  Includes index.
  ISBN 0–412–99121–7 (pbk. : alk, paper)
  1. Management—Handbooks, manuals, etc.
HD38.15.S595   1995
658—dc20   95–45287
                                                       CIP

British Library Cataloguing in Publication Data available

To order this or any other Chapman & Hall book, please contact **International Thomson Publishing, 7625 Empire Drive, Florence, KY 41042.** Phone: (606) 525-6600 or 1-800-842-3636. Fax: (606) 525-7778, e-mail: order@chaphall.com.

For a complete listing of Chapman & Hall's titles, send your requests to
Chapman & Hall, Dept. BC, 115 Fifth Avenue, New York, NY 10003.

Contents

Chapter 13: Costs and Measurement: What Do I Need as a Manager?

Preface

Although I suppose that I should be used to the idea of ongoing change in management concepts, I still find it a bit unnerving to read new research and discover novel theories and practices that tend to modify the ideas of the past. This requires that I continually update the materials that I teach in the many seminars on technical management delivered yearly for practicing professionals and students in academia. This edition is one result of that updating process. It is an updated edition of a prior work entitled "The Technical Managers Survival Book."

In almost every one of my classroom experiences, there are always a few students whose reactions are somewhat similar to mine since they remind me how very disconcerting it must be to study management when one has obtained an original training in the sciences. In mathematics, physics or chemistry, there were formulas and data that could be relied upon. In that original training, the "absolute" rules predicted the correct answer to a particular problem. For example with a knowledge of calculus, it was easy to integrate the formula for a curve to determine the area under it. Similarly, knowing Newton's law of gravity, one could find the gravitational attraction between two different bodies using his straightforward formula. But even in the sciences times change. Now we know that at the atomic level, the "basic" so-called unchanging laws of many sciences no longer apply. The simpler Newtonian, predictable world becomes more complex and less predictable. It can be very confusing. In many ways, management is like that. It seems that as we learn more about it, the more the "absolute" rules begin to fail. Some of those "absolutes" such as a limited management "Span of Control" or responding to only "One Boss" have become obsolete and have been superseded by newer concepts such as "participation" or "Benchmarking". And it is probably reasonable to assume that even these concepts will be superseded by others in the future.

Some students have said that if it was possible to place the various concepts, theories and applications about management into neat compartments, life would be so much easier. Then, the manager could quickly categorize each problem, look into the handy "Management Encyclopedia", find the answer, apply it and everything would be solved. But, of course, that is not what happens. Management seems now to involve an almost infinite number of variables and solving the immediate problem is only a general guideline for solving problems in the future. There are several reasons for this.

First, we ourselves are part of both the problem and the solution. Our own decision-making processes are not always logical and consistent. We are living, changing, emotional beings and how we approach a problem today is not the same way that we approach it tomorrow. Tomorrow, we will gain some additional experience or learn another problem-solving method and that new knowledge will change the way we view the world. Our perceptions, and in turn, our thinking processes will change. In some cases, if there is a major occurrence, the effect upon ourselves will be evident but most times, the changes are minor and slow. But even so, there has been a change.

Second, we deal with other people and there are other kinds of changes probably happening to them. Therefore while we, as managers, may have general rules that might help us to determine what motivates the next person, we never really know if that general rule will apply to the specific individual. The specific individual is always changing.

Finally, there are interactions affecting management that we may not be aware of and over which we have no control. We now live and work in a one-world economy. Social changes, politics and culture in other parts of the globe may directly affect how we work. Our competitors, for example, may now be in different countries or our brand-new (and improved) products may be rendered obsolete overnight by someone who is better attuned to the world market place.

Therefore, it is vital that we, as technically trained "Knowledge Workers" recognize that our educational processes will never end. The effects of change on ourselves, our co-workers and our organizations will always have to be considered. Graduating from a technical school or a university is now only one step on the never-ending learning process that we need if we are to improve our ability to manage change. We will always be students and this edition is intended to assist you in that process. Good luck!

People make choices during their lives. They make choices about the kind of work they do, the companies they will do it in, and how long they will stay at it. Sometimes these choices appear to be limited and at other times almost unlimited; but in all cases, there are always choices. Often, the choice that is made depends upon the normative or predictive expectancy of the person(s) making it. This means that there is an expectation that the alternative chosen (from those alternatives perceived to be possible at the time of the choosing) will result in the best future situation. Therefore, the choosing process depends upon the personal theory of the chooser, and that personal theory in turn is based to a great extent upon the chooser's past (education, experience, etc.) and how that past is coupled with a subjective evaluation of the future. It's not a completely logical process; emotion is involved. There is no such thing as complete objectivity where people are involved; personal values and assumptions determine which "facts" that one sees.

Much of our education and training as technical managers, engineers and scientists attempts to minimize this subjectivity by proposing a relatively value-free, objective view of the world. Those attempts can be only partially successful, since it is obvious that this education process too has a hidden value: value-free data and decisions are best. That "best" approach can work only if no humans are involved. Values, ethics, and prejudices are vital attributes of each individual's personality. These attributes are often prime contributors to our success or failure as managers, yet they are often overlooked in management books.

That is not the case here. This book is the result of many choices that I have made over the years in dealing with human attributes in technical organizations. These choices were based partly on extensive management experience in industry and academia and partly on the same kind of "value-free" training that all engineers are supposed to receive. Therefore, they included partially biased selections of various theoretical points of view, literature reviewed, analyses made and recommendations offered. The major bias behind these choices is my belief, based on experience, that managers of technical operations are potentially among the more influential managers in any organization. Another bias is that these technical managers are often poorly equipped to handle that influence because their training provides insufficient background about the complex (and less-than-objective) human interactions within all organizations. Technical success is usually based on observation and implementation of the relatively fixed, logical relationships found in nature. Management success requires this and the more elusive ability to respond to and use the relatively flexible human relationships of people. This book, therefore, results from my choices about these ideas. As you read it, compare your own choices and theories with those that I suggest, and be aware of the similarities and differences between them.

Before we begin I would like to stress one important point: this book is intended to benefit you. It is intended to help you improve the choices that all managers must make. You and I have a common goal here: to help you survive as a technical manager and to optimize your position now and in the future. This survive and optimize approach is almost unprecedented in industrial organizations. Technical people now can choose the quality and quantity of work to be done and can therefore directly affect the future of their organization. This kind of individual power or influence at work is relatively new.

WHAT IS THIS BOOK ABOUT?

In our modern, complex organizations producing technical products and services, I believe that the decision makers (managers, technical staff, etc.) determine the direction of the company through the decisions that they make. This decision making is often decentralized. It is the technical workers who design and develop innovative products who keep the company alive and growing. Typically, these workers operate in small groups or teams that actually do the work. The groups may be organized by project or by function (e.g., by "Blue Turbine" project or "Hydropack" project or by engineering design, new products development, or quality assurance). However organized, the outputs of these groups are vital to the company. They are managed by highly skilled, independent, technically qualified leaders, and it is really these leader-managers who direct the productive and growth capacity of the company, and who are responsible for technical achievements and continued organizational good health.

This book is about the power of managers to make decisions, the power that they (and you, as one of them) inherently possess because of their technical competency, and the control over company growth that this power gains for managers. This book is intended to help managers optimize this power. First we will review recently developed management techniques that strengthen often underemphasized management areas. These areas include psychology, sociology, anthropology, information science, economics, and finance, topics that often seem to be missing in technical manager's backgrounds. We deal with these areas and more in order to develop and improve the quality of your decision making as a technical manager. This development and improvement should result in at least an equivalent improvement for the company. Being better equipped to deal with uncertainties in decision making in both technical and human problems can only result in a general improvement for everyone concerned. There are many ways, however, to improve. Since we all have different strengths and ways of learning, I have tried to make the approach eclectic. We will review appropriate management readings, compare them with applicable practice and theory, and prescribe some uses for them. Your responsibility is to select the prescriptions that you think will best fit your situation. You will be making the final choices; you will develop your own management methods.

It is necessary to develop your own methods because technical managers in different organizations are involved in different problems that are situationally dependent. These managers rarely have the luxury of solving problems in uniform ways from organization to organization or even within different departments of the same organization. They are required to be as creative and flexible as the people they manage, and different amounts of creativity and flexibility are required in different situations. When nonrepetitive problems must be solved, creativity and flexibility are strengths. Conversely, in repetitive situations, they could be weaknesses. A problem that is solved once should not be handled again if it reappears. That would be a waste of your own invaluable thinking assets. Solved problems are placed in books or in organizational policy manuals. Then, if the problem happens again, the solution is at hand.

Another reason managerial responses or choices should be unique is that technical groups occasionally must respond to unusual crosscurrents in directions received from others: e.g., "We need some new creative ideas around here; but be sure that everyone adheres to company dress codes and working-hour standards in addition to keeping time sheets. That's company policy and everyone has to do it." Responding to that kind of direction requires the adroitness of a management Houdini, and certainly requires a unique management theory. However, even unique theories do not preclude consistent personal guidelines. New situations generally have some old, familiar parts that have been faced before. Without some consistent personal management guidelines, decisions might just as well be made at random.

This book consequently addresses the problem of developing a consistent personal theory of technical operations management (e.g. treating people as individuals). It begins by showing you, the technical manager, the inherent advantages you have in your organization. It describes some of the better methods you can use in improving your personal management techniques and assists you in optimizing your position. Your decisions as a technical manager affect the growth of your organization, and when your decisions improve, both you and the organization gain. But first and foremost, this book is for you, to help you improve your knowledge and abilities. The organization will automatically profit if you do.

FOR WHOM IS THIS BOOK MEANT?

This book is intended for the manager or would-be manager who makes the decisions in the technical departments of the company—those decisions that eventually determine the company's future viability. Although I have tried to make this book easy to read, I have also tried to avoid simplification of the material to the point where it becomes just another "management cookbook." I believe that you, as the reader, want more than that. Therefore, I have tried to keep my simplifications and interpretations of the materials to the minimum needed to coordinate and explain them.

These materials are among the most complex that we have; they involve human beings and their behaviors. The book, therefore, might be a bit difficult to read and absorb in one sitting. I suggest a more reflective pace. Read one part or chapter at a time, and integrate it thoroughly in your mind before moving on to the next part. Examine it well and ask yourself as you go along, "How can I use this in my job?." In my opinion, management in general, and technical management in particular, is not learned in "ten easy lessons." The process involves understanding difficult situations and people, learning about what others have done in similar settings, setting up the theories that are intended to work in your own situation, and then testing them on the job. The process never ends, since management improvement never stops.

As a technical manager, you are among the more influential managers of the modern technical organization. When you started your professional life as an engineer, technician, or scientist, the variables you considered in problem solving and decision making were relatively limited, since the laws of physics, chemistry, metallurgy, and similar fields seemed to be quite stable. They were very predictable and in general, applicable to all the technical situations you encountered. However, as a manager you find that some rules no longer hold uniformly and the number of variables has increased geometrically. The complex psychological, social, and economic relationships are not as predictable as relationships in physical laws, and applicability across different situations is quite limited. For example, you can consider the politics that affect the use of your company's products and service. There are many more political variables in the product user's environment than the relatively few variables of pressure, temperature, and so on in the product designer's environment. You, as the technical manager, must now consider more sets of variables when new products are on the drawing board. Additionally, the situations that involve human variables are not always predictable; therefore, the more structured thinking of technicians, engineers, or scientists—who used to be concerned primarily with the product's function and cost—now is forced to deal with social variables that might have been irrelevant not-too-many years ago. These new management situations require new understanding, different modes of analysis, and the development of unique management responses. They must be unique since you (a unique creature) are not only responsible for managing the situation but are also part of it!

Your success in technical management therefore depends upon a multivariable, very involved process. That process of management includes both the variables of the technical background that you learned in the past and a new, more flexible, set of variables. When these new variables (and their changing relationships) are mastered, your personal opportunities are greatly increased. If you can survive (and then optimize) in your own situation, your organization will gain by being able to respond successfully to the changing demands of its environment. And when it does, some of that success should redound to you and your further advantage.

WHAT MAKES THIS BOOK DIFFERENT?

In the past, many texts dealt with management as if it were a static activity, and the recommendations in the texts seemed to apply to a generalized "one best way" to plan, to organize, and to control. These "best ways" have followed styles in management just like the styles in manners or clothing. For example, earlier in this century scientific management theory was popular. This theory relied on a triangular, classically hierarchical, organization structure in which the manager was to have a limited span of control (six to eight people), report to one boss, and send information up, with directions coming down the organizational structure. I've worked in that kind of an organization and I know that *sometimes it works*. On the other hand, *sometimes it doesn't*.

But styles change. There was a period during which decentralization was a popular recommendation for organization design, and the model of the large multidivisional corporation was used to show how well this particular concept worked. For quite a while that also seemed to work. However, times changed. More recently, foreign auto builders have used quite different management models and seem to be doing very well, thank you.

Obviously, other factors, even in the limited area of organizational design, helped determine organizational growth, and these were still not defined. Within very recent times, the development of a relatively new branch of management theory and practice that is *situationally* determined seems to have produced ideas and practices that include many of the positive aspects of prior theories and yet is able to overcome many of their deficiencies. This relatively new set of theories provides the flexibility to organize and manage differently to meet the different needs of complex technical organizations. As an example, it seems to be able to explain why and when both rigid, centralized structures and flexible, decentralized structures are appropriate. It can also handle both functional and project-matrix management structures. It can even resolve why all these structures might be necessary within the same company—because different situations may give rise to different problems at different times and places. We will explore those alternatives. Since few, if any, books apply these relatively new and powerful set of theories to the unique problems and opportunities of technical managers (from their own viewpoints), this book could be a very practical tool for managers in technical organizations who wish to optimize their situations.

HOW IS THIS BOOK ORGANIZED?

This book is organized into many parts, each of which is intended to follow a familiar sequence of explanation, analysis, and synthesis. The first part is concerned with developing backgrounds, explanations, and a general model or operating hypothesis for the organization against which you can test your perceptions of your own organization. It deals with applicable theory in two general ways: first by description of applicable concepts and then by synthesis or prescriptions for the use of these concepts in your own situation. The second part disassembles the general model into its major components of people, structures, and technology and shows how you can modify and use these components in building your best management "style." Other parts deal with the special problems of communications systems and leadership in technical organizations and shows how to develop systems that provide you with the data that you need and how to respond to the changing leadership needs. It also summarizes some current data on developing change processes needed to implement your theories.

WHY IS IT NECESSARY?

For many years, first as an operating technical manager and then as an educator and consultant, I have felt that there was a need for a flexible and yet effective set of management concepts that could be adapted to fit most situations in technical management. Existing texts and systems seemed to be oriented either top-down, with some mythical "top management" making major decisions that everyone else was supposed to implement, or bottom-up, with the "all-knowing" lower managers sending up their needs and requirements to be satisfied by a cooperative company. Neither of these extremes reflected the typically varied behaviors of successful managers in the fast-changing economic environments associated with technical products and services. There might be an occasional similarity, but situations changed and successful behaviors, systems, and products always seemed to be in a state of response to this change.

This did not match the static approach many management texts seemed to recommend. An example is the often-found recommendation for constant leadership behavior (which is usually to be both supportive and participative). If one followed this advice, the successful supportive, nondirective "theory Y" leader of a technical group would invariably fail when external economic conditions became difficult. A different and more flexible leadership style would then be needed. Conversely, the successful directive, results-oriented "theory X" leader of a similar technical group would fail if changed economic conditions require advancement in the state of the art. Leadership needs change with changed situations, and the technical manager who is aware of these changed needs may be able to respond better if he or she is also aware of some of the various responses that can be used. The manager may not be able to respond completely (it's very difficult, if not impossible, to change our personalities), but he or she will be able to recognize the changes in the situation and will therefore have an increased ability to modify some situationally dysfunctional behaviors.

In addition to leadership needs, there are other management needs and responses that are similarly variable. These also depend in part upon the situation. If we can make an obvious assumption that managers, as individuals, are unique, it should follow that the manager-situation interaction is also unique. The ideas in this book attempt to assist you in handling these unique situations. Many of the ideas have been developed over many years and have been tested against the perceptions of hundreds of successful technical managers in the diverse organizations with which I have worked. These ideas do work, but they also require cooperation on the part of the user. This is a different kind of "how-to" book; it requires you to take the components and concepts offered here and build your own management framework, a framework that can be modified as the situation changes.

ACKNOWLEDGMENTS

As an engineer, a manager, and a consultant, I was always aware that few of us accomplish major achievements alone. Interactions with others are vital elements, and I could not have written this book without those supportive interactions. My appreciation and thanks are therefore extended to all who have been involved in it. I particularly wish to thank those who were kind enough to read innumerable (and deadening) drafts of the manuscript and offer the constructive, supportive criticism that extended beyond the call of duty. They include Drs. William Engbretson, Michael Grimes, Leslie Kanuk, Harvey Levine, Michael O'Neal, Frank Riessman, A. E. Russo, Herb Shepard, and Alfons Van Wijk. The continuing moral support of my family (especially Liz) helped me get over many "writer's dry-up" days. Finally, the many professional associates and attendees at my continuing education seminars who thoroughly tested the book's concepts and case studies before they were included were invaluable.

Now a word about the case studies. Those are fictionalized accounts based upon both my experiences and those of professional associates. They are not intended to represent any existing people or organizations.

Some artwork in this book is from Lotus SmartPics for Windows. © 1992. Lotus Development Corporation, an IBM subsidiary. Lotus and SmartPics are registered trademarks in the United States of Lotus Development Corporation.

There are other graphics developed from software called "Snapgrafx," Version 2.0, © 1994 the product of Micrografx Corp. Richardson, Texas.

A final word before you begin this book: I freely acknowledge any errors, and I would like to hear from you about any that you find. I'm sure that they exist. I would also like to receive any suggestions for improving the ideas that I present here or anything else that you feel is important. Chapman & Hall will forward your comments to me. I hope that you enjoy reading the book.

Melvin Silverman

About the Author

Melvin Silverman has a varied and extensive background as an engineer, operating executive, and university instructor and is a well-known consultant on project and engineering management. Dr. Silverman is the author of many books and numerous articles for the technical and management press. He is currently a Managing Partner in the consulting firm of Atrium Associates, Inc., in Cliffside Park, New Jersey. His academic background includes a B.S.M.E., an M.S.I.E., an M.B.A., an M.A. in clinical psychology, and a Ph.D. in industrial-social psychology. He is a registered Professional Engineer. With over twenty years of direct technical experience as a manager in, and as a consultant to, a broad spectrum of industrial organizations, he is able to combine an extensive training in applicable management theory with the seasoned approach of a practicing professional.

CHAPTER 1

Technical Management:
Who and What We Are

1.0 OVERVIEW AND INTRODUCTION:

Congratulations—you have been promoted. Now what? The part of the learning process that involves testing (behavior) should be concerned with development of a management style (i.e., observed behavior by others) that reflects the new duties and responsibilities of the technical management job and not the duties and responsibilities of the job from which you were promoted. Most of the positive behaviors of technical personnel that result in their promotion into management can become negative once those people have gotten those promotions. (This is generally true for everyone no matter how high up in the organizational structure a persons rises. If you continue to do your old job after you have been promoted, you'll soon be demoted back to it.) A successful equipment designer who still likes to check the drawings after being promoted to chief engineer is not performing as a chief engineer. The new job is no longer primarily concerned with the adequacy of the design; it is now centrally concerned with the adequacy of the designer (possibly the one who replaced the engineer who was promoted) and the design group itself. When this is understood, the problem of: How can I manage people who are more qualified technically than I am? becomes unimportant. The job is no longer that of managing technology; it is now managing the more complex and varied rules about people, about organizational structures and the impact of the environment on the company.

Therefore, the *first task* of that new job is to develop a different set of cognitive processes and behaviors—a new theory that will work for you in your new situation. That theory and the hypotheses that you try to make operational will help you to develop your own management style. The *second task* is to modify and sharpen problem-solving methods, as the number of variables to be considered has increased dramatically and the amount of data to be processed through the "human computer" has also increased.

(A "problem" is some situation that has occurred that does not match your predictions and is considered to be negative in its impact. For example, an increase in returns of goods by customers because of poor quality is a problem. A decrease in returns because of high quality is not. In both cases, however, there is a difference between the situation and your predictions. Problems are negative.)

To continue, processing that additional data accurately and rapidly to extract the relevant facts upon which to act is part of the manager's professional expertise. It is very different from the technical expertise of the engineer or scientist because of the greater number of variables and the incomplete (and greater quantity of seemingly irrelevant) data to be processed mentally. Managers, similar to physicians and attorneys, never get enough data to act absolutely correctly every time. But there comes a time when it costs more to wait for corroborating or more data than the possible gain from waiting. The general who waits to be absolutely sure of the enemy's strength before attacking will probably lose the battle. By the time that his forces are committed, the enemy may have changed positions, gained reinforcements, or done a number of things to make the general's data obsolete. Therefore, if this analogy holds, the data upon which most management behaviors rest is usually incomplete.

Art is built on innate human capabilities, and these capabilities can always be improved and thereby become more valuable. Therefore, both your increasing value and the unsatisfied organizational demand for effective technical managers should provide you with the opportunities to try new (and possibly better) concepts to improve your management abilities. However, learning and experimenting (or testing) occasionally do lead to setbacks. In your organizational situation, these setbacks should not be considered failures; they are chances to revise and improve your management model and are to be used as guides for obtaining better solutions to the problems raised in the future.

1.1 The Important Technical Manager

You, the technical manager, are the key executive in today's rapidly changing industrial and governmental organizations. You have the responsibility for both the current economic health of the company and its future growth because you manage the two most important groups in any organization. These two groups, *functions* and *projects*, have different but equally important roles in the company. The function-oriented groups, such as production, design, quality and maintenance, are concerned with *today*. They work to improve the products and services being currently produced. On the other hand, the project-oriented groups such as research, development, and special project-matrix teams, are concerned with *tomorrow*. They develop the new products and services for the future. When these two groups cooperate and support each other in the use of the organization's people, finances and facilities, that organization stays economically healthy today and able to grow tomorrow. When they don't assist each other, the organization will surely suffer and eventually fail. Achieving this internal management cooperation is vital because both function- and project-oriented groups must reorganize themselves as the company itself changes in response to varying external pressures. These come from demanding economic, social and political environments and they are constantly changing. The ongoing functional groups must be able to continually decrease current costs while increasing quality and decreasing delivery time. The more or less temporary teams must be able to continually innovate to produce future products and services to suit changing customer requirements. It never stops. The company has to simultaneously produce faster, better and less expensive products and services for today while developing new and better products or services for tomorrow.

Our world has changed. Products may be designed in one continent, manufactured in another and sold around the world. For example, computer hardware can be designed in the United States and built in Taiwan, while the software is diagrammed in Europe and programmed in Asia. Within the last century, gains in science, transportation, and communications have eliminated many older market constraints, thereby producing the gains of global markets and the problems of global competitors.

There has been a parallel change in management systems with a steady obsolescence of many older management concepts that are replaced with an outpouring of new concepts, systems, fads, and techniques. These are often quite confusing as they seem to promise universal solutions to all management problems. This confusion is mirrored in the management literature. Each new management system seems to offer the only solution that will ever be needed.

For example, as the traditional large-scale organizational frameworks of older, more traditional companies such as IBM and General Motors are being successfully challenged by smaller, more agile competitors, many smaller companies are simultaneously forging semi-permanent relationships with other small companies in order to gain size and market share. As one newspaper reports:

> New technology has spread around the world, trade barriers have come down, financial markets have been deregulated and consumer tastes have converged across borders. All these changes were once expected to give big firms even more scope to flex their muscles. Instead they have granted business opportunities to thousands of small and medium sized companies and shown the bodies of many corporate behemoths to be mostly flab. . . .
>
> As the advantages of corporate gigantism diminish, its long ignored costs are becoming painfully evident. Many large firms are scrambling to reduce these, scrapping layers of middle managers, cutting overheads and reorganizing themselves into "federations" of autonomous business units—that is, they are trying to become more like their smaller rivals. They are also increasingly wary of becoming yet bigger, choosing joint ventures and alliances as the way into new markets, rather than following the earlier path of buying other firms. (*New York Times*, April 17, 1993)

2.0 THE WORLD AND ORGANIZATIONS ARE CHANGING

Therefore, large companies want to get smaller and small companies want to get bigger. The situation becomes even more confusing when we read about attempts to become leaner and more aggressive. Sometimes this works and sometimes

> The lean production model relies on centralized coordination and performance indicators. . . . At the moment, "lean" managers are obsessed with "re-engineering" their businesses in order to cut overheads and reduce cycle times. The team model relies on workers to take decisions and produce innovations. . . .
>
> Is history repeating itself? Talk to managers . . . and you will learn that they are all "reinventing the workplace." Visit those workplaces, and you often find

that the changes are marginal, introduced out of faddishness rather than conviction. (*The Economist*, 9 April 1994, p. 76)

Our more or less familiar traditional organizational model resembles the military structure with its multiple, redundant layers of management. That military structure is necessary in wartime because those "managers" may suddenly be eliminated and "backups" must be ready to take their place. A centralized, redundant framework ensures that top-down management orders are carried out. But, in many modern technical organizations, the decision-making process is dispersed so that these organizations can respond more easily to the customer. Because they can no longer afford the military structure, their attempts to become leaner involve shedding many middle managers.

Unfortunately, even with dispersed decision making and downsizing of large structures to become leaner and more responsive, not all these attempts at improving organizational effectiveness work. For example, the recent management technique of re-engineering (see Hammer and Champy, 1992) is aimed at redesigning internal business procedures, to accelerate responses to both market changes and organizational performance. This has often resulted in widespread elimination of many middle management jobs as a part of shrinking or downsizing the company. But shrinking without equivalent internal improvements can lead to problems. Consider the following lead sentence stated in a recent news article in *The Wall Street Journal*: "Downsizing - job cutting medicine fails to remedy productivity ills at many companies." (*The Wall Street Journal*, 7 June 1994, p. a2) The conclusion of this news article is that success requires major positive changes in management thinking and operations and a single solution is not complete. (What a surprise!)

Re-engineering reviewed
    The report says that the two biggest obstacles to change are fear and turf protection: managers worry that decades of experience will count for nothing in the new organization and everybody worries that they will lose their job. As a result, the "soft" side of re-engineering (winning over the workers) is even more important. (*The Economist*, July 2, 1994, p. 66)

For specific details on re-engineering, see Teng, Grover, and Fiedler (1994).

Therefore, responding to the changing world environment by downsizing to change the organizational structure or forming alliances to gain market share constitutes necessary but incomplete solutions. There must be other changes in the work situation.

This book suggests that most of those changes concern the kinds of workers in the organization today and the work they do. These workers no longer include the hourly wage earners who would produce products and services with their hands. In technical organizations, that kind of work is minimal. In those organizations, it is the "knowledge" workers who produce products with their heads that constitute the work force. This different kind of worker and different kind of work are a challenge to management because no one can see it happening.

2.1 The Workers Are Changing

Physical activity can be predicted and measured. Mental activity is relatively unpredictable and is difficult to control. Although productivity during the past 125 years has provided most of us with a standard of living that could be enjoyed only by the upper classes before that time, a great deal of that productivity was based on the contributions of a few technical workers (i.e., engineers, scientists, physicists, chemists) and a major investment in manufacturing capital. Today, however, the investment in manufacturing capital is significantly less than the investment made in the recruitment, maintenance and ongoing training of the "knowledge" worker. The almost daily increase in technical knowledge requires continual updating of one's learning in order to combat technical obsolescence. There is research that indicates that the half-life is approximately five years for engineering know-how. In order to keep current, a part-time continual graduate study program is necessary (Thamhain 1992, p. 3).

Because the world, the technical organizational structure, and the workers in that structure are changing, it follows that the individual manager must also change.

2.2 Management Is Changing

Technical management used to be concerned primarily with engineering the optimum application of machines and facilities. As for people, they could easily be trained to operate machines. This is no longer true. Management is now concerned primarily with people as unique contributors. Management requires different skills than those needed in engineering, physics, or chemistry. As successful technical contributors, we were able continually to update our knowledge and technique by learning new science. We worked in teams. We had a minimum knowledge of budgets and reports. And we applied our knowledge very well, which is probably why we were promoted into management.

A report by the Carnegie Foundation noted: "The most important fact brought out by . . . surveys over a period of thirty years is that more than 60 percent of those persons who earned (engineering) degrees in the United States, either became managers of some kind within ten or fifteen years or left the engineering profession entirely to enter various kinds of business ventures." (Ackoff 1974, p. 80)

But management is very different from engineering. In engineering we deal primarily with objectivity. In management, we deal primarily with people and they are not always objective. It's a major change.

> The engineer does not believe in black magic, voodoo, or rain dances. The engineer believes in scientific truth, that is, truth that can be verified by experiment. (Florman 1987, p. 68.)
>
> As engineers we are pledged not to engage in merely wishful thinking. We are not the grasshopper; we are the ant who knows that winter is coming. We are the grumpy little pig who builds his house out of brick while his friends play and sing. (Florman 1987, p. 69)

Although people are not necessarily objective, we do have some objective similarities to each other. We all have gender, race, are born with binocular vision, and so forth, but at the same time, it's obvious that each of us is different. It is those differences that are the major concern of management. But although we deal here with some of the similarities of people, just as there are some similarities in general management, we emphasize those aspects that are different in people as well as those parts of management that are particular to technical operations. For example, technical management is differentiated form all other types in three areas: the manager, the subordinate, and the world.

1. The Manager—Changing our thinking processes: If we have a technical background, we have been trained to think logically and consistently. "Mother Nature" is our prime opponent and she provides the same answers to everyone who asks the same questions in the same way. But ask questions of a person, and the answers are bound to vary. Those answers could depend upon any number of variables; how one slept last night, problems at home, position in the organizational hierarchy, and so forth.

One analogy could be to think of people as if they had a computer software program in which both the program and the data bank were constantly changing. How people think and respond depends on their perceptions. Their perceptions affect the amount and type of data they receive, and that in turn may change the way they think. It's interactive. Therefore, the success we may have enjoyed as technicians dealing with observable and consistent phenomena may no longer apply in management. We manage other people who are producing improvement in the company's goods and services. Much of that production occurs in their thinking patterns and is unobservable. We must now deal with many more variables than we dealt with as technicians. We have a new career where intuition and communication skills are needed more than technical abilities. We are no longer judged by technical competence but by managerial skills involving business,human relations, finance and administration.

2. The subordinate—Loss of technical expertise: Technical expertise becomes obsolete quickly and in many cases managers do not have the time to be continually upgraded. Therefore, it is possible that our subordinates can become more technically knowledgeable than we are. In effect, we can now only "test" consistencies in data reported to us instead of understanding the details of the whole design process. We must delegate technical decisions that we used to be able to make ourselves.

3. The world—Personal liability: Society seems to think that we, as technical managers, should always know exactly what we are doing and what the eventual effects of our designs will be. Therefore, we can be held personally liable for the technical decisions that we make ourselves, because we are technically qualified, and for those made by our subordinates, because we are responsible for them. If that were really so, we wouldn't need safety factors in product design, and research and development departments would never have any output failures.

Although the technical manager is always concerned with product safety, it is possible when technical designs are developed that a faulty design will slip through; there is never complete surety. In many industrialized countries, if this happens it is possible for the technically qualified manager to be held personally liable even though that manager is a corporate employee. Technical people consequently have a more direct connection to the customer than any other type of employee and should act appropriately to ensure the customer's safety. In some cases, top management decisions have affected that relationship negatively.

From an organizational point of view, the management of technology is critical. As noted before, rapid changes in technology demand appropriate company responses and, because technical obsolescence never ends, technical management training must be continual in order to optimize the work of the innovative knowledge worker.

To summarize, technical management is different because, to paraphrase R.E. Shannon (1980):

Technical operations are associated always with creating something new; it is a onetime activity.

The costs of these onetime activities are very difficult to estimate because there is no organizational history to rely upon and therefore they are incorrect many times.

There are differences in expediting work. Production can be expedited when additional resources are applied. Not so in technical operations. In fact, additional resources may even slow it down.

The ability to measure performance is different. We are rarely concerned with what has been expended as when the task will be completed and the expected total expenditure at that point. Interim measurements may not be relevant. (Adapted from Shannon 1980, pp. 1–7)

2.3 Changing the Technical Contributor into the Manager

Much of the content of engineering and technical curriculums of universities and other teaching institutions is oriented toward very objective and measurable goals. If we lived in an orderly world, based on the relatively predictable laws of nature, goals would be more easily achieved.

But our world in management is a mix of logical predictability and all those variations that characterize the human condition. Rarely are products designed wholly according to logic based on the laws of nature. Both the designer and the customers, consciously or not, are emotional beings. Success in managing technical operations is really determined by the ability to handle human differences and that includes the unseen drivers of human behavior: the emotions. Unfortunately, knowledge about these drivers of behavior seldom gets enough attention in the academic training of technical managers.

As an example, the technical student quickly learns that the current in an electrical circuit flows according to immutable, predictable rules. When the wires are connected correctly in the lab, the circuit performs its task. If not, the circuit breakers interrupt the current (if the student is lucky), and a prompt learning process is completed. That is exactly as it should be—Mother Nature is a strict and attentive teacher who exacts retribution immediately when excessive "creativity" intrudes. The rules are clear and unchanging and the attentive student can understand them quite easily. Repeated inputs of some specific type of variable always result in the same type of repeated output. With Mother Nature, logic leads to success.

3.0 PEOPLE AND IRRATIONALITY

But when the technically trained, logically oriented person becomes a manager, the rational framework of thinking that contributed to success in the past no longer serves . The relatively simplistic rationality of the design of the machine or the electrical circuit no longer applies. We now must deal with the more complex (occasionally irrational) thinking process and emotions of individuals and also with the intricate interpersonal relationships of groups of people who work together. When managers direct individuals and/or teams of technical contributors and achieve success one time, the same managerial approach may not lead to another success or output the next time. Although the direction is intended to be the same, it may result in a different human output, no output, or any combination of outputs. The people (and the manager) have changed.

Nature is very complex, but those complexities are understandable when one learns the relatively logical mathematics and analytic processes of the "hard" or natural sciences. Unlike physics or chemistry, where the rules and results are straightforward and constant, humans are affected by new learning every day and consequently their thinking and behavior can change. Predicting some of the simpler drives behind human behavior is fairly straightforward. It is obvious that starving workers will be very motivated by food and not very interested in complex mental tasks. However, most of the drives or bases of human behavior are not that simple.

We know that people are more than consuming, moving, reproducing machines, whose parts are diagrammed in biology texts. The source for new, improved products and services is science plus human creativity. They have to exist together. Human creativity (which is of course based in emotion) is not necessarily rational but it is as necessary for technical innovation as the physical laws of Mother Nature. Most of us who have had some technical training are familiar with the more predictable hard sciences and laws, but the equally important aspects of the "soft" sciences, such as psychology, sociology, anthropology, and the other "—ologies" often require strengthening. This book is directed to that strengthening. Objectivity and creativity are codependent.

4.0 USING THIS BOOK: A PROCESS MODEL

In this book there will be many references to the company, the group, or the worker, but those references should be understood as merely generalities. They can never apply exactly to your situation. Current developments in changing management and organizational situation prevent the recommendation of any overarching, grand organizational strategic concepts. As Carroll points out,

> A less visible compositional shift has occurred within particular industries where small specialist organizations have emerged at the same time that overall market concentration is increasing.
>
> These developments have presented managers with a host of new and difficult challenges. . . . All in all, it makes little sense to speak generally (as much as the current management literature does) about a single organizational solution for the new downsized industrial world. (Carroll 1994, p. 39)

We need practical ideas and concepts first that you, as the reader, can use. Consequently, we'll concentrate primarily on optimizing or improving your personal management capabilities within your perception of your organizational situation. There will be no panaceas, but rather specific pragmatic recommendations to improve your management skills in a technical environment. A word of caution is appropriate now. Those recommendations cannot be totally adopted without modification to fit your interpretation of the situation. No two organizations or situations are identical (as there are no two identical people in those situations), and you will have to select and modify any ideas to be applicable. New learning must be modified to fit your view of the situation. Then that view should be tested on the job.

This book will use a *process model* for learning. That process model (or learning how to change) will include a description of the typical organizational components, then analysis to determine applicability and finally a synthesis into a more effective and newer management thinking mode. You should then be able to test that mode against your interpretation of your situation. The sequence to be followed in this process model of analysis and synthesis

1. *Description and explanation:* (Thinking and learning) What have others done? How do their actions relate to technical management, and what do those actions mean to me? What general prescriptions can I use?

2. *Prediction and testing:* (Testing and applying) How can I specifically use this? When it is used, what happens, and how effective is it when applied to my situation?

The first step in the process requires cognition (thinking) and behavior (testing or managing). Your cognitive and your behavioral changes are interrelated in the learning process. As we'll be dealing first with attempted changes in cognition, a word of caution is in order. Those changes depend to some extent on how the new data (such as in this book) is interpreted by you. Even in the hard sciences, the same information has been interpreted in very different ways (Gould 1981, p. 8).

For example, Galileo insisted that planetary orbits were circular, even after Kepler had provided meticulous observations showing that they were not. (As you know, they are ellipses.) And we all know about the trouble that occurred several hundred years ago when Galileo stated that the earth was not at the center of the universe. (Within the past few years, the Catholic church finally agreed with him.) So the interpretations of data may not match the objective results. The same can be true of your own concepts. If the data reported do not match your predetermined concepts, you have two basic alternatives: disregard or discard the data (e.g., Galileo's discovery that the earth was not the center of the universe was data that was "discarded," as was Galileo himself when the Church imprisoned him for a while) or recognize that you have an incomplete set of concepts and that those concepts have to be changed. (Most of us now agree that the earth is not the center of the universe.) Therefore as you read, you might want to mentally question whether this book and/or the authors referred to in the references are being objective, according to your own criteria. A healthy skepticism is very useful.

To help you to learn how to deal with the human side of managing technical organizations we'll use a tool that is fairly familiar. That tool is the *scientific method*, summarized as follows:

1. Define the terms and develop the theory.

2. Set up the hypothesis that tests the theory.

3. Gather data and compare results against the predictions of the hypothesis.

4. When the variance between predictions and results is equal to or less than expected, accept the hypothesis. Otherwise, go back to Step 1 and re-evaluate.

5.0 DEFINING THE TERMS

We start by defining *theory*. There are many definitions. One that I like is that it is a tool that prevents the person from being confused or dazzled by the complexity of perceived data and/or natural events. It is also a general explanation for a class of events. Therefore theoretical statements should help you to distinguish the "important" from the "unimportant," support categorization and classification, and communicate clearly.
    Consider this:

    . . . a theory is a set of related concepts or best guesses about what is going on in a given area. . . .
    By beginning with theory we develop the basis for processes of -
    1. Inference or conclusion based on a guess or hypothesis (deduction), or expanding a known datum into a generalization (induction),
    2. Test of the inference (what is the right guess?); and
    3. Correction, based on feedback from behavior.
    A theory, therefore, is chiefly a mode of organizing one's thinking. (Levinson 1976, p. 3)

    That last point is not always self-evident, but it is always true that the ways in which we organize our thinking direct the consequent behaviors. The behaviors in turn affect our thinking. It's an interactive, never-ending process. This becomes more obvious as we consider attempting to solve management problems. The selection of the appropriate definition of variables, parameters, and constants as part of a theory or philosophy followed by a hypothesis prepares the mind to learn to see things differently. Perceived data is always evaluated with respect to some personal theory. We see what we are prepared to see. Just the act of defining the terms and developing the plan (theory and hypothesis development) affects our perception and prepares us for learning in a predetermined way.

Without this kind of preparation, the world would be a random pattern. The following quotation says this a bit more elegantly.

> . . . reason has insight only into that which it produces after a plan of its own, and that it must not allow itself to be kept, as it were, on nature's leading strings, but must itself show the way with principles of judgment based upon fixed laws, constraining nature to give answers to questions of reason's own determining. . . . Out of the profusion of words uttered by nature, the scientist selects those that he can understand. (Kant 1958, Preface)

Being immersed in our own organizations, we always hear the "profusion of words" uttered, but we must try to limit that profusion by defining our terms, building theory, and limiting our hypotheses closely. This process, of course, is not followed only one time. When there are differences between our predictions and what actually happens, they can be seen as errors, problems, or opportunities. When these differences do occur, the process is repeated with new data to confirm that what we found was not unusual. This book can assist by providing analyses of selected research findings (what others have done) that seem to be applicable to your own theories and hypotheses. You are the theory builder or the "scientist who selects those words that can be understood."

The research findings presented here are based on academic research summaries, and tested by personal experience that was collected over the years. Admittedly, this is not completely rigorous testing. There are no predefined controls or statistically valid experimental sample sizes here—those are difficult to achieve in the real world of industrial or governmental technical organizations. Time cannot be turned back to repeat an experiment to determine its reliability. But experience and common sense aren't so bad either. In any event, you have to decide if the data presented in this book can assist you in limiting the confusion as definition and analysis concerning the organizational environment develops. Unfortunately, some of the data will repeat as it appears in various sections of the book. Although there are many simultaneous interactions among the people in a company, books that describe these interactions are written (and read) linearly. As an example, the topics of motivation, structure, and the technical group's technology may all be interacting at the same time but they appear and reappear for discussions several times under different headings.

Let us start with the first description or definition of the place in which we find the technical organization: the economic/political/social environment.

5.1 Definition: The Organization's Environment

The environment is defined as those important interactions recognized by the organizational participants as being *outside* the organizational structure. This definition is not exact (in fact, it's a bit circular), but that inexactness happens because there is little agreement on exactly where the organization stops and the environment begins. [Psychologists call this type of definition an operational definition; see Skinner (1953).] For example, do the organizational boundaries include stockholders and customers? What about unions and regulatory bodies, or families of employees? This general definition is not as exact as it could be if we were dealing with one specific organization and could make some assumptions about its multiple internal and external relationships, but inexact measurements do not preclude useful results.

For example, consider the builders of the great Pyramid at Gizeh in Egypt. They lived before the development of the accurate, precise surveying techniques that could achieve the absolute flatness of the foundation needed to carry the millions of limestone blocks, which were to be placed without mortar. Any unevenness in the foundation would probably result in an eventual sideward thrust that would cause the blocks to move and the pyramid to collapse. In spite of this lack of exact definitions of flatness, they developed a pyramid that, after thousands of years, still stands square and vertical: good enough not to need cement to hold the blocks together. The squareness and verticality were not perfect, but they proved to be entirely adequate for these several thousand years and that's good enough. Similarly, our "good enough" definition of the organizational environment should be adequate for our purposes. We'll now move on to a "good enough" definition of the people in the organization and then determine why these people in the technical groups have become so important within recent times.

5.2 Definitions: The Organization's People and the Organization's Environment

The people in the organization are those who interact *regularly* with it, *draw* economic and/or social payments from it, *contribute* physical and mental inputs to it, and are *primarily* employed in the creation or production of the organization's goods and services. Their tasks within the organization are affected and defined by changes in the organization's external economic, social, and political environment. One example of this change is the recent general movement from a manufacturing to a service economy. We are no longer primarily a manufacturing economy, if that type of economy is defined as one in which production of physical goods is the controlling force. Now our economy is concerned mainly with the production of services such as communications, finance, and information. In the United States manufacturing labor accounts for about 20% of the total work force (Ginzberg and Vojta 1981). Our labor force has changed from the manual labor of prior years to the mental "knowledge work" of today. Because the organization's worldwide communications have shown less developed sections of the world what more affluent societies can provide, there are consequent social movements in these less developed nations toward greater industrialization. That changes life styles and customer demands and affects the services and products provided by a company. Obviously, other requirements such as antipollution statutes and health/safety rules are political changes that also affect the organization.

Other changes include the explosion of educational opportunities and scientific knowledge that supports rapid improvements and expansions in technology (the way we do things). In the past, all of these changes would have been driven by investments in machines and plants. The modern organization is driven by investments in training its knowledge workers. The individual technical worker has become the human capital that supports current economic growth. Although machines and plant facilities are necessary, they are really a secondary kind of capital. Now the capital that fuels the increase in real wealth comes from the skills and minds of professional and technical workers.

The change from a primarily goods-producing to a services-producing economy is fairly conclusive. Recent demographic research concerning producer services, which includes engineering and technical functions, indicates that the value being added by producer services now approximates that added by manufacturing in the U.S. economy (Ginzberg Vojta 1981), and this value is growing.

In other words, the production of wealth by technical (or knowledge) workers is surpassing the contributions of manual workers and physical capital of the manufacturing sectors. If this trend continues, our primary production source "is the . . . knowledge worker . . . who puts to work what he has learned in systematic education, that is, concepts, ideas and theories, rather than the man who puts to work manual skills or muscle." (Drucker 1973, p. 32.)

This knowledge worker is a relatively new kind of worker. In some superficial ways, he or she has a resemblance to the artisan of the past who had highly developed skills, but there are major differences. Artisans apply their skills in a fairly logical and repetitive fashion within an extensive but still limited repertoire, whereas the technical professional must add unique cognitive, creative contributions to those craftsman-like skills to develop the new products and innovations upon which the organization depends. The present-day technical professional has a higher level of education upon entering the work force than his predecessors of twenty or thirty years ago and that educational level is constantly being added to. The technical professional's studies never stop.

The definition of the people, or the knowledge workers, in our modern organizations has also been affected by the changes in the size of the organizations themselves. The growth and proliferation of the typically large diversified industrial organizations of thirty years ago has slowed down. The extensive multinational conglomerate is not a new phenomenon. During the last two centuries, the Hudson's Bay Company controlled much of North America, the East India Company ruled most of India, and the Dutch India Corporation ruled Indonesia and many Pacific Islands. These large enterprises controlled the lives of thousands of employees, vendors, and customers, and even the course of governments. Some of these organizations had grown into huge conglomerates with the power to redistribute work and resources in ways that many nations of the last century were unable to do. For example, the huge Krupp companies, which produced steel and coal in Germany during the late 1930s, supported the growth of militarism in Europe. Without a dependable source of oversized weaponry, it is difficult to pursue a large-scale war. In the United States, there are many legal restraints that are designed to preclude this type of direct interference in political processes by industrial organizations. These are the typical legal limitations on corporate political contributions and monopolies. (There is always some discussion as to the adequacy of these legal restraints.)

But large, centralized multinationals are being limited today less by political considerations and more by economic considerations. The growth of the large manufacturing line-and-staff organization is over. It has been stopped by the ability of smaller, more responsive organizations to serve changing customer requirements quickly and efficiently. But serving these requirements often requires innovative specialized technology, and that in turn comes from the knowledge workers. Large industry, with its stable, restrictive structures, has difficulty competing in recruiting knowledge workers when smaller companies offer growing structures, stock options, and promotion. This resulting competition for the services of the knowledge worker has begun to equalize the previously unequal power relationship between the organization and these workers. The workers have become quite mobile and because of their easily transferable skills (and equally transferable benefit plans), they have fewer loyalties or economic ties to any one organization. They move to other jobs very easily if dissatisfied or pressured in their present jobs, which often leads to loosened organizational pressure for behavioral conformity.

5.3 Definitions: The Concerns of the Technical Manager

In addition to dealing with problems concerning the organizational environment, highly mobile, more independent knowledge workers, and changing organizations, the technical manager faces other problems trying to manage the required special blend of logic and creativity. One element of a creative mind is an independent capacity for judgment and a resistance to the conformity of group pressure. We know that, both as individuals and in groups, people do not function as consistently and predictably as the laws of nature. Creative people are, if possible, even less consistent than the general population. They do, however, have a major asset in their ability to see patterns in data that others have not seen before. That ability often results in innovative products and services—a central concern of the technical manager.

It is unfortunate that the consistent, logical thinking (based on the models of nature's laws) that is taught very well in the technical and engineering schools is not supplemented by an equal emphasis on the flexible thinking of the creative innovator, which provides personal and organizational growth. This flexible thinking must often be learned on the job if, indeed, it is ever learned at all. It is rarely, if ever, taught in academia. In many respects, managing technical functions or organizations is closer to the innovative process than it is to a disciplined, logical process. The data being used by the technical manager are extremely variable, including, as they do, the behaviors of very independent people, changing organizational structures and expansions of technology.

5.4 Definitions: Science and Art

If we can continue to use "good enough" definitions, we can move on to *science*.

1. Science is a process that is open, objective, and observable by anyone with adequate training to understand.

2. The results are independent of the scientific researcher (otherwise it's magic, not science).

3. The results can be replicated by anyone with the same or equivalent equipment.

We find that *management* does not fit this definition. For example,

1. The process is not open. It occurs in the thinking of the manager. All that we can observe is the behavioral outputs (and sometimes only the results of that behavior).

2. The results are dependent upon how other people perceive and interpret that manager's behavior.

3. The results can never be replicated because the situation always changes. (It is impossible to go back in time and do it over if we want to be sure that the results are reliable. It's also impossible to run a "controlled" experiment in management. Laboratory experiments are merely poor simulations of actual management situations.)

If it is obvious that management is not a science, then what is it? It is an art, if we use the following "good enough" definition for art.

1. An art is a subjective process that is dependent on the artist.

2. The worth of an art is dependent on the value that the social environment

places on it. (During van Gogh's lifetime, his paintings couldn't be sold. Today, they're worth millions.)

3. It is not replicable, even though some of the techniques can be learned by others. However, even when practiced by others, the output is never exactly the same. (This, of course, does not include plagiarism, which is not art but theft.)

Accepting this definition, management is an art and the value of that art depends on where and how it is practiced and valued by others. A superbly trained manager is useless unless that manager works in a situation that can respond to the excellent management process that is appropriate to it. We need both the manager and the appropriate situation. The typical situation in the modern technical organization is one of rapid product and environmental change. That requires extraordinary artistic ability to direct and optimize the work of the technical organization's creative contributors: the knowledge workers. Rapid change requires equally rapid (and intuitive) management responses.

As examples we have the changes discussed before: the growth in the importance of the knowledge workers and the proliferation of more flexible, rapidly adaptable, smaller organizations that make changing jobs easier. As recruiting demands increase, the knowledge worker's independence increases.

These changes are very new. The adequacy of the manger's art and adequate response in dealing with changes such as these often will determine if the company survives and grows or if it dies. Speed is essential. On the other hand, when dealing with science, actions are generally slower and more deliberate because one proposes questions to Mother Nature very carefully—her responses to wrong questions can be very painful. With rapid change in the economic environment or in the organization, response time becomes critical and a management reaction becomes almost instinctive—in many ways like the artist who "knows" what is good or bad in an artistic situation. One definition of the manager that seems to fit best is that of the *professional*.

1. The professional is required to undergo extensive training and that training is certified by some external body.

2. The professional's knowledge is tested by his or her peers.

3. The competency of the professional involves both science and art and cannot easily be judged by those who have not had the same or equivalent training. It requires peer evaluation.

Typically, physicians, attorneys and professional engineers, among others, qualify for this definition.

5.5 Becoming A Manager: Building Your Own Theory

But developing artistic or professional capabilities requires more than reading books (even this one). Although this book and others presents a general *process* model of learning and personal change (e.g., the scientific method noted before) and a general *tangible* or static model of the organization (see **Chapter 4** for a static model), both the method of learning that you finally use and your organizational model may be quite different from those suggested here.

Try to recognize these differences, test the hypotheses presented, use the parts of the models presented that you can, and retain the rest for possible future use. Build your own theory by selecting the parts that fit your perception of your situation to support your own artful and successful management style, as no two situations or organizations (or managers) are exactly alike.

Of course, no style can ever be static; the world and you are always changing. Therefore, provide for an iterative testing of your style and consequent modification based on new data received. This is not a simple process, and under the best conditions might require you to learn in a calm and reasonable environment, which the working organization may not have.

Research shows that the activities of managers during a "normal" day rarely reflect any description of that calm, reasonable environment in which to learn about technical management. One popular myth is that the manager is a careful, reflective planner and if he/she is managing effectively, has no regular duties to perform. The regular tasks are carried out by competent subordinates and the manager is concerned primarily with forward planning.According to this fantasy, management is, or at least is quickly becoming, a science and not a profession. If it is, it still has a long way to go. In my experience, many management courses in academia seem to be taught as if it actually were a science. It's not. According to research findings, the real-life situation is a bit different.

> . . . the job of managing does not breed reflective planners; the manager is a real-time responder to stimuli. . . . The managers' programs to schedule time, process information, make decisions, and so on remain locked deep within their brains. Thus, to describe these programs, we rely on words like judgment and intuition, seldom stopping to realize that they are merely labels for our own ignorance. (Mintzberg 1979, p. 110)

Building the model of the art (or the professional management style) that works for you is a process that involves off-the-job thinking (there's that calm, reasonable environment), then hypothesizing, and finally on-the-job testing of these hypotheses. Simply reading about models of managing and using what others have done as a basis for changing your manager's mental programs is insufficient by itself. In other words,

Cognitive learning is detached and informational . . . But cognitive learning no more makes a manager than it does a swimmer The latter will drown the first time he jumps into the water if his coach never takes him out of the lecture hall, gets him wet and gives him feedback on his performance. (Mintzberg 1979, p. 122)

Professional physicians, attorneys and engineers follow an action-oriented process model of learning. After they complete their formal schooling, there is a practicum or internship that they must successfully pass. Learning management requires study plus extensive experience. This book is part of the "study" part. You have to try it by testing on the job to gain the "experience." But that "experience" can be partially vicarious. One can learn by watching how others do it and by studying appropriately.

If knowledge and skills could be acquired only through direct experience, the process of human development would be greatly retarded, not to mention exceedingly tedious, costly and hazardous. Fortunately people can expand their knowledge and skills on the basis of information conveyed by modeling influences. Indeed, virtually all learning phenomena resulting from direct experience can occur vicariously by observing people's behavior and the consequences of it. (Wood and Bandura 1989, p. 362)

But even study and experience may not result in an optimum management style for you. For example, assuming that the samples of management behavior are entirely adequate for your present situation, they may not prepare you for any surprising changes that may come from other parts of the company or from the outside environment. On-the-job training is inadequate by itself because it has several implicit assumptions: that your trainer is demonstrating the best model, that you can duplicate what you see, that only action triggers mental programs, and that you're ready for anything once you've finished your course of training.

The fallacies of these assumptions become obvious once they have become explicit. Management, as art superimposed upon logic, can be learned through practice, but that practice has to be corrected and predicting practice, based on solidly researched theory and the successful experiences of others. Your trainer (and the training situation) has to be able to provide this type of practice. It is also obviously important to recognize that your trainer's perceptions and behaviors are not yours. (Only Picasso painted like Picasso; the imitators could never duplicate his work exactly. But you knew that because Picasso—and you—are artists!)

Fortunately, exact duplication is neither necessary nor desirable because situations change rapidly. The task is learning to use the variables of flexibility and foresight in order to be able to set up your own successful solutions to situations that have not occurred before. In addition to the variables in your state of mind and in environmental change, other variables affecting the situation include the organizational structures and the technical processes used. Consider the possibility that a "Picasso" of management in a well-structured organization producing standard electrical parts who is transferred to the relatively unstructured matrix of a defense plant. In that unstructured organization, the Picasso would probably attempt to behave the way that had succeeded in the closely structured company. However, the new structure is different, the technology is different, and the people are undoubtedly different. Now success depends upon the two factors that we discussed: education and corrected testing on the job. Education means learning those theories and practical applications that have already been developed successfully. (Why reinvent the basic wheel?) Corrected on-the-job testing of these educational materials means ensuring a fit to the current situation (i.e., adapting the basic wheel to fit the wagon you're working with).

6.0 REVIEW AND PRACTICAL SURVIVAL GROWTH TIPS

The technical manager is faced with an environment and an organization in rapid change. A manager who has been successful in other situations may not expect to immediately succeed in a different situation.

A manager doesn't have to be a psychologist or a behavioral scientist in order to be an effective manager of engineers and scientists. However, in order to become a knowledgeable and practical engineer or scientist, one must first become well acquainted with theories and laws. Likewise, in order to become an effective manager of human resources, one must first become familiar with some of the theories that attempt to define and interpret human behavior and personalities, formation of habit patterns, needs, motivational drives, and direction (or redirection) of human energies in positive and productive directions. (Wortman 1981, p. 12)

The first step is to learn, then do. There are opportunities in these changes for the technical manager to learn and grow. Interpreting the environment or the world has never been an entirely objective process. Therefore, you need a personal theory to aid you in subjectively organizing your thinking. Defining terms and selecting hypotheses based upon some theory helps you to find potential answers. One definition of an intelligent person (i.e., manager) is one who possess the ability

To respond to situations very flexibly; To take advantage of fortuitous circumstances; To make sense out of ambiguous or contradictory messages; To find similarities between situations despite differences which may separate them; To draw distinctions between situations despite similarities which may link them; To synthesize new concepts by taking old concepts and putting them together in new ways; To come up with ideas which are novel. (Gomez-Mejia and Lawless 1992, p.21)

One persistent problem is named "organizational inertia," that is the tried and proven ways of doing things that rule organizational life. Typically this inertia is manifested when in troubled times people try to do the same things that they've always done, but do it harder, or faster, or longer. If it doesn't work the first time, do it again, faster and more carefully. I prefer a different suggestion: if at first you don't succeed, don't do the same thing again. Do something else.

The technical contributors to the organization have become its human capital, and that human capital is responsible for the production of more income than the physical capital. There has been an increase in technical education of "knowledge" workers. These workers have become more mobile and are less dependent upon any particular company for continued employment.

There has been a change from a manufacturing- to a service-oriented society.

There are many ways to approach the problem of learning how to manage technical operations, as one must consider both the participants and the situation; neither are fixed for long.

Successful technical management requires creativity and logic or, in other words, art combined with science.

Understanding what others have done in management is an initial step in developing your own art, but you are cautioned to test that understanding in your own situation. If the results of your tests meet the goals you selected, use the hypothesis behind the test as part of your behavioral repertoire; otherwise, go back to the starting point and take another look at both what you have been doing and the research of others for improvements. This learning and testing, with on-the-job feedback from peers, subordinates, and superiors, is my suggestion for a guided leaning pattern that can produce the positive cognitive changes you need.

Although the situation on the job generally does not support the quiet, introspective learning that is usually associated with cognitive change because there is still the demand for rapid, almost instinctive, but accurate responses to difficult management situations, it is necessary that any learning that is done both off and on the job, be done very well: until your responses are almost automatic. This type of rapid response requires both on-the-job training and introspective learning. Consequently, a recommended process would start with introspection (i.e., reading and study), development of a theory and hypothesis for action (i.e., a plan), and testing that hypothesis on the job (i.e., on-the-job behavior). Success as a technical manager will be yours only after the results of those tests are used to modify your future behavior.

You, as the technical manager, are a tremendously valuable asset to your organization, and that knowledge should provide you with the confidence to test your hypotheses. (Organizations don't discard valuable assets if there is an occasional loss, and you can't make progress if you don't stick your neck out once in a while.) Those hypotheses that do not work out as well as expected may have to be modified to cope with the changing situation. (We also learn from experiments that fail.) Finally, when you are able to develop effective and useful management hypotheses (as measured by your success as a manager), you have learned the basics of combining the logic of technical requirements with the art of human requirements.

7.0 "ONE-SENTENCE" SUGGESTIONS

1. The "human element" introduces uncertainties that require the "art" of management.

2. Each manager needs to discover/create his or her own unique management approach.

3. Different management styles are appropriate to different situations.

4. Managers can change their styles as the situations change.

5. The power to make and test independent hypotheses is perhaps the most important concept in management.

6. This power can be used to enhance the growth of the company.

7. Theories can be molded to assist managers in developing their personalized approaches to management and leadership.

and finally, on the one hand,

No problem can stand the assault of sustained thinking. (Voltaire) and on the other, remember that even under the most rigorously controlled experimental conditions, the organism will do as it damn well pleases. (anon.)

In the next chapters, we will explore more of the "how to" that might be used to further and expand that success and begin the outline of a straightforward decision-making model.

Case Study:
The Case of the Quality Program

Cast

Tom Berry: President of Thorax Medical Products. His father started the business. Tom took over ten years ago after working in various training positions throughout the company following his graduation from Alpha Technical University.

George Beardsley: Vice president, sales. Had been with the company for fifteen years and was primarily responsible for the growth of sales over that time.

Marvin Loren: Chief engineer. Had been with the company for more than twenty years. He had started as the company's first drafter and had been promoted as the company grew.

Bob Spelvin: Chief, quality control. With the company six months. Had been recruited from an aerospace company to start a formal quality program.

Thorax Medical Products had been manufacturing high-quality stethoscopes and other mechanical instruments for hospitals and physicians for more than fifty years. Its products were considered to be the best in the field. Within the last several years, sales had begun to drop when one of their competitors, Atlas Medical, came out with a series of remote reading electronic stethoscopes.

Tom Berry, the president of Thorax, had immediately instituted a product redesign program, and that program had been in progress for two years. During that time, the engineering design group had developed several concepts that seemed to be ahead of anything else on the market. Some of these products included a microcomputer that could take vital measurements at any bed in a hospital automatically, and a diagnostic chair for physician's offices that could take a patient's blood pressure, temperature, heart signs, and other data automatically when the patient sat in it.

George Beardsley, vice president, sales, had been pressing Tom to get the factory going because his salesmen had been out with preliminary brochures describing the new products for several weeks and tremendous interest had been shown by key customers, even though delivery was not promised for another six months. Marvin Loren, chief engineer, had assured Tom that these products were well designed, and even though the company had not done this before, he felt that the products could be manufactured by subcontracting out major parts to several computer manufacturers. Mechanical parts and assembly were to be completed in Thorax's own plant. Bob Spelvin, chief, quality control, had sent a memo to Tom a week before, with copies to George and Marvin, flatly disagreeing with Marvin. He had recommended that an extensive program of vendor quality inspections and in-house quality audits be instituted. Tom didn't even understand some of Bob's language. What was a quality audit, anyhow? With pressures building up because of dropping sales, Tom called a meeting in his office the following Monday morning. At 8:30 A.M. everyone was there.

Tom: Well, you're all aware of our situation. This meeting was called so that we could thoroughly air our differences of opinion and decide what to do next. Who wants to start?

George: Well, I'm sure everyone knows what our position in the market is.

It's deteriorating, and unless we do something fast, we'll be in serious trouble. Our field representatives have been screaming for something to beat the competition with and we have to move on it or we'll lose our market and the best group of field reps that any company ever had. Those people are independent and they won't stick with a loser, especially as they work on commission.

Marvin: I agree with you. Some of the ideas that we have come up with are really fantastic. Why, I personally spent my vacation right here checking the detail drawings and assemblies for our new microcomputer products. And Bob here has been a great help in developing life tests for the equipment, getting it past the Underwriter's tests, and making sure that every component going in was thoroughly checked out. Frankly, I don't understand the reason for this meeting. We have great products and they work beautifully, both here in our labs and in the field tests.

Bob: Well, Mr. Berry, as I see it, the computer chips that we are buying and the other electronic components are unlike anything this company has handled before. Incoming inspection cannot do an effective job of checking them and we cannot afford to have any field failures of our equipment. Some of those components may deteriorate over time and I'm trying to prevent a field repair program that could ruin our reputation. As far as our own plants are concerned, I'm not sure that quality control is the answer anymore. I'd like to place the responsibility for quality with the plant operations office and start a quality assurance program that would prevent problems before they happened, rather than quality control, which would find the problems afterward.

Tom: So far, Bob, I haven't heard anything to disagree with, but why should this affect our new product introduction schedule? Engineering and sales have agreed, so why not go ahead and produce?

Marvin: Well, I've heard some things that I really don't understand. Bob, why can't incoming inspection check those components? The prints are very clear about the specifications, and our plants have been running under quality control systems for years. Why should we change now? It is just an additional frill. Why raise our costs for something that could happen but has never happened before?

Bob: You're right, we can check components being received, but computer chips cannot be tested 100% and therefore our present testing is not effective enough. I suggest that we develop a vendor quality evaluation program in which we check the way our vendors design and manufacture our parts. In that way, we'll have a higher confidence level that we are getting what we are supposed to. As far as our plants are concerned, they should be working to the same procedures as our vendors. Let them produce quality parts and we, in the quality department, will just be checking their methods and procedures. If those methods are right, the product will be right.

George: Bob, how long will this take and how much will it cost?

Bob: I don't have the final details but I would guess about six months and an increased budget to provide four more quality engineers.

Marvin: Bob, you're going overboard on this. We're not Amalgamated Aerospace, where you came from. We can always fix things and there are always a few bugs in any new product. Everybody knows that and the customers have always accepted it. I vote for getting into production now.

Tom: Fellows, the decision to move ahead with production has to be looked at very carefully. I think that we should sleep on it. Let's meet again tomorrow.

QUESTIONS

1. What actions did Marvin take, and what should he have done?
2. If you were Bob Spelvin how would you have been prepared before this meeting?
3. What should George have done before the meeting? What should his recommendations be tomorrow?
4. What should Tom Berry do before the next meeting?
5. What are the implicit theories and hypotheses that each of them is using, and how are they different?

Before reading my answers, take a minute to reflect on the contents of this chapter and your own experience. Try to present your own answers first and then compare them with my suggestions.

SUGGESTED ANSWERS

1. It seems to me that Marvin had not determined that his role as a technical manager is different from that of a designer. His personal theory is out of date because it applies to his previous job with the company. He mentioned that he had spent time on the drawing board determining that all the components would work well. That behavior is more appropriate for the designer that he used to be. To set up his theory, he should understand that there have been changes in technology and in the company, which is moving into an entirely different kind of market and manufacturing process. He should be concerned with whether the product design goals can be met in the marketplace, because the product has been drastically modified from a mechanical to an electronic device. The product operations and controls are different. Because the product works differently, it might not be as easy to define product acceptance tests.
2. Without objective data to back them up, a person's statements can only be evaluated subjectively, and Bob Spelvin had not been with the company long

enough for that to work for him. Therefore, data were required. For example, in a worst-case scenario, what would the company's liability be if one of the new diagnostic chairs were to fail? What would the costs of the quality assurance programs be to minimize that potential field failure? In effect, Bob should have come in with a cost benefit set of data.

3. George seemed to be as unprepared for this meeting as the rest of the group. He certainly was aware of the problems (and opportunities) in producing the new products. George should have supplied data for the meeting that were obtained from his field representatives showing how serious (in a quantifiable measurement such as lost sales) they thought the delay would be. Or he could have offered an experimental hypothesis with one obvious cost-effective alternative that could have prevented possible trouble in the field. For example, after development of final acceptance tests, the company could provide the diagnostic devices to several teaching hospitals on an experimental basis through selected field representatives. Data obtained from the devices would then be checked at the hospitals by more conventional means, giving some indication of product effectiveness. Meanwhile, the field representatives would be able to maintain contact with their markets (i.e., the hospitals) and would be able to get the sales credit when the superiority of the new products was established.

4. Define the terms. (Set up an agenda and require each attendee at the meeting to provide specific recommendations with justifications for them before the meeting so that they can all understand each other's position.) Propose a theory. (The company needs new products because. . . .) Develop a hypothesis to be tested at the meeting. (Take either Bob or Marvin's position and, independently, determine what the potential effect would be.) Then use that working hypothesis as a standard against which to measure the recommendations of others.

5. The implicit or mental theories of the participants seem to be:

Marvin: The future is almost an extension of the past. If that's the way we did it in the past, that's the best way to do it in the future.

George: Sales is where it's at. It doesn't matter if the technology has changed. Selling complex products is just the same as selling simple ones. Just get out there and sell.

Bob: The solution to a similar problem in my last organization was to set up a vendor quality program. This organization (and, implicitly, the situation) is the same, so the answer should be the same.

Tom: Have no theories or hypotheses of your own. Depend on the advice of "experts." How can you manage people who are more qualified technically than you are?

QUESTIONS FOR YOU TO ANSWER

1. Do you agree or disagree with these interpretations? Why? Do you see how they are related to the chapter?
2. Is this situation similar to those in which you have been involved? What happened?
3. If it were possible to "re-live" the situation noted above in Question 2, what would you do that is different?

REFERENCES

Ackoff, R. L. **Redesigning the Future: A Systems Approach to Societal Problems.** J. Wiley & Sons, New York. 1974. p. 80

Carroll, G. R. "Organizations . . . The Smaller They Get." **California Management Review.** Fall 1994. 37(1):28–41.

Drucker, P. F. **Management Tasks, Responsibilities, Practices.** Harper & Row, New York, 1973.

Florman, S. C. **The Civilized Engineer.** St. Martin's Press, New York. 1987

Ginzberg, Eli and Vojta, George J. The Service Sector of the U.S. Economy, **Scientific American**, March 1981 244(3), pp. 48–55.

Gomez-Meija, L. R. and Lawless, M. W. (eds.). **Advances in Global High-Technology Management.** JAI Press, Greenwich, CT. 1992.

Gould, P. "So Human a Science." **The Sciences** (May–June 1981); p 6–30.

Hammer, M. and Champy, J. **Reengineering the Corporation: A Manifesto for Business Revolution**, Harper Business, New York. 1992.

Kant, I. **Critique of Pure Reason** (2d ed.), Smith N. K. (trans.). New York Modern Library, New York. 1958.

Levinson, H. *Psychological Man.* The Levinson Institute, Cambridge, MA. 1976.

Mintzberg, H. "The Manager's Job Folklore and Fact" **On Human Relations,** Harper & Row, New York, 1979. pp. 104–124. Reprinted by permission of the *Harvard Business Review*. Excerpts from "The Manager's Job: Folklore and Fact," by Henry Mintzberg (*HBR*, January–February 1975). Copyright © 1975 by the President and Fellows of Harvard College. All rights reserved.

Shannon, R. E. **Engineering Management.** John Wiley & Sons, New York. 1980. pp. 1–7.

Skinner, B. F. "The Operational Analysis of Psychological Terms." **Readings in the Philosophy of Science,** Feigl, H. T. and Brodbeck, M. (eds.). Appleton Century Crofts, New York. 1953. pp. 104–124.

Teng, J. T. C.; Grover, V.; and Fiedler, K. D. "Business Processing Reengineering: Charting a Strategic Path for the Information Age." **California Management Review.** Spring 1994. 36(3):9–31.

Thamhain, H. **Engineering Management.** John Wiley & Sons, New York. 1992. p. 3.

Wood, R. and Bandura, A. "Social Cognitive Theory of Organizational Management." **Academy of Management Review.** 1989. 14(3):361–384.

Wortman, L. A. **Effective Management for Engineers.** John Wiley & Sons, New York. 1981.

————, **The Economist** (9 April 1994): 76.

————, **New York Times** (April 17, 1994).

————, **Wall Street Journal** (7 June 1994): A2.

CHAPTER 2

The Manager: Dealing with Yourself—The Most Important Individual

Case Study:
The Ordinary Day

Cast

Leona Russo: Chief engineer
Secretary: (Leona's)
Bill Watson: Head, quality assurance
Sam Smith: Controller
Arnold Mitch: Design engineer

Leona kept the car well within the speed limits as she drove toward the plant on a sunny Wednesday morning. There was still plenty of time to get to her desk, review the capital budget for next year, and make some last minute adjustments before her presentation at the board meeting scheduled for 10:00 A.M. that day. She went over her morning's schedule in her mind:

1. Review the budget.
2. Approve the final design review notes on the latest turbine pump.
3. Find out what happened during the last interview that Arnold Mitch had with the graduating engineer they had recruited from State University.
4. Check out the vendor assessment that Bill Watson had promised her on the potential hydraulic motor vendor.

She had everything all set up in her mind by the time she wheeled into the parking lot. As she approached her office, promptly at 8:30 A.M. her secretary was waiting for her with a worried expression.

Secretary: Good morning, Leona, the Board of Directors' secretary just called and said that the meeting had to be moved up to 9:00 A.M. Can you be ready? Here are some other phone messages that just came in.

Leona quickly checked over the phone messages. One bothered her. It was from Sam Smith. She decided to return his call first because he rarely called unless there was a major problem. She picked up the phone and dialed.

Leona: Hello, Sam, what's up?

Sam: Leona, I think that you're in for a bit of a problem at the board meeting. George Wishley asked me for a summary review of project costs on the last three capital expenditures and I had to give it to him. You remember, those were the installations that finally ended up costing about twice what the appropriation had allowed. I know that there were extenuating circumstances, but you might have a little explaining to do this morning. I called you late yesterday afternoon but you had left for the day.

Leona: Thanks, Sam. Yes, I have the data on those. I'll talk to you later.

As she hung up, Bill Watson poked his head in the door.

Bill: Hi, Leona, got a minute? (Without waiting for an answer, he draped himself over one of the office chairs and continued.) The material review board has been approving rejected materials from Appleby Valves for the last three months. Your engineers don't want to change the drawing and we've "bought" ostensibly "unacceptable" materials every time that my inspectors rejected them at incoming inspection. Look, we either have to really reject that stuff the next time or, if it's really OK, let's change the tolerances on the drawing and maybe negotiate a better price with Appleby Valves. We just can't go on wasting time like this. You know that I'm shorthanded and those board review meetings take up a lot of time. I'm sure that your review engineers have other things to do too.

Just then, Arnold Mitch walked by; he saw Bill in Leona's office and waved at them both. Bill got up and walked over to the door to speak to Arnold.

Bill: Hi, Arnold. If you have a second, I'd like to get your opinion on the latest lot of Appleby Valves.

Arnold: Well, if you ask me, those guys at Appleby are getting away with murder. They swore up and down, when we did the vendor review at their plant, that they could meet our tolerances, and the evaluations of our vendor review team tend to support that. Now they ship late, and we always have to waive our requirements because we need their valves, and another thing . . .

Leona: (Interrupting) Look, people, we've had these discussions time and time again. Both the design team here at the plant and the vendor review team that visited Appleby agree that we need those valves and it was cheaper and more effective to have Appleby supply them. There doesn't seem to be too much that can be gained by going over the same old ground. What would you like to do today? This problem doesn't seem to be going away. I'll work on it and lay out some policy by the end of the week. Meanwhile, Mitch, what happened at the design review meeting on the turbine pump?

As Mitch launched into a rambling discourse on the meeting, Leona thought about the rest of her schedule that morning. Mitch was interrupted by a call from Leona's secretary.

Secretary: The data processing center just called and said that the central computer is down and would be unavailable for at least two hours.

Leona was visibly annoyed. She had requested her own computer system for the engineering department for the past two years, but had never been able to justify spending the money to the Board of Directors. The central computer was down on the average of once a month and always when there seemed to be an impending disaster. At that point Leona noticed the time. It was five minutes to nine. She told Bill and Arnold that she had to leave and would talk to them later. They went back to their offices. She gathered up her notes from the previous night, and headed down the hall toward the board meeting.

Questions

1. What management techniques should Leona have used that would have helped her to handle the situation?
2. How should she resolve the engineering quality dispute on the Appleby valves?

3. How is she resolving the problems of handling uncertainty and what would you do?
4. How should she handle any problems about the capital overruns if George brings them up at the board meeting?

My suggested answers for these questions will be found at the end of this chapter. Why not answer these questions before you read my suggestions? Remember, there are no wrong answers, just different ones.

1.0 OVERVIEW AND INTRODUCTION

Chapter 1 described new and recent changes in the various organizational environments, such as those caused by improvements in global communications resulting in worldwide competition, the downgrading of large, slow-response organizational structures in favor of smaller, more responsive ones and finally, increases in mobility of the knowledge workers. Those changes increase the need for new thinking and modified internal management procedures. This chapter describes the actual process of managing in response to these changes and prescribes the relationship between the management job and the individual who occupies it. We start with the company and then the job.

2.0 COMPANY SIZE AND THE INDIVIDUAL

On the surface it would seem to be logical that larger companies should be more successful because they can develop overall corporate policies that reduce unintended deviations in human behaviors. When those behaviors are standardized, the products that are produced should have fewer deviations. This is needed when customers demand a steady stream of unchanging products. But that's not the case in today's environments. These environments change quickly and the organization must have the agility to respond easily to these changes. Overall corporate policies that apply rigidly to every part of the company can often lead to disaster. Fast response usually requires innovation, which is usually easier to achieve when policies are flexible and are different for different departments of the company. For example, whereas the standards engineering department procedures should be strictly codified and enforced to ensure the use of pre-tested, interchangeable components, the research and development (R&D) department should have minimal procedures to eliminate any constraints upon creativity. In this example, one might actually observe that these two departments are really two small, and different, organizations. Innovation would not be acceptable in standards, but it would be a necessity in R&D.

It is almost a truism that successful innovation seems to be greater in smaller companies or in independent divisions with separate identities within larger companies because there are fewer constraints. The underlying idea is that there are advantages to being structured in smaller sizes to allow innovation and creativity (i.e., individual emotion).

Organizational procedures that tightly restrain the individual can be the enemy of innovation because it is the individual, not the group, who provides the primary creative force. Groups may assist the individual by testing ideas or by presenting agendas to guide general boundaries to innovation (i.e., "We need improvements in our rotating electrical motors, so let's drop any innovations that would lead us into designing submarines."). But, in any event, the individual is the original and mature innovator, no longer expected to follow a predetermined, all-encompassing set of company rules.

Creativity functions best when there are few restrictive procedures. And as there is an ongoing worldwide shortage of qualified, innovative knowledge workers, it behooves the organization to retain those people by maximizing job security and minimizing administrative controls. In too many cases, rapid changes in economic, social, or political environments result in across-the-board, short-term, cost-cutting exercises and even unjustified downsizing. This doesn't provide a supportive environment for innovation. Neither does a company internal environment that changes personnel as they get older and have more experience with the company.

Even though current research (Economist - 1995) seems to indicate that the widely held perception that the American work force is becoming increasing mobile is not true, there is almost a daily progression of newspaper articles that contradict this through the highlighting of down-sizing of the technical work force. While overall job stability of all workers may not show a decline, specific job tenure in engineering and technical areas seem to be decreasing rapidly because of management decisions about the possible increased obsolescence of the organization as the technology changes. Technical companies need the state of the art in order to succeed and if they refuse to gain that by funding internal training of their technical employees, they try to gain it by replacing them. When that happens, the organization loses a great deal. Insecurity does not support innovation. It depresses it.

Management occasionally forgets that the experienced knowledge workers are the human productive capital of the company. When company policies that are intended to decrease costs and increase efficiency by downsizing or releasing experienced personnel are put into effect, the expected positive results may not result. People need some security for creativity, and those who remain are thinking about surviving, not creating and the new employees quickly learn to think that way too.

3.0 THINKING ABOUT MANAGING, PROMOTION, AND LOGIC

The academic kind of logic is often justifiably emphasized in technical schools and universities, when the output is expected to be scientists and engineers, not technical managers. However, although logic is needed to be technically qualified, as noted before, art is needed to manage those technical persons who bring that academic logic to the company. To understand the reports of technical subordinates and test their validity, the manager should be familiar with scientific concepts. Promotion without this adequate scientific knowledge doesn't provide the new manager with the tools to understand. On the other hand, promotion based on technical competency without adequate management training can be a double loss to the organization. There is a loss of a competent engineer/scientist and a resulting incompetent manager. The decision to promote is often inadequately thought through. The new manager may have been a successful technical contributor, but technical competence, although necessary, is an insufficient basis for management success.

Engineers reach a permanent decision about their future usually between three to seven years after graduation with an engineering degree. They find themselves in need of choosing between a technical specialty, a technical management position, or a nontechnical management job. There is a major reason for this: "The reward system in industry, by and large, seems to favor individuals moving through the management, rather than the technical specialty ladder." (Cleland and Kocaoglu 1981, p. 8).

Although management competency, like creativity, is not gained easily, with study and practice, it can be improved. As noted in the previous chapter, one way is by using a typical process model called the scientific method to build personal management theory and test applicable hypotheses that flow from that theory. The testing is done on the job.

4.0 THE PROCESS MODEL

When we developed the process model (noted in the previous chapter), it was stated that there were requirements for both cognitive change (such as that represented by reading this book) and behavioral change (represented by testing your hypotheses in your organization). To briefly review, let us begin with definitions of the terms to be used in your possible cognitive change. We then describe the methods of theory building (a theory is a general explanation for a class of events), develop testable hypotheses, gather test data, and, if necessary, restructure the theory to fit the data received more closely and try again.

4.1 The Reluctance to Change

That description of the process model seems clear enough, but it doesn't equip us to deal with those theories that we carried with us before we began the process. Those implicit theories we bring with us are not always easy to change, especially when the data upon which to base our new theories seems to conflict radically with our previous notions. Reluctance to change can occur even when the new data and the results seem to be irrefutable. One example of this occurred in the theoretical world of physics when Albert Michelson and Edward Morley obtained experimental results that conflicted with Newtonian mechanics. Zukav describes it well.

> The problem is that no matter what the circumstances of the measurement, no matter what the motion of the observer, the speed of light always measures 186,000 miles per second (in a vacuum) . . . suppose the light bulb is standing still, and . . . we are moving away from it at 100,000 miles per second. What will the velocity of the photons measure now? . . . 86,000 miles per second? . . . the speed of light minus our speed as we move away from the approaching photons? wrong! . . . it should, but it doesn't. The speed of the photons still measures 186,000 miles per second. Two American physicists, Albert Michelson and Edward Morley . . . completed an experiment (in the late 1800s) which seems to show that the speed of light is constant, regardless of the state of motion of the observer. (Zukav 1979, pp. 149–151)

This crucial experiment contradicted most of the previously closely held theories of physics of the day, but it eventually led to important modifications to all our theories about the universe. (One of those modifications resulted in the mixed blessing of atomic energy.) Yet in spite of the disconfirmation of Newtonian physics, the theories still are usable (good enough) on a macro level. They are now considered to be a subset of the larger theories of modern physics that this experiment showed existed. In another example, there was a tremendous uproar caused by Darwin's theories of evolution when they were first published. Darwinian theory conflicted with closely held religious theories of the Victorian era (in some cases, even today, the uproar has still not died completely away). The unwillingness to change the closely held theories that we carry into the learning process can be a difficult psychological stumbling block.

It is especially difficult when there is a move from technical expert into management because the technical knowledge which caused our promotion, is based in apparent logic and consistency. But now the job has changed radically from being primarily technically oriented to being people oriented. The "management" part of the new title requires that the past sources of success may not be enough for future success. We have to be a bit less structured (less rational, if you will) in order to deal with the emotional inputs of people. The closely held theories that may have seemed to be right for you because they led to your present position as a manager must become more fluid as you become more attuned to people. This new theory has a larger framework that now includes both people and Mother Nature. The ability to change one's personal theories as new data is received is necessary to learn the art of technical management.

Michelson and Morley showed that Newton's ideas were a subset of larger theories, but they also showed that larger theories existed. Newton's laws still worked, but within a more limited context. Changing closely held personal theories is therefore recommended as you learn or as the situation changes, but it doesn't always require completely ignoring previous successes. Consider that those successes will be only a subset of a new and larger theory and there always has to be one. With no theory of your own, any management action may be as correct as any other, and that rarely is satisfactory. Random answers rarely teach you anything.

For example, assume that some type of change or improvement is proposed and a person affected by it says, "But that won't work here because we're different." That "difference" may be due to a very limited theory held by the person making the statement. Possibly there is a more general theory that supports some common understanding upon which we all could agree. Otherwise, we could never learn to change and apply management lessons from one situation to another. Children often react either by resisting change or else by accepting every possible change because they have little experience or personal theory to work from. Everything is entirely new to them. We, on the other hand, must always have some minimal common ground from which we can gather data to expand our personal theory. Maybe an idea didn't work before because the situation at that time was different.

5.0 DEFINITIONS

Our process model begins with definitions. Some of the classical definitions of management that appear in the texts are not very specific.

Management is defined here as the accomplishment of desired objectives by establishing an environment favorable to performance by people operating in organized groups. Each of the managerial functions . . . is analyzed and described. (Koontz and O'Donnell 1964, p. 1)

. . . management can be defined as a process, that is, a series of actions, activities or operations which can lead to some end. (Gibson, et al. 1976, p. 30)

Management can be defined as involving the coordination and integration of all resources (both human and technical) to accomplish various specific results. (Scanlon and Keys 1979, p. 7)

The common ideas seem to be that management is a process that results in some positive accomplishment and that is associated with a group of people or an organization. These definitions do not show a direct relationship between the management process itself and either specific organizations or specific jobs within organizations. One might assume from them that management is independent of the situation.

These definitions provide no directions for personal cognitive or behavioral adaptation. Therefore, they appear to be universally applicable, except that we now know that can't really be so. Company environments differ. Technical management in a high tech computer manufacturing company is not the same as in an insurance company. And jobs within companies also vary. The president of the high tech corporation is a manager and so is the chief engineer but they have very different jobs and responsibilities. Obviously we need better guidance to begin the development of a personal theory. We need a definition that *connects the manager with the situation*.

I suggest that an initial definition of management is that it is a *decision-making* process that is an inferred mental procedure that is goal oriented, within an organizational framework, applied primarily to modify the behavior of people in the organization. It is also nonrepetitive and is applied differently for different jobs by absorbing uncertainty.

This defines the What is it? and the How is it done? of management. Although at this point, the definition doesn't differentiate among companies, it does do so for different jobs within any company. It proposes that uncertainty is directly related to the position. Because the actual process of decision making is unseen, it happens in the manager's head. It is only the result that can be observed and even these are interpreted by the observer.

5.1 Inferred Mental Process

For example, the Chief Technical Officer (C.T.O.) in the "perfect" organization spends the day "thinking" about data received from the rest of the organization. She or he then makes a decision and announces it.

"The technical group will set up a separate organization consisting of project operations. It will have its own personnel, policies and facilities, thereby ensuring a stream of new and improved "widgets" that should cost 25% less to produce and be made in three colors; Blue, Red and Green to suit our world-wide customer preferences. The initial designs have to ready in ten months."

The C.T.O. then requests the rest of the organization to produce the detailed development plan within one month to get that decision implemented.

In this "perfect" situation, it is impossible to determine how that C.T.O. made that decision since we can't access those mental processes.

5.2 Goal Oriented

All decisions have to have some tangible goal, otherwise they become generalities that nobody pays any attention to. In this example, the goal of that decision was to reduce costs by 25% and increase market share through product diversification into three colors.

5.3 Within an Organizational Framework

The technical group has been given this assignment. It is understood that this group manager and the team managers will provide a plan to implement within the ten month schedule or else provide an acceptable alternative.

5.4 Applied Primarily to People in the Organizations

It is pointless to make decisions about science since it has a fixed, unchanging basis. (Science doesn't change, just our perceptions about it does. Remember Galileo and Darwin.) Of course, science is used by people in the company but making decisions, for example, that gravity will no longer apply on the Earth's surface doesn't affect the force of gravity. Only those decisions that affect the people in the organization can be applied.

5.5 Nonrepetitive

It's a waste of resources for mangers to make repetitive decisions. That is the reason for developing "standard operating procedures" for basically repetitive decisions that have become successful in the past or those required by law. These kinds of internal decisions make up the "decision matrix" (see Figure 2-4). On the other hand, examples of repetitive decisions that deal with the external environment are found in the Information Systems surrounding our organizational model. See Figure 2-1. Typically, these could be the processes to purchase materials, hire personnel, and sell company products. We issue purchase orders, personnel requisitions, or sales documents. After those processes are developed, they are documented so the manager doesn't have to repeat them each time. [The organizational model (Figure 2-1) is noted here merely for continuity. It is dealt with in more detail in following chapters. Briefly, this model suggests that the total organization can be represented by an inner core of three interrelated parts: people, structure, and technology. This inner core is maintained and supported by the nonrepetitive decisions of the manager-leader. This maintenance-supportive set of decisions is in turn buffered from the organization's external

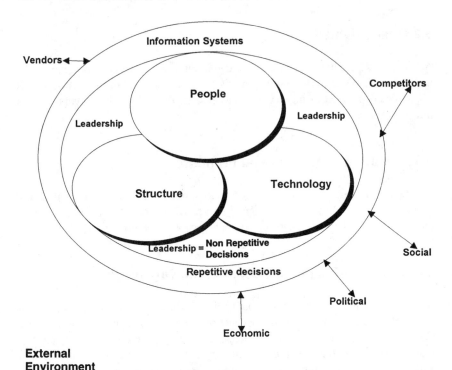

**External
Environment**

FIGURE 2-1: Organizational Model

environment by a set of external policies-procedures or repetitive decisions. More about this later; this chapter deals with *you*. The following chapters deal with the *situation* and that's where we describe/prescribe the organizational model.]

5.6 By Absorbing Uncertainty

In our "perfect" organization, that CTO makes very few decisions in any year, but when they are made, they shake the entire organization. See Figure 2-2. Those decisions should always be unique and concerned with predicted future events. Although there may be inputs such as recommendations from the staff upon which to base a decision, the responsibility stops at C.T.O.'s desk.

The uniqueness of the problem to be solved, its magnitude, and the period of time in the future with which it is concerned should always be related to the level of management in the organization. Even though the ideas supporting differential decision making (i.e., absorbing uncertainty) are the same throughout the company, the *amount of uncertainty absorbed* is supposed to vary directly with the management level. The higher the level, the higher the amount of uncertainty absorbed.

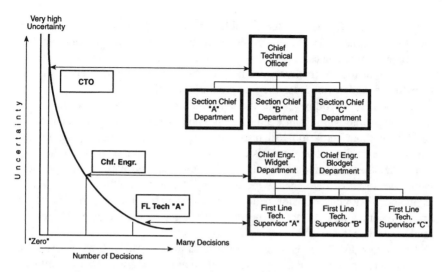

FIGURE 2-2: Organizational Position vs. Decision Making and "The Job"

Consider the decision making of the First Line Technical supervisor for the "A" line of products (FLT "A") in the shop for that same "perfect" organization. See Figure 2-2. He or she also arrives at the office each morning and also spends the time thinking. But that thinking period is usually quite short. Decisions are made quickly because they are concerned with less momentous situations and shorter time periods in the future and therefore have less of an impact on the total company. Instead of deciding how to restructure the entire technical group, there is the problem of several vendor components that may not be exactly to specifications. In order to minimize delay in production, the decision must be promptly made to reject, repair, or replace those components. If this situation reoccurs, the manager may decide to document the prior decision to guide subordinates. And so it goes all day long—decision after decision.

In both examples, we have eliminated the repetitive decisions that plague managers in less than perfect organizations and allowed both the CTO and the FLT "A" to concentrate on the peculiarly human skill they both have, that of making decisions by absorbing uncertainty.

It is now appropriate to define the term uncertainty, because it is at the core of decision making (which in turn is the reason for management to exist). Decisions can be classified under three general categories (modified from March and Simon 1958, p. 137).

Certainty: There is complete and accurate knowledge of the consequences that will follow on each alternative available to the decision maker.

Risk: There is correct knowledge of a probability distribution of the consequences of each alternative and the decision maker has that knowledge.

Uncertainty: There is a set of consequences for each alternative which belongs to some lesser set within all the possible consequences, but it is impossible for the decision maker to assign definite probabilities to the occurrences of the particular consequences. The probability distribution is there, but the decision maker cannot use it. If there were no probability distribution, decisions could be made at random, because they all would be equally acceptable.

Decisions must follow one of the above criteria. They cannot be made randomly by people—in fact, it's almost impossible for anybody to make a random decision without some tool, such as a coin to be flipped. We're programmed with memories that influence decisions and keep them from being truly random.

On the other hand, if effective decisions could really be made randomly, there would be no need for managers because any decision would be equally likely as another. With any decision being equally acceptable, there is no personal or even any other kind of theory needed, just a set of honest dice. Therefore, random decisions are out.

One major difference between the two examples above of the CTO and the FLT supervisor is the amount of uncertainty that each absorbs in order to make decisions. Another is the frequency of those decisions. Considering the curve shown next to the organization chart in Figure 2-2, it seems reasonable that there is an inverse relationship between uncertainty and the frequency of decisions. In my opinion, it would probably be too simplistic to suggest that this relationship is linear. Organizations are much too complex for that. That's why I feel that the curve becomes steeper as it approaches the top levels of the organization. I suggest this: The higher the organizational position, the fewer the decisions that are made, but the greater effect they have.

According to this model, the values of Uncertainty (vertical axis) versus Number of Decisions (horizontal axis) would probably decrease exponentially from left to right (i.e., minimum decisions at the maximum uncertainty to minimum uncertainty at the maximum number of decisions).

As noted before, completely maximum uncertainty might be treated as randomness, in which case almost any decision would be equally correct. Thus, there is an assumption that both randomness (randomness means that when $x = 0$, $y = $ infinity) and complete certainty ($x = $ infinity and $y = 0$) are not in our type of decision making. The CTO is required to make a few decisions, and the FLT makes many minor decisions. (See the curve in Figure 2-2.)

6.0 TOOLS FOR DECISION MAKING: CERTAINTY AND RISK

Management decision making, as discussed before, can take place under conditions of certainty, risk, and uncertainty. Making decisions under certainty is a fairly straightforward process. A completely rational path in that process would be as follows:

> In the certainty decision model the decision between alternate strategies is relatively obvious—namely select the strategy whose payoff is largest or smallest depending

upon whether the decision-maker is maximizing or minimizing. (Archer 1967, p. 455)

This is self-evident because all the decision maker has to do is to evaluate the available information (which is assumed to be complete and available) and select the optimum alternative. It's quite logical and consistent, similar to a computer program or the correct solving of a mathematics problem. Decision making under conditions of certainty does not require management. It needs only a formula. Complete rationality works invariably only with Mother Nature. But there is a process that can minimize unknowns, if not eliminating all of them.

1. Investigate the situation: Define the problem in terms of the organizational objectives. As an example, exactly what is the difference between the desired end result and the end result that will occur or has occurred right now? What are the underlying causes that make this decision important? Causes may not be obvious. They are similar to symptoms of an illness. The problem is not that the patient has chills, aching bones and a loss of appetite—those are symptoms. The underlying disease or cause (such as influenza) has to be diagnosed.

2. Identify the objectives of any decision and develop alternatives: What would an effective solution or decision be? Is there more than one alternative? Is there an acceptable method to compare these alternatives? What advantages would the selected decision have over all others?

3. Implement and monitor: Giving directions is not sufficient in implementing decisions. Procedures for measurement of progress have to be established and responsibilities fixed with specific individuals. Occasionally there is a tendency to feel that the decision that was made is being carried out well, but every decision is subject to re-examination as implementation progresses. Typical questions are:

Has this proven to be the right decision based on what we know now?

Is it being implemented correctly as it had been planned to be?

Are the budgets and schedules being followed?

There are some quantitative tools that can also be used to minimize the unknowns. Some of these tools have been described in the literature on operations research. For example,

Algorithms and heuristic processes; An algorithm is a process for solving a problem which guarantees a solution in a finite number of steps if the problem has a solution. (Example: finding the maximum of a function for which the equation is known. Take the first derivative, set it to zero, solve for x, etc.). . . .

A heuristic is a process for solving a problem which may aid in the solution, but offers no guarantee of doing so. . . . A familiar and widely employed heuristic is the use of analogy. Look for an analogy between the situation with which you are attempting to deal and some other situation with which you have successfully dealt in the past.

> A Means - End Analysis compares what you have with what you wish to obtain; identify a difference between the two; find and carry out an operation which may reduce the difference; repeat the procedure until the problem is solved. (Taylor 1965, p. 73)

Algorithms are extremely useful in relatively low-level organizational decision making, where the rules are clear and unchanging. In those very rare instances when this does occur, decisions are made under certainty conditions, almost as if they were programmed into a computer program. For example: Is George entitled to participate in the corporate profit-sharing plan or is he not? The profit-sharing algorithm has everything that is needed for this decision, such as the regulations on length of service, compensation, and the other relevant considerations. These types of decisions, with very limited circumstances, use algorithms best. Heuristics are helpful, but they are not guaranteed to work. They are not algorithms.

Decision making under conditions of risk is a bit different than using algorithms.

> Under risk . . . review the expected value or the sum of the payoff each multiplied by its respective probability of occurrence. . . . Select the strategy that optimizes expected values (Archer 1967, p. 455)

This process requires us to gather more information about expected frequencies in addition to expected values. It is easy to describe: just determine what the frequency of the occurrence or answers will be, multiply each frequency by its expected value to the decision maker, and choose the best answer.

But this process can never be completed because either the frequency or the expected answers are usually missing. When that happens, the missing data is filled in by the decision maker and as that data is human based, it relies partially on human emotions and judgments. It is therefore less exact, because the required objectivity of an answer is diminished, if not eliminated entirely. We will get a range of answers and we're not sure which is absolutely correct. We define the expected values and the frequencies of their occurrence subjectively. Even though calculating those expected values of the decision under various conditions of risk is more complicated than determining a single value under conditions of certainty, we can do it. But we must recognize that we have included a subjective, nonrational, nonmeasurable mental evaluation process as part of the calculation of risk. It's "This is what I think the probabilities are," rather than "This is the right answer."

But even then, the answers do not spring forth in their complete and full beauty. The reasons for this limitation lie in the fallibility of the human thought processes. We do not have the mental ability to search our mental and physical environment totally and provide the complete data that exists. We just cannot perceive or think about it all. Without this optimum data input, it is impossible to get an optimum output. There is probably another and maybe better answer out there somewhere, but we rarely find it. We don't know where it is and, even if we knew where it was, we probably would not have the time to look for it. Because our decision making is usually based on the data provided through human relationships (and they are somewhat nonrational), our search for the mathematical perfection of the definite and optimum answer is bound to fail when deciding under risk conditions.

Furthermore the limitations of our senses to receive and process data prevent our being aware of all the factors that could be relevant to our particular problem of the moment.

As an example, consider the oil company forecasters of the early 1970s who were aware of the repeated demands of the raw petroleum producers in the Mideast for increased crude oil prices but were unable (or unwilling?) to think about the consequences of a possible concerted action to raise prices on a world-wide basis. The problem is not that some managers don't try to optimize, but that they are limited and must be satisfied with a less than best answer to temporarily solve the problem. This type of decision making is called "satisficing" rather than optimizing. This was first pointed this out many years ago. For example, this was defined as

> The theory of intended and bounded rationality, of the behavior of human beings who satisfice because they have not the wits to maximize. (Simon 1965, p. 13)

More recent research amplified this important idea.

> Human information processing carries with it four major consequences. First, the individual's perception of information is not comprehensive but selective. Second, people tend to process information sequentially because they can only integrate small portions simultaneously. Third, one's memory depends upon operations that simplify judgmental tasks and reduce intellectual effort. Finally, and most importantly, overall human memory capacity is severely limited. . . . People attach more weight and importance to information and data that they consider to be causally related to a target outcome. (Gattiker 1990, p. 35)

6.1 Decision Making with Incomplete Data

Another reason for the failure of a search for mathematical certainty in decision making is intuitively obvious when we make decisions about other people. It is evident that we can never know exactly what others are thinking and therefore can never totally predict their mental processes and consequent reactions when that decision concerns them. About the best that can be done is to observe the other's behavior and (depending upon our own criteria) infer what that behavior means in terms of that other person's thinking processes. In other words, because we never really know what's happening in somebody else's head, we use our own interpretations of what we see. And because what we perceive comes partially from our own thinking and past experience, we may be somewhat "wrong" in our interpretations of the behavior observed.

We do know, however, that decisions are defined within an organizational framework by two sets of data (see Figure 2-3):

1. The hierarchical position that provides the formal authority and responsibility, and

2. The Personal attributes of the decision maker that provides the informal support for the decisions.

One obvious problem with our inference is that we don't know how much of the other person's behavior results from the present circumstances, from past circumstances remembered, or from the person's expectations of the future. Present behavior is always affected by these three elements. A valued future reward by that other person can spur positive working behavior, just as an evaluation that there will either be no positive reward or negative future consequences can slow it down or even stop it. Because our observations do not supply all the reasons for the other person's behavior, our inferences are limited. However, even though we never have all the data about other people, we still must make decisions.

There are other interpretations of decision making in the literature. Drucker (1973) gives us these two: A decision is a judgment. It is a choice between alternatives. It is rarely a choice between "right" and "wrong." It is at best a choice between "almost right" and "almost wrong." (p. 470) People do not start out with a search for facts They start out with an opinion. (p. 471)

These are not the rational processes of the physical sciences. But why do these descriptions of decision making, which include both nonrational and emotional

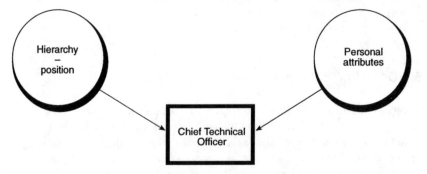

FIGURE 2-3: Inputs to the Decision Position and Personal Attributes

thinking resulting in "satisficing," rather than optimizing, seem to be simultaneously correct and incorrect? It is because they refer to a type of decision making that is really the central concern that we have as technical managers: that of uncertainty in which no one answer or even range of answers can be completely evaluated, analyzed, and used objectively to optimize the decision making process.

But decisions have to be made and we often base them on data that are (a) inadequate, (b) inaccurate, and even (c) nonexistent. We always do this. We've been doing that all our lives. What we do to minimize this "data" problem is create information or extrapolate from experience. Examples are all around us; we select university programs, marry, have children, take or leave a job and make many other major life decisions without all the data. If all the data were available beforehand in organizational decision making, a computer could replace managers.

Even though we have discussed "satisficing" as if it were a single process, it covers many different processes, (perhaps as many as there are managers), and untangling them to select the best process for ourselves is a very difficult task. There is no one "best" specific decision-making process, but fortunately the overall processes of decision making are the same as those of problem solving ". . .processes important in problem solving are also important in decision making and creative thinking" (Taylor 1965, p. 48).

This simplifies the selection of our decision-making processes. Novel decisions and novel problem solutions that involve other people require creativity. Problem solving (and decision making) in technical organizations includes both rational and nonrational (or undefinable) processes, and one of those processes is very familiar to us: intuition.

Using one's intuition means making a choice that relies on unconscious processes that have been modified by experience. "These processes (which are probably non rational since they are subjective) cannot be used easily without an established frame of reference" (Levinson 1981, p. 18).

If it is possible to develop an established frame of reference for intuition, that same frame of reference may let us handle other, less accessible, processes as well. For example, one researcher suggests that using variables such as speed and multiple inputs in decision making (using more intuition) is preferable to going slowly.

. . . slow decision makers become bogged down by the fruitless search for information, excessive development of alternatives, and paralysis in the face of conflict and uncertainty.

. . . Slow decision makers rely on planning and futuristic information. They spend time tracking the likely path of technologies, markets or competitor actions, and then develop plans. . . . the current fast decision makers look to real time information—information about current operations and current environment which is reported with little or no time lag.

. . . They prefer indicators such as bookings, backlog, margins, engineering mile-stones, cash, scrap, and work in process to more refined, accounting-based indica-tors such as profitability.

. . . fast decision makers keep the key financial managers close to operations, and not in a watch-dog, staff role.

. . . fast decision makers emphasize frequent operational meetings—2 or 3 meetings per week are not unusual. . . . these meetings cover "what's happening" with sales, engineering schedules, releases, or whatever comprises the critical operating information of the organization (p. 41.

Research on the development of intuition suggests that the basis of intuition is experience.

. . . managers who track real time information are actually developing their intu-ition. . . . they can then react quickly and accurately to changing events. (p. 45) (Eisenhardt 1990)

In addition to speed and multiple inputs to support fast, accurate decision making, the following steps can be helpful as a beginning of your decision-making framework.

1. Look for an analogy between the situation with which you are attempting to deal and some other situation with which you have dealt in the past.

2. Set up a self-developed process that will support repetition of your past success in the present and in the future, e.g., a document that outlines what happened.

3. Continually revise the framework as both the situation and you change.

This may sound like practicing until you get it right, but that would not be entirely correct. Practice does not make perfect: *corrected* practice does. Cor-rected practice begins with some type of ongoing development of the unique, personal management practices that matches perceptions of and expectations with the responses that the situation produces. Developing a continuously modified decision-making process is always based on your changing your hypotheses as better ways to make decisions begin to appear.

Define the situation. If it is in your experience or mental data bank, possibly there are elements that are recognizable and useful. With these as bases, develop and modify your management hypotheses and processes as the test data come in. That adaptation should be in accordance with successful practice, but the world being what it is, it might also be modified by some less than successful practice.

A creative executive must be able to plan a program, try it out, tear it apart and then start over again before settling on a system that works. Many executives feel such pressures to get things done that few examine closely what they are actually doing. They do not see the implications of their actions and often prematurely close on an incomplete system which must be redone later. (Levinson 1981, p. 221).

Assume the following:

- You have avoided the trap of being blind to the implications of your choice and actions before you act.
- You have been able to build a preliminary decision-making framework against which you will measure the results of your actions.
- You are aware that you cannot emulate other situations exactly but you can use your observations of these other situations to contribute to the changing of your own framework.
- You can evaluate the data you receive from testing your hypotheses and modify your processes. In effect, all the data received from such things as on-the-job experiences, nonrational thought processes, and learning through cognitive change (e.g., reading books) can become just another part of personal development and growth in your management job.

You then have a foundation for management success. Usually scientists, engineers, or technicians enter the organization in jobs that are fairly well defined and therefore need objective, logical, repetitive thinking. (I remember that one of the first jobs I had as a neophyte engineer in training was to check material sheets to be sure that nothing was missing and all the parts were appropriately listed. This sort of job is almost obsolete today, because computers do it so much better than human beings. It emphasized objectivity, logic, and painstaking accuracy. I hated it.)

Becoming a successful manager requires a different approach. There is less need for the mind-deadening logic of some repetitive technical jobs. The approach now is primarily intended to support innovation, improvement, and change. Our new organizational tasks require individuality. The approach is therefore more emotional and creative (i.e., nonrational) and we in turn must develop personal management models that include those aspects if we are to optimize our management decision making.

6.2 Decision Making by Absorbing Uncertainty

There is a typical process of making decisions by absorbing uncertainty that can be another starting point for the creation of your own framework-theory-hypotheses. It's not the only one but it seems to be a reasonable start. In the "perfect" organization described before, uncertainty was a major variable that distinguished the various levels of organizational responsibility. Maximum uncertainty resided at the top and diminishing levels of uncertainty followed downward with the decreases in organizational responsibility. Uncertainty occurred on every level because the decision maker at every level could not assign a complete probability to all the alternatives that existed. It was impossible to perceive them all as the information needed for evaluation might not even be available, be incomplete, or even be wrong. In most cases, the decision maker accepted a nonrational choice to stop searching for more data and was "satisfied" with what was available. Obviously, we cannot deal here with any specific individual's reasoning in decision making because of people's variability and the limitations of this book, but it is suggested (and prescribed) that you use this less than perfect but still practical general approach as a beginning. You can adapt it later to fit your specific requirements.

One initial step is to categorize all problems requiring decisions into repetitive and nonrepetitive classifications after you have decided that the problem is your responsibility. The repetitive problems can be disposed of quickly under conditions of certainty. With repetitive problems there should be some type of organizational decision matrix that produces an answer every time, as with decision making under conditions of certainty using algorithms. These matrixes exist in every company, although not in the same detail. Typically, they might include such decisions as employee benefits, insurance programs, design standards, or corporate operating procedures. Creating them is one of management's first responsibilities, as they represent major decisions that have dealt with prior uncertainties. They have now become decisions under conditions of certainty. Sometimes they're called company policy. These repetitive decisions are the answers that can be produced by algorithms (e.g., a computer program). As noted before, company policy, in a general sense, includes the decision matrix (the internal decisions) and information systems (the connections to the external environment).

The next task of management is handling nonrepetitive problems under all other conditions. In this model, both risk and uncertainty can be subsumed under the heading of uncertainty because the assigning of an array of possible values in decision making under risk seems to involve many of the same nonrational choices as making decisions under conditions of uncertainty. The dividing line between risk and uncertainty rarely exists outside of textbooks. Before discussing selection of a personal theory later on in this book, you might consider the following prescription as another alternative to use in decision making. See Figure 2-4.

1. Determine if you have (or even wish to assume) the responsibility to make a decision. If it has not been made before, it seems to be a major decision, and you cannot make it, then push it upward to the next level of responsibility since you have determined that the problem is a symptom of a potentially negative situation that you believe can be solved or avoided without your intervention.

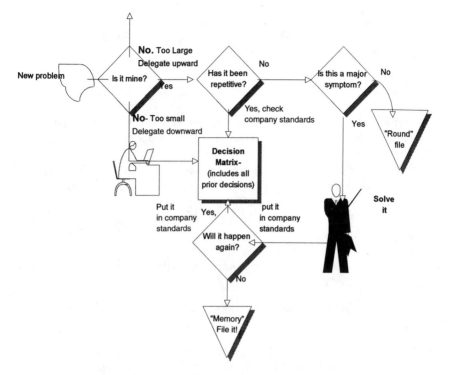

FIGURE 2-4: Suggested Decision Process

If that situation is likely, delegate it to someone else at a higher management level within your area.

If the problem is minor in your opinion and can be dealt with by a trusted subordinate, delegate it downward. That delegation should result in a decision that might occur again in the future. This means the result should be included into the "decision matrix."

The "decision matrix" includes the formal predetermined decisions such as standard operating procedures, company policies, and formal organization charts in addition to all the informal or cultural operations that the company uses. (In my opinion, it doesn't take as long to learn the company's "culture" as it does to learn the formal procedures.)

However, if you can or wish to accept the responsibility, go to Step 2, because you have now arrived at the point in the decision-making process that requires your efforts.

2.   Next, *decide* if it is a repetitive or nonrepetitive problem. If it is repetitive, find the similar decision-solution in the "decision matrix." If this is done, stop and let the matrix provide the answers (don't keep solving repetitive problems).

When it cannot be delegated, but you believe that it is not important at the moment, forget it. (Sometimes decisions just go away.) Finally, if it seems to require your intervention, go to Step 3.

3. Then *absorb* the uncertainty required at your level of authority/responsibility and solve the problem or make the decision. When the problem is solved (i.e., the decision is made), the next point is to decide if it will happen again in the future. If so, document the decision and have it included in the company "decision matrix." If not, the solution just becomes part of your experience.

Now you are ready to test the next problem using this general method for decision making. Even with no personal theory adopted yet, it is possible to test this process using part of the scientific method.

1. Establish your hypothesis.

2. Make your decision with this process (logically, intuitively, or any other consistent way that you feel fits the situation).

3. Determine how closely the end results of that decision matched your predictions.

4. If the variance between predicted and actual data is within your tolerances, stop. Why modify something that seems to be working well enough? Paraphrasing an old New England saying, "If it's not broken, don't fix it." On the other hand, if the variance exceeds your tolerances and those variance limits are still valid, modify your process-framework-theory and hypotheses, beginning with the largest variances first, and try again.

This proposed problem solving (or decision-making) process assumes that there is no one better equipped to solve that particular problem than the person to whom it is finally assigned in the organization. In other words, the amount of uncertainty depends in part on the decision maker's organizational level and how much uncertainty that the decision maker is willing, or able, to absorb. The initial decision was to determine within whose area of uncertainty the problem rightly belonged. That obvious statement does not provide objective guidelines for absorbing uncertainty, but there are interesting, and equally obvious, statements that might assist in setting these guidelines.

For example, no one in the organization knows more about your responsibilities and your job than you do. You're the immediate expert on your job and you always will be because you're the one occupying it! Organization charts and job descriptions cover no more than an insignificant part of any management job, and in many organizations, the basic reason the structure even functions at all is because managers (and, in fact, all participants) do things they feel need to be done. If they did only what they were supposed to do, the company would probably stop. These extra tasks are never covered by the job description.

In most cases, the formal organization is not even aware of these actions, but they are necessary actions. Every technical organization would quickly come to a grinding halt if participants did *only* what they were told to do. Therefore, no person in the organization is better qualified than the manager to determine when, and if, the organizational level requires her or him to make a decision. If the decisions are made in accordance with your own prescriptive model, problems will tend to be placed at the lowest organizational level for solution. And that's the optimum.

6.2.1 Step 1: Uncertainty and Delegation.

For example, if a lower-level manager finds that he or she does not have the resources, or the authority, to solve a problem at that level,the problem should be pushed one level upward *with a proposed solution* (delegate upward). ("I can't solve this because it involves areas beyond my responsibilities but if given these additional resources, this what I propose.") The next upper-level manager must either then absorb uncertainty and make a decision or decide that it is also beyond his or her responsibilities, in which case it is sent upward another level, and so on until it is finally resolved. Conversely, if the problem can be delegated downward, that should be done, but the delegation must include the resources or authority for a successful implementation of the eventual solution.

There are, of course, difficulties in learning how to use this prescriptive decision-making process because it includes many nonrational concepts that are not easily described. Intuition (based on personal experience) may be one of those processes because you have developed a method of corrected practice that supports improvement of that intuition. Documentation is important for this corrected practice. For example, it might be useful to jot down all the reasons for supporting a particular decision (i.e., documenting the data behind your limited hypothesis). When the results come back, check them against your predictions and correct for the next time. If this sounds like the previous prescription for using the scientific model for personal decision making, you're right. As noted before, it is not practice that makes perfect, but corrected practice. In many cases, intuition can be improved as practical experience supports it.

6.2.1.1 Practical tips on delegation. Returning to the decision-making model and the process of absorbing uncertainty, there are several operational questions that may come up. For example:

How can problems be delegated upward? That might involve a risk to my position as a manager.

Pushing problems upward because they are beyond your capacity to absorb the uncertainty is relatively straightforward provided the rules for doing so are well defined beforehand. Those rules should be defined and discussed with upper management levels before they are implemented. Start by looking for similarities between particular past kinds of problems and their solutions and the present situation. How were the problems solved successfully in the past? What kinds of extrapolations or modifications could be accepted? As the expert in your particular management job, you are best qualified to determine your level of uncertainty and to suggest when that level has been exceeded. This should result in uncertainty guidelines. Developing these guidelines with upper management beforehand tends to minimize surprises, which are generally regarded unfavorably. Always delegate upward *with a recommended solution*. Usually that recommendation will include an extension of your authority or resources.

When appropriate suggestions for decision making accompany a problem upward, only two alternatives are possible: acceptance or rejection. If your suggestions are accepted, you have either explicitly or implicitly been given the responsibility to make similar decisions. In that case, your level of authority and consequent uncertainty have been raised. If those suggestions are rejected or returned to you for modification, you have a better concept of the limits of uncertainty (and the kinds of decisions that will be made) at the next upper level of management. Because you will be modifying your personal management processes as you receive new data (such as the data on the level of uncertainty), your processes will be that much better next time. The process is not one of taking a risk but of learning how to manage both yourself and others, including the boss. Delegating upward was quite straightforward. The decision exceeded your level of uncertainty. You provided a solution and requested approval.

How do I know when and how much to delegate downward?

Delegating downward is more difficult because you are responsible for that delegation even though a subordinate has been given the job. Delegating downward is not the same as abdicating. When delegating downward, you are relinquishing the authority to get the job done. Authority is digital. You either have it or you don't. Therefore, when a task is delegated, the authority that goes with it is gone. Nevertheless, the responsibility for getting that delegated task completed is not digital; it is infinitely expandable. Even though you have delegated the responsibility, you still have the same amount! Therefore, delegating downward is more complex and must be measured.

6.2.1.2 Measuring delegation. The measurement process can be fairly straightforward. As you are the delegator, you are the one who has to determine how much to delegate.

That delegation process is measured by the amount of error or mistakes that you estimate you can accept if those errors are made by the person to whom you delegate.

Remember, as the delegator, you are still responsible for the end results and you can be held liable for any errors in those results. The amount of delegation is therefore dependent upon the amount of time and resources that might be wasted if the delegated task is not completed correctly. In some ways, it is similar to calculating the insurance premium that increases with a greater potential loss. The greater the amount of time between your act of delegation and the feedback or response from the person to whom you delegated the task, the greater the amount of delegation (assuming all other things equal, of course). The amount of delegation is not initially related to the overall size or complexity of the task being delegated but to the *frequency of feedback* on progress. That is what you, as the delegator, must establish when the job is delegated downward to some individual. When *you* think that uncertainty is high, you might want feedback every week. When *you* think it is low, a monthly or even quarterly report on progress might be satisfactory.

Another point is that you never delegate a task to a group, only to an individual. If a "committee" is assigned to a task, the delegate is really the chairperson of that committee.

There is another decision to consider: that of even deciding to delegate at all. The cost of the delegation process itself (which is measured by the time you took to explain what you want) must be less than the potential value of the job delegated. If it takes as much time to explain what you want to have done to someone as it does to get the job done, you either don't know how to express yourself well or else the job is too complex to delegate to someone else. As noted before, it is the similarity to the premium (i.e., time to explain) on an insurance policy, which must be less than the potential loss (the complexity of the job that the policy covers) if the task delegated is not taken care of as expected.

In other words, there must be a relationship between the amount of delegation and the potential managerial loss to you if the delegated task is not completed quickly and effectively. Defining this relationship in terms of time or cost is an excellent process to follow in order to develop your subjective (nonrational?) delegation standards. The definition process that you establish in order to decide about delegation doesn't have to be formalized. Documenting it with notes on a scratch pad can be quite effective. Learning may be achieved just as well from your notes in a daily personal log as from a formally developed corporate policy.

Although mentally measuring the potential loss to the delegator is difficult, it is easier to determine than the next decision: what to delegate.

You should delegate what you know.

Most of the time, as managers, we find the most difficult problems are those that we have not prepared for. It's the things we don't know, not the things we know, that can be bothersome. For example, if your expertise is in, say, "mechanical structures" and you have little knowledge of "electronics," and there are two jobs of equal size, cost and complexity to be delegated: one in "mechanical structures" and one in "electronics." I suggest that you delegate the "mechanical structures" job. The reason would be that you, as the delegator, could more easily detect small problems in the structures job than in the electronics task.

Therefore, your exposure to error would be less even though both tasks would be equal in size. In other words, the signal-to-noise ratio for problems that might happen in "mechanical structures" would be smaller than those in "electronics." This leads to the conclusion that you, as manager, should concentrate your efforts in areas that you are less familiar with. Then, you would always be getting more feedback and consequently would be learning about areas with which you are not immediately knowledgeable.

"What you know" also affects "who" you would delegate to. The longer that you associate with a subordinate, the more you know about that person. The more you know about that person's capabilities, the more you can delegate (or decide *not* to delegate). Consequently, there are two parts to what you know:

1. Your personal technical competence (e.g., structures vs. electronics)

2. Personal knowledge (who is best qualified in your opinion, as the delegator?).

For example, assume that the problem to be resolved involves accepting or rejecting a new vendor as qualified to supply your organization. If you recognize this as a repetitive problem, You follow Step 1 and determine if this your decision to make. If it is, the next step is to determine if there is an existing policy for this kind of decision. If a company policy (i.e., decision matrix) exists, follow it (remember, this problem is repetitive) and the decision will probably be automatic. If no matrix exists and the problem is yours to solve, now is a good time to consider designing a new decision matrix for vendor evaluations, and use this vendor to test it. If you are not responsible for this problem being solved (and remember, you are the only one who can really make this decision) but the problem does have to be solved, push it upward to the next level of management *with appropriate suggestions* for action. Typically this could be the design of a new vendor evaluation process.

Conversely, if this problem has not been solved before, you decide that it is important, and you feel that it is your responsibility, the next step is to determine who is to handle it. Do you want to tackle it yourself or delegate it?

If it is to be delegated, what is the potential loss to you, as the manager? This assists you in determining whether to assign this particular task to the neophyte quality engineer in your department or the departmental veteran who has handled many similar jobs well. Assuming that you were the expert on this vendor's products, you can easily determine small errors in judgment before they escalated into disasters and therefore might give the problem to the neophyte as a training exercise. On the other hand, if you have no expertise on this product, you give it to the veteran, but you follow up closely because you are facing greater uncertainty and therefore a greater potential loss (and you also want to learn something from observing how the veteran does it).

Some might argue that the veteran to whom you delegated it would be able to decrease the potential loss in delegation, but that is not relevant here unless your expertise is at least equal to that of the veteran. We are concerned here with the potential loss as measured by *your own level of technical knowledge*, not someone else's. As noted before, this method eventually teaches you about the areas with which you are less familiar because that's where you spend most of your time in follow-up. It is part of the never ending, self management education process.

Of course, this example is deliberately simplified to illustrate the decision-making process that is correctly matched to the level of uncertainty; but simplified or not, most of the time spent in managing can be wasted if either this or another type of procedure that you have developed because it is more applicable to your situation is not developed, corrected, and practiced continuously.

To emphasize another point about delegation, it is not often intuitively obvious but it is logical to delegate the greatest amount about the things you know best. As another example, the chief hydraulics designer should be able to delegate the hydraulics designs very easily when he or she is promoted to head of the mechanical design group. With his or her expertise in hydraulics, the potential loss in delegation is minimized for this area, since he or she can spot potential errors quickly when they are relatively minor. Therefore, in this example, there is a lower potential loss in hydraulics compared to other areas. It also follows that the less you know about a subject, the less should be delegated until you learn enough to be comfortable with it. Increasing the feedback schedule can increase your comfort level.

As noted in Chapter 1, many of us tend to try to maintain our technical expertise when we change management jobs, and that could be an error. There is the example of the sales manager who outsells many of the salespeople and is not doing the job of the sales manager. The new management job requires delegating almost all of the actual selling, concentrating instead on less familiar areas, such as developing better administrative procedures or communications between customers and internal order clerks.

6.2.2 Step 2: Is This Decision (or Solution) Repetitive?

Have you ever seen this problem before? Are there any policies, guidelines, or procedures that have been used before that can affect this decision? If so, consult the decision matrix, which includes both the formal organizational procedures and the informal ones. Sometimes an informal relationship with someone else, or searching the culture of the company, will supply an answer when there is nothing formally documented. If you decide that there is no precedent, the next step is determine if the problem is important to you. If in your opinion, it's not, put it on hold. If it is, solve it.

6.2.2.1 Which repetitive problems should be checked out first?   One mechanism is to keep a histogram or frequency distribution of problems on a

regular basis over a specific period of time. The x axis should number each separate class of problem or decision and the y axis the frequency of that problem or decision. Using a type of risk analysis, you should subjectively assign a potential loss factor to the frequency (i.e., how much that problem cost you each time it occurred). Following the classical description of risk analysis, the highest product of frequency and loss defines the highest priority, the next highest product determines the next priority, and so on. With this priority sequence in hand, you can determine if the problem is repetitive and what the cost would be to develop a decision matrix (e.g., design an algorithm for it). That cost would depend on who is to develop the algorithm, you or someone to whom you delegate it, and the expected development time.

6.2.3 Step 3: Is It Important?

This is perhaps the easiest step of all since you are the only one that can determine this answer.

Making decisions can be very difficult unless there is a process that you have developed to categorize them. Decision making is neither a random procedure nor is it a fixed, unvarying one. It is dependent upon how you view changes in the organizational situation. For example, if it is a repetitive problem, you might feel that you know generally how to handle it. Delegate it downward to someone (if there is someone available) with instructions about what you see as the differences between this problem and all the others already in the existing decision matrices. Then pass unusual or major nonrepetitive problems upward (with appropriate recommendations, of course). Now how would you handle the nonrepetitive problems that are left?

Having gotten this far, you have probably found that you are making fewer but more important decisions. Some problems just go away by themselves and you don't want to make a major contribution to a minor problem.

Problems that don't go away could be important. But because you are the expert in your job and are best qualified to absorb uncertainty at your level, now is the time (and I'll say it again) for combining your evaluations of successful management patterns of others in your organization with your reading and continuing education, developing your personal processes, setting up your hypotheses that forecast specific results, and solving the problem or making decisions in accordance with those hypotheses. Then measure the variance between the result and your prediction, and if they are acceptable, stop. If not, you have some data that does not agree with your hypotheses, so you either have to discard those data or modify the hypotheses for the next time. In any event, as there was no one better qualified to make the decision at the time, your existential management framework was the best that existed, and that framework can only get better as you modify it.

7.0 REVIEW AND PRACTICAL SURVIVAL TIPS

Management is an art, not a science. Therefore, it depends upon the "artists" interacting with the organizational situation. The primary purpose of management is make decisions, which are contingent upon the decision maker's organizational position, experience, and the organization's culture. Usually, the higher in the organizational framework, the fewer objective criteria exist upon which to base these decisions. It then follows that if the organization has a culture of pushing day-to-day decisions downward, that is a strong delegation process. The higher in the organization that the manager is, the fewer the decisions that should be made, but those fewer decisions will have a greater impact. The organizational level of the manager (or decision maker) is directly related to the amount of "uncertainty" that is absorbed in decision making. We need people as decision makers (managers) because people can make decisions with inadequate, inaccurate and missing data. As people, we "fill in" the missing pieces, as required. There are many obvious examples that can be taken from everyday life. The decisions to select a trade, a life partner, or a job are invariably made without having all the data. But these types of decisions are always made and as shown in Figure 2-3, they are always based on two sets of data: the situation (the job) and the decision maker's attributes.

People cannot make decisions at random. No matter how we try, there is always some hidden bias in our decisions that springs from our past history. Random decisions require some tool such as dice or flipping an honest coin. Therefore, all human decision making is based on some thought and some memory or past experience. One way to use that experience is to develop an initial decision-making mode that we can modify as we gain more data. Eventually, modification results in decision-making processes that are unique to the individual. One of the methods that can be used to start development of our own unique decision making process is the scientific model. Although it is very basic, it is a powerful tool. It assists decision making because it helps to clarify the "missing pieces" that we don't know about and then to create them. Human beings rarely have the time, resources, or capacity to gather and evaluate all the data upon which to base a decision. They therefore "satisfice" rather than optimize. Managing involves "satisficing." The scientific model helps you to learn from your experiences and begin to create the educational- and experience-based model that you have modified to suit your perception of the situation.

In other words, each of us is limited. We can make decisions by starting with our education and experience and use the scientific model to help us initially to solve problems. As we learn more, we modify that model or even build our own models to suit our own situations.

Now that we know that a manager is a decision maker and have suggested the relatively straightforward scientific model as a beginning decision-making model, in the next chapter we can analyze and build some general models (or theories) to help develop our thinking processes, upon which to base decision making.

8.0 "ONE-SENTENCE" SUGGESTIONS AND QUESTIONS

SUGGESTIONS

1. We always make decisions with inadequate, inaccurate and/or missing data, so don't worry about what you can't possibly know.
2. Managers are decision makers and should have a "model" or theory upon which to base their decisions. Randomness is not acceptable.
3. The decision-making "model" used is unique to the decision maker but it changes in response to changes in the situation.
4. Delegation is measured by the amount of "trouble" the delegator can get into.
5. Delegate what you know, downward.
6. Always delegate upward *with* an answer.

and finally:

All business proceeds on beliefs or judgements of probabilities; and not on certainties. (C. W. Eliot)

and remember it is unfortunate but true that if you only use experience as a teacher, the test comes first and the lessons afterward.

QUESTIONS

These questions are not intended to test your retention of the contents of this chapter but to help you to think about and assimilate any applicable ideas and apply them to your own situation.

1. Which of the conditions certainty, risk, or uncertainty dominated your decisions before you became a manager? Did anything change when you were promoted? Has anything changed recently? What is or was responsible for that change? What would you like to do about it? How? When? Why?
2. How many times have you made the same decision in the past six months? What was it? Why did you do that? How can you change the results, considering your present evaluation of your situation?

3. In terms of your past management strategy in your present situation, have you received any data that are intuitively reasonable and yet seem to conflict at present with your own and/or company policy? Why? What should you do now when data do not agree with existing frameworks? How?
4. Can you really refuse to make a decision because it is beyond the limits of your uncertainty? If so, how would you do it? If not, what are the constrictions upon you and should you do anything about them?
5. Can we define the limits of our uncertainty? How? Isn't that a contradiction, defining what we don't know?

———————————

Your answers will always be the right ones for you, so why not write them down? It's reasonable to suggest that behavior can change thinking, and just the taking of notes does help many of us to learn. Try putting the notes aside for a few days and then take another look at them. Would you want to change some of those notes? Why? How? What happened?

———————————

SUGGESTED ANSWERS TO CASE QUESTIONS

1. She started out well by reviewing the things that had to be done within the near future, but she did not stick to her plans. There was no valuing of the many problems that were dumped on her that day. She had not categorized the problems into those that had to be resolved immediately (for the meeting) and those that could be put off (such as the problem of the Appleby valves). She had not calculated her potential losses if she had delegated properly, nor set up her own decision process. See the first decision box: Is it mine? in Figure 2-4.
2. This seems to be a repetitive problem requiring a general answer. Possibilities include:
   (a) When engineering has accepted a product that fails to meet specifications twice, the prints shall automatically be changed through the issuance of an engineering change order and purchasing shall be required to negotiate a different price with the vendor.
   (b) All products that are rejected twice in sequence shall be removed from the accepted vendor list and a new source obtained by purchasing.
   (c) Fill in your own answer. The repetitive answer is not as relevant as the fact that an answer has to be given. See the decision box: Has it been repetitive? in Figure 2-4.

3. She is not handling uncertainty at all. On a more limited scale, she could have listed all her tasks of the day and categorized them as A (to be done first), B (to be done this week), and C (to be done when I have time). There are other ways. Have you any of your own?

4. In line with my suggestions for handling uncertainty, I would advise Leona to point out to George that the item was not on the agenda and (diplomatically, if possible) suggest that it be put on the next agenda so that everyone attending the meeting would be prepared. My feeling is that you do not have to know everything, and a functional or problem-solving meeting should be run in order to minimize uncertainty.

Do you agree with these answers? Do you think that you could use them in your own situations Why and how?

REFERENCES

Archer, S. H. "The Structure of Management Theory." **Readings in Organization Theory,** Hill, W. A. and Egan, D. (eds.). Allyn & Bacon, Boston, MA. 1967. pp. 440–466.

Cleland, D. I. and Kocaoglu, D. F. **Engineering Management.** McGraw Hill, New York. 1981

Drucker, P. F. **Management: Tasks, Responsibilities, Practices.** Harper & Row, New York. 1973.

Eisenhardt, K. M. "Speed and Strategic Choice: How Managers Accelerate Decision Making." **California Management Review** (Spring 1990):39–54.

Gattiker, U. E. **Technology Management in Organizations,** Sage, Newbury Park, CA. 1990.

Gibson, J. I.; Ivancevitch, J. M.; and Donnelly, J. H., Jr. **Organizations.** Business Publications, Dallas, TX. 1976.

Koontz, H., and O'Donnell, C. **Principles of Management** (3d ed.) McGraw Hill, New York. 1964.

Levinson, H. **Executive.** Harvard Univ. Press, Cambridge, MA. 1981.

March, J. G., and Simon, H. A. **Organizations.** J.Wiley & Sons, New York. 1958.

Scanlon, B., and Keys, J. B. **Management and Organizational Behavior.** John Wiley & Sons, New York. 1979.

Simon, H. A. **Administrative Behavior** (2d ed.). Free Press, New York. 1965.

Taylor, D. W. "Decision Making and Problem Solving." **Handbook of Organizations,** March, J. C. (ed.). Rand McNally, Chicago. 1965. pp. 48–86.

Zukav, G. **The Dancing Wu Li Masters.** William Morrow, New York. 1979.

———, **The Economist,** Jan 28, 1995, p. 25.

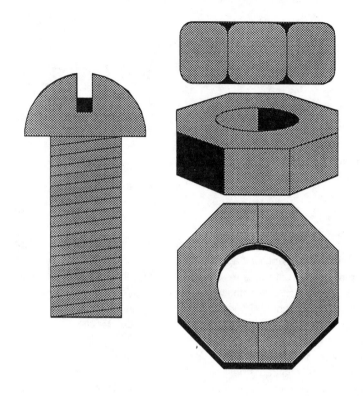

CHAPTER 3

Beginning: Building Your Own Model/Theory

Case Study:

The "Model" That Failed

Cast

Tony Neelson: Chief engineer, B.D. Machinery Corp.
Jon Koroton: Project engineer
Shelby Index: Electrical engineer/designer
Ed Masur: President

The sun emerged from a cloud and shone brightly through the blinds just as Tony Neelson entered his office early on Monday morning. He didn't notice it. He had spent the weekend at home reviewing everything that had happened at the company during the last six months when he had tried to improve technical operations by installing the latest management techniques and processes. He slumped into his chair and again began to review the situation. Maybe he'd think of some answers now.

He had joined the company, B.D. Machinery, seven years ago when he graduated as a mechanical engineer from one of the best technical universities in the country. His first job was as a junior engineer in the mechanical design department. At that time the company only had about ten employees. With time, the company grew and the engineering department expanded. There were always new and exciting projects to complete and Tony was an enthusiastic and prolific designer. Four years after joining B.D. Machinery, he was promoted to be the head of a much larger mechanical engineering department. The company then sent him for a month-long management training course at the local university.

After the course, Tony arranged his mechanical engineering department into small teams of designers to handle the various mechanical requirements of each of the new machines that the company developed. For over two years, this management model using small teams seemed to work well. The work was varied and exciting. Then business began to slack off. The company lost several large contracts because its quoted prices were too high. Rumors began to circulate about a potential layoff in the future. Tony's department could be affected.

Then tragedy struck the company. During a flight in the company plane to another state, the plane ran into a storm and crashed. The chief engineer was severely injured and hospitalized. The forecast was that he would be out for at least six months if not longer. A month after the crash, Tony was promoted to acting chief engineer of the company. Now, all the electrical, mechanical, and project engineers reported to him. Tony remembered the day, several months ago, when Ed Mazur called him into the front office to tell Tony he was being temporarily promoted.

Ed: Hello, Tony, come on in. I have some news for you. As you know, our chief engineer was hurt in the crash last week and we're not sure when he'll be able to come back. That's an important job for this company because whoever has it is responsible for coordinating the company's technical and design work. I've always said that engineering is the most important part of this company because there's more money made and lost during design than there ever is in sales or manufacturing. I'd like you to take that job over until he gets better. Will you do it?

Tony: Well, it's a big responsibility and I'm not sure that I can handle it.

Ed: Your mechanical design department seems to run very well. Your idea to organize the teams gets people to work together. The work itself is done well and on time. It may take a few weeks to get adjusted, but I feel confident that you can handle this temporary promotion. Who knows—if the recuperation is not as fast as we think or if he decides to really retire this time, because he's talked about it for awhile now, you could probably be the next chief engineer of B.D. Machinery.

Tony: Well, OK, I'll do the best I can.

Ed: Now that that's settled, there's something else I want to talk to you about. I'm sure that you know that we haven't been as successful in the past year as in prior years. Our competitors are undercutting us on price and delivery. Our customers are demanding higher quality levels. Our sales have been down and, as engineering is one of our strong points, I'd like you to think about restructuring the way we do our design and development work.

We have to get better. Our competitors are no longer in this state or even this country. We compete on a worldwide basis and we have to get smarter if we want to survive. You went to that management course awhile ago so you should have some ideas as to what to do. Your mechanical engineering department works well so that should be a help. We need to do something else on a company-wide basis and I'm counting on you to help.

Tony: What resources do I have?

Ed: Tony, you're a logical engineer, you can have whatever you need. How soon can I see your plans?

Tony: I'll have a proposal for you within four to six weeks.

Ed: Great! And congratulations on your promotion.

Tony presented his plan to Ed, as promised. He would develop cooperative teams to work across all the engineering groups. He had learned how to develop teams in his management course and installed them in his mechanical engineering department. It had worked well and he felt that extending those ideas to all the engineering groups would be just an extension of his past management successes. He would select a team leader for each new job. The team leader would select his staff from each department, supervise any purchasing, coordinate the manufacturing and test of the machine, and maintain contact with the customer throughout the whole process. Each team would have objectives such as cost, delivery, and customer satisfaction. According to the management course that Tony had attended, it was the best way to get things done. Ed agreed with everything Tony proposed.

Tony then called a meeting of all the group engineers. There were six people in that group. He explained his plan and told them that they would be team leaders rather than group engineers as before. They would become managers primarily and technical contributors secondarily. Then Jon Koroton, (a group engineer) raised his hand.

Jon: Tony, I like this idea but there's some problems you may not know about. You've been managing just one department. If we use your idea, as the team leader, I'll still have to answer all the questions and information requests that the customer throws at us. In addition, I'll have to be sure that the technical details are correct during preliminary design. Because I'll also be checking the electrical designs and the mechanical aspects, I just won't have time to supervise purchasing and keep an eye on manufacturing.

Tony: Jon, from now on, you have to start to think like a manager. Learn to delegate. Just pick your staff, give them direction, and help them when they need it. You're not supposed to oversee every last technical detail as you did before. Now, you have to be sure that other people do that job.

There were no further comments from the group. After they left his office, Tony issued the general procedures to be followed. He left the specifics to each team leader but he referred back to his management course quite a bit to be sure he hadn't missed anything.

After six months, unforeseen things began to appear. It wasn't like his management books at all. There were many arguments among the team leaders themselves and between various team leaders and the manager of electrical design and the newly promoted manager who took Tony's place as head of mechanical design. There were numerous complaints from all of them. For example, the design managers complained that the team leaders never provided the right information, or that it was late or it was changed after the departments had started work. Sometimes Tony felt as if he were in a debating club rather than a machinery manufacturing organization. The peculiar thing was that he really couldn't figure out how to solve the arguments. There were always different reasons. It all came together one Friday afternoon.

Shelby Index phoned him and asked for an appointment. She said that she was considering leaving the company and she wanted to talk to him about it. Tony was surprised. Shelby was an excellent engineer and she even had several patent applications to her credit since she had joined the company a year ago. She arrived at the appointed time.

Tony: Hi, Shelby, come on in. What's this all about?

Shelby: Tony, I can't work with Jon. Since he became a team leader under your new plans, there has been nothing but arguing between my boss and him. Jon tells me what and how to design. He tells me what my objectives are supposed to be and then when I start on his job, my boss tells me to work on something else that is a major problem. Naturally, I do what my boss wants. Later, Jon checks on my work and starts complaining that I'm not meeting the objectives. He tells my boss that I'm not doing what he wants me to do. My boss then has me go back on Jon's job until the next hot problem comes along. This back-and-forth business has me about worn out. Jon argues with me, and my boss blames me for not being cooperative. I'm ready to leave. Something has to be done—I'm not the only one who feels this way.

By the way, I've heard there's the same kind of problems between the team leaders and the head of purchasing and the manager of manufacturing. Nobody is sure of what they are doing. It's a real mess.

Tony: Shelby, let me look into this. Don't make any major decisions now. Try to leave things alone for the next few days and we'll talk again at the end of next week.

Tony didn't want Shelby to leave the company. Therefore, after Shelby left his office, Tony asked Jon to come into his office. Jon showed up in fifteen minutes.

Tony: Jon, how is the new team leader system coming along? Are you having any trouble managing your staff?

Jon: I'm glad you brought that up. I used to think that the electrical people were really experts, but I'm having some second thoughts now. For example, almost every time I tell Shelby in electrical design to do something, it's almost impossible to get her to commit to a realistic delivery schedule. Then when she finally agrees to a delivery date, she's always late and has some excuse about being put on other jobs. You know how the rules were set up, Tony. When anyone is on my staff, they report to me. She thinks that just because she's in the electrical department, she can stop my job whenever her boss says so. She thinks that she doesn't have to tell me anything. I'm about to recommend that she be taken off all my projects. I can't work with her.

The same kings of things are happening in purchasing and manufacturing. No matter how many times I check on those people, they always seem to be doing things that don't apply to my projects. The managers of those groups don't give me any reasons for that happening. It's almost as if my directions don't count. How can I be responsible for the whole project and meet my objectives if nobody does what I tell them?

Tony: Well, Jon, we may have to change some things around here. Before you submit a memo suggesting that Shelby be removed from your projects, let's think about it over the weekend and see if we can't come up with less drastic answers.

By the way, who is answering all the requests for information that the customers send in, now that you are the Team Leader?

Jon: Oh, I still do that. If the people in the company don't do what I want in the engineering departments, how can I delegate this very important customer-interaction part of the job? No, I still do that and I still check all the details and I still do a lot of the preliminary design. I told you six months ago that this idea wouldn't work and now we can see that I was right.

Tony: Jon, I have another meeting in a few minutes so we'll have to continue this on Monday. See you then.

Now it was another day. The sun was shining bright, but Tony felt that his office was in deep gloom. What should he do? He had tried to repeat the success he had achieved by developing teams just he had done when he was head of mechanical engineering. The team approach was developed as an extension of his past success. However, the company teams weren't working as his mechanical design teams had. The textbooks in his management course had nothing about changing their recommended systems to fit new situations. They presented only one answer. His changes were supposed to be an improvement, but the B.D. Machinery engineering organization was beginning to look like a battlefield rather than a smoothly functioning technical group. Now what?

QUESTIONS:

1. What is the central underlying problem?
   (a) Why? What are the "symptoms"?
2. What should be done about Shelby's problems?
   (a) How should that answer be implemented?
   (b) Why?
3. What should be done about Jon's situation?
   (a) What are the alternatives?
   (b) How would they be evaluated?
4. What should be done to support success in installing any change?
   (a) Do you feel that your answer will work in your company?
   (b) Why?
5. Is this a familiar situation?
   (a) What happened? Was it resolved? How?

1.0 OVERVIEW AND INTRODUCTION

In the last chapter, we defined the job of the manager as making decisions by absorbing uncertainty and suggested that the higher in the organizational structure the job is, the greater the amount of uncertainty to be absorbed. With adequate delegation, there would be fewer decisions at upper levels but they would be very important. In other words, upper-level jobs make a very few but highly uncertain decisions affect the whole organization. Because decision making is dependent upon the individual, as well as upon the organizational position that is occupied, it is an art, not a science. That "art form" changes if the person occupying that job changes. Art can be learned, but no two people will practice it exactly the same way. In addition, every situation is different and is never repeated in exactly the same way; there are always changes. But it is possible to use some generalized learning tools to improve our decision making artistic abilities and then try to apply those tools to our own unique situations. Several suggestions for problem-solving or decision-making tools were given in the preceding chapter. This chapter is intended to start the foundation for your own unique management model that begins after you have categorized the decision as one you should handle.

2.0 BUILDING OUR OWN ART

Personal theory is intended to assist you in making decisions that will optimize your position (first) and the efforts of the organization (second). Without you as the manager, there is no organization. The idea of absorbing uncertainty as a function of the job you hold defines the inferred primary management task of managing or decision making, within your organization. A next step would be learning how that could be done in your unique situation. We begin by using the by-now-familiar scientific method. First we review applicable research (why reinvent the wheel?) and then develop applicable hypotheses and test them in our work situations. We follow up by determining how closely the data received as a result of the hypotheses tests match what was predicted. If necessary, we then build improved hypotheses. With several workable hypotheses, we can have the beginnings of successful personal management theory.

The workable hypotheses will always be based on personal attributes such as experience. Because the situation is in almost continual flux, your theory will necessarily be in a somewhat equivalent change and improvement process. The development process is fairly straightforward to describe, but the challenges of using and improving personal theories can be difficult. You need discipline and concentration. You can apply the processes of the scientific model in building your own unique art form as follows.

Among the first tasks, produce the general definition: Management is an inferred process of optimizing your decision-making efforts, thereby positively affecting the efforts of an organization.

Then outline the theory: Managing is an art form that is dependent upon the interaction between you, the manager, and the organizational situation. We initially selected a process defined as effective decision making about people in organizations that is goal directed and is concerned with absorbing uncertainty. (These first two steps of theory and definition are interrelated; their sequence, therefore, is not important.)

Now select a testable hypothesis: Evaluate applicable research results and practices against your special needs as a technical manager. These are generally described in this and other books. Then select the hypothesis that fits both you and your perception of the *situation.*

Finally test the hypothesis and, if necessary, correct it: Try it within your limited areas of responsibility and predict the results. Compare the data that result against your predictions. If the data do not match the predictions of the hypothesis (within predetermined limits that you had set), change your theory and hypothesis or disregard the data as unacceptable, erroneous, or in conflict with your preconceived notions. (Books like this help in the beginning of cognitive development, but feedback from associates at work guide the correlated practice of improved management behaviors.)

This generalized definition and theory building, cognitive change, and behavioral practice is a learning process that never really ends. There will always be a need for updating your management skills as new findings in both research and practice become available, as your education and experience expand, and as the organization itself changes. We are all becoming lifelong students, irrespective of the management position we have in the organizational structure. Accepting this can be a problem for insecure, second-rate people, but those who accept it and find no difficulty in learning from others are probably not even aware that it might be a problem (Boettinger 1979, p. 199).

We have many competent teachers. Those that are quoted from or are listed in the bibliographies are just a small sample of those available. To be an effective manager is to be a student for life.

3.0 CHOOSING DEFINITIONS: THEORIES AND HYPOTHESES:

But the learning or educational process should have some kind of structure because there often seems to be just too much information in the world. Accordingly, some type of classification scheme must be used to select applicable data. Otherwise, we have no way to select out random, nonapplicable inputs resulting in few guidelines for future use. An appropriate classification scheme to use for selection and testing of ideas is "modeling."

> A model is some representation, qualitative and/or quantitative, of a process or situation that shows what we think the significant factors are for the purposes we have chosen.

> Scientists and engineers use models as exploration plans to demystify nature. With no models, all facts would be equally relevant. Nevertheless, models always have some degree of subjectivity because they include past experience. When experience is coupled with our predictions of the future, we have the beginnings of a model.

> Observation and experience can and must drastically restrict the range of admissible scientific belief, else there would be no science. But they cannot alone determine a particular body of such belief. An apparently arbitrary element, compounded of personal and historical accident, is always a formative ingredient of the beliefs espoused by a given scientific community at a given time. (Kuhn 1970, p. 4)

Unconsciously, we limit ourselves because what we perceive is limited by our experience and expectations and that affects the range of our beliefs. If we accept new information, we change how we think and that in turn changes our perceptions. In management, models are based on both objective (research data) and subjective (researchers' opinions) entities, modified by our prophecies of the future.

The simplest management model is organizational experience codified in some sort of design standard. This model assumes that the future will merely be an extension of the past. To some extent, the establishment of the design standards in every technical company fits this model. As an illustrative example, consider all the alternative designs and materials available for simple cap screws that are available to the machine designer and therefore can be considered before an assembly drawing can be completed. If the decision to select screws was based solely on one designer's wishes, the likelihood of selecting the best screw could be low. To minimize uncertainty, someone in the organization develops standards that have been proven to be successful in the past. Some novel and creative cap screws are probably eliminated, but the model has accomplished its purpose. This "standards" model has limited alternatives and directed thinking by incorporating the organizational past experience. That experience may not be correct because it is a model, but it seems to work. Similarly, the models selected for this book may have both the same flaws and the same assets. They may not be all inclusive but they do work.

This example is almost trivial, but it is illustrative of the process of building and using models. That process requires some careful thinking about what the important variables are and determining how those variables can be controlled to gain some desired result. If the results do not match our predetermined results, we change them. For example, the American automobile industry's model of believing that they were able to build superior cars was severely shaken by the market penetration of higher-quality Japanese cars. Management models do change. Typically, "knowledge about the process being modeled starts fairly low, then increases as understanding is obtained and tapers off to a high value at the end" (Chestnut 1965, p 130).

3.1 Using Models: Advantages and Disadvantages

When we assume that we have selected the relevant variables of whatever we wish to understand and have been able to develop the relationships among them, we can develop a relatively consistent and powerful thinking process that allows us to hold parts of the model (or the situation, in this case) constant while we change the other parts. For example, if we hold an organizational structure fixed, we might be able to predict the effects on our model of total organizational effectiveness when we replace a supportive, people-oriented manager with one who is task or only bottom line results oriented. The prediction might not be perfect, but at the least, we could build a testable (i.e., predictable) hypothesis before the change was made and the organizational situation irrevocably altered. Therefore, model building assists in understanding and predicting the central processes of the situation that are supposed to be represented. It involves abstraction and is intended to represent what we think are the essential properties of whatever we are studying.

A disadvantage includes the obvious restrictive thinking that might be to the variables and relationships that were selected to use in the model. Some variables that are not considered may be quite vital. Conversely, some that are considered, might have justifiably been neglected. This disadvantage of limiting thinking could be particularly important to you as a technical manager because, as noted before, we must change the past success patterns of the technical contributor to the future (and different) success patterns of the manager. The academic training that we all received while becoming engineers, physicists, or whatever was often quite structured and provided little reward for creativity. That mental model must now change to support more flexible thinking.

Engineering and technical curricula stress logic and consistency. This is quite reasonable, as it is the basics of natural science that must be learned first. Structural engineers, for example, do not initially design for maximum novelty or creativity. They first learn to develop foundations and then the supporting framework for a building before even attempting to create the aesthetically pleasing outer building walls that will be so universally admired by the editors of the architectural magazines. Creativity in management goes further because there are many more variables in dealing with people than there are in dealing with anodized aluminum curtain walls.

Creativity is almost impossible to teach, but it can be nurtured whenever possible, and our management models must therefore allow a great deal of freedom and uncertainty in our subordinates. Our peers, subordinates, and superiors change, and learning how to deal with that is very difficult.

3.2 Changing Models: Functions and Projects

The models do change and that change process can be troublesome because it might seem to be due to irrational occurrences. Most of our technical management (i.e., artistic) models include parts of both the natural and the social sciences. In the natural sciences, when new and conflicting data occur, they kill an outdated theory. Because the social sciences do not have the developed, more objective methodology of the natural sciences, it is not as easy for new data to alter old social theories. In many cases, the assumption is that the new data merely extends current social theory and doesn't displace it. The social sciences are too young to have developed the integrating processes or the superordinate theories that the more mature natural sciences have. Therefore, early ideas such as hierarchical, top-down directed, organizational structures may live quite comfortably in the same organization as more modern structures, such as project and matrix management.

Therefore, when measured by the standards of the natural sciences, most general management models are relatively inexact. In addition to inexactness, management models dealing with technical operations have some special problems of their own. Technical operations have two diametrically opposed organizational requirements placed upon them, "functions" that emphasize continuity and survival and "projects" that emphasize creativity and innovation. It is almost as if there were two opposing lists of needs that had to be satisfied simultaneously. One list labeled survival, consistency, and control includes the functional organizational needs that typically stress design standards, drawing controls, growth through product improvements, and fixed procedures that promote organizational stability. An opposing list might be labeled innovation, uniqueness, and creativity and these are needs that typically stress change or radically new products and processes. Functions promote stability. Projects, by definition, promote organizational change or instability.

When these differences are organizationally separated, they have different labels such as the engineering department as opposed to the research and development group. However, in my opinion, when these responsibilities are combined into one technical area, the result is a poorly managed organization. The differences between them should be emphasized, *not* subsumed into one department. Then the technical manager is required to try to maintain peace between two often-warring, but necessary technical activities. To be really effective, technical managers must develop two distinctly different processes and structures to suit these differences. That task is not an easy one.

4.0 WHICH MANAGEMENT MODEL TO BUILD ON? SITUATIONAL THEORY

The management hypothesis (or theoretical subset) that seems to be adaptable to these two areas of functions and projects fits within a general theory set called situational theory. That theory implicitly states that when there is a defined cause, there is a *recommended* action to be taken. It defines management actions as being contingent on or related to interactions either within the company or outside of it. But these interactions always affect the company. Because of the limits of this book, we will deal with internal operations only, starting with the development of various overall hypotheses. These are the three interrelated areas at the core of the technical organization. See Figure 3-1.

More detailed hypotheses follow to outline three internal interdependent areas: the people (Chapters 6 and 7), the structure or situation (Chapters 8 and 9), and the technology (Chapter 10). All these hypotheses are held together by leadership or nonrepetitive management (Chapter 11). Finally, there are repetitive decisions surrounding the internal organization that are relatively standardized (Chapters 12 and 13). Typically, these could include repetitive operations such as purchasing, selling and accounting.

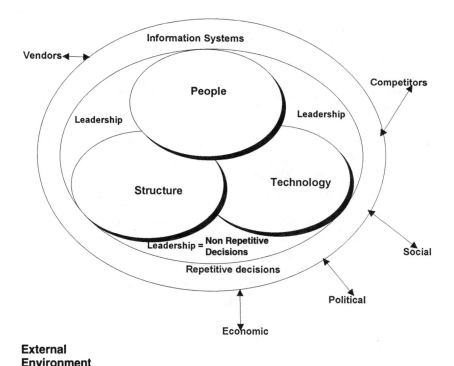

FIGURE 3-1: Organizational Model

With all these interactions, there are changes occurring continuously within the company. Therefore, personal management hypotheses (or models) must be flexible and will always be in an improvement or modification mode. As noted before, this is like a computer program in which the operating system, the applications and the data are always being altered. You can't expect to duplicate organizational situations exactly because the situation includes both the environment and the people. Those environments are rarely fixed and people gain new experiences all the time.

The subset of situational theory that we will use deals initially with interactions in three areas: the people, structure and technology. These areas are surrounded by nonrepetitive decision making defined as leadership-managing, and the total organization is separated from the external environment by repetitive or standard organizational procedures.

Because this is intended to be a manager's "model," we'll begin by extending our discussion of the manager before taking up the organizational model. We start our study of this subset of situational theory with the first major area, *you* as the technical manager. This approach follows those of the physical sciences, in which the investigations usually start with the most independent variable, which you are. In the following chapters we will deal with the more dependent or malleable variable, the situation.

5.0 THE IMPORTANT INDEPENDENT VARIABLE: THE MANAGER

We begin with some general ideas about human behavior before dealing with specifics. Human behavior is a special concern of psychology and, in my opinion, is much more challenging than concerns about the physical factors in technical operations. Among other things, psychology is also obviously interested in those apparently nonrational (but very human) aspects such as creativity. Individuals are always unique and therefore exhibit more behavioral diversity than groups of people. Groups can have average behaviors and the larger the group, the more predictable overall human behaviors become.

It follows then that unless we know a lot about someone or have intimate knowledge about that person's past behavior, experiences and expectations, it is more difficult to predict an individual's responses than those of a group.

> Behaviorists contend that, the bigger the group, the more successful the manager can be in his predictions. Perhaps here we have a partial explanation of the success some managers exhibit as they move upward in the accountability for work groups, seeming to become more competent as the head count of their sections, departments, or divisions is increased. (Wortman 1981, p. 13)

Therefore, we will describe some general ideas about decision making because it's easier. Then you can develop the more difficult, typical individually oriented, prescriptive action-oriented model intended to support your unique decision-making hypotheses, by absorbing uncertainty. That basic process, general description followed by individual prescription, is an appropriate method for learning new ideas. One idea that has been mentioned before is dealing with other people.

> It came as quite a shock to me to realize that my success as a manager was so dependent on what other people could and would do and how effective I was in working with them. It wasn't this way when I was an engineer; then my success depended on how effective I was as an individual working alone. Being a manager is so much different from being a technical person; one must not only be able to see the big picture but see how the parts and people relate and fit into the big picture.
>
> My training as an engineer taught me to look closely at the technical details of the work. This was a form of myopia; my particular engineering discipline helped me to develop a form of tunnel vision. This presented no problem as long as I remained in engineering work.

I was an outstanding engineer—in fact so outstanding that I was promoted to be a manager of engineers. Now my future is in someone else's hands. Now if I don't produce by working through my staff I will be a failure. This is not a comfortable feeling. (Cleland and Kocaoglu 1981, p. 45)

Discomfort can also occur when there is a fixed set or notions about "human nature" as applied to yourself or others. Even the following general categories of "human nature" may not completely apply to yourself or to specific individuals in your organization. We deal with very small groups of people, and generalities don't always work. Be introspective about your own thinking and the applicability of the findings of this researcher. This researcher creates four classifications, but concludes by stating that within physical limits, people are almost infinitely malleable. That means changing one's management model when required.

For example, how do you think about "people"? Do your mental categories fit? Are these real categories that relate to your situation? As an example of categories, you might consider the following.

(It is possible) to create four . . . categories of human nature:

Democratic Man—In this eighteenth century model of humans, human nature is a creation of natural law. It is relatively constant in the sense that it is believed to exist prior to socialization and does not change much due to socialization.

Modern Man—This view, developed in the nineteenth century, dominates social and psychological theorizing today. Humans are assumed to be bound by law-like external forces, but their ability and propensity to learn makes them malleable. . . .

Totalitarian Man . . . Humankind is viewed as constantly irrational, emotional, self-bound, willful, and random. Whatever social there is in institutions and organizations only can come about through elite coercion or propagandizing focused on channeling emotional discharges.

Hermeneutical Man . . A human is defined as a self-bound creature which creates its own nature in the course of its interactions. Interactions mold that nature. (p. 536)

(Modern man) encompasses most management theories . . . because it stresses what they imply: humankind has a nature bound by laws and rules in such a loose manner as to make humans almost infinitely malleable.

The modernist view does not reject the democratic view that humans are law-bound; it only rejects the non-malleable nature of humankind. (Sullivan 1986, p. 539)

6.0 THEORY CLASSIFICATION: DESCRIPTIVE AND PRESCRIPTIVE

As noted before, there are two approaches that are sequential: explanatory or descriptive followed by normative or prescriptive. We will discuss the descriptive or explanatory concepts first, and then move to the more complicated normative or prescriptive approach. This is reasonable because we can use the descriptive ideas as a basis for the prescriptions for our own personal theories. Most of us accept prescriptions better when we understand the reasons behind them, the potential benefits others have achieved and finally the problems in using them. We will use generic theories, then narrow down to more specific and directly applicable hypotheses to be tested.

The data collected from the tests are the stuff of learning experiences or empirical information that improves us as managers. If the hypotheses and data fit your situation, I would prescribe using it as is. But it is quite rare to find an exact fit; thus, a more useful process might be to develop the specific hypotheses and then use these specific prescriptions as further raw materials to be adapted as required, in a repetitive or iterative way. Our first prescriptions are concerned with the primary responsibility of managers: that of making decisions and some practical advice about the realities of decision making and managing in most organizations. Once you know what you want and can interact with others, it's appropriate to evaluate your relationships with peers, subordinates and the boss. It's important to step back mentally and try to understand the environment, your place in it, where you want to go and when you want to get there. Then there is—

> In dealing with bosses, learn your boss's style; keep your boss up to date. Don't be surprised if the boss doesn't give an answer or a precise direction, often they don't know themselves. (Herman 1994, condensed and annotated)

7.0 REVIEW AND PRACTICAL SURVIVAL TIPS

The problems associated with technical management have some similarities to but many differences from those faced by other managers in the organization. There are few other operations within the organization that are continually and simultaneously responsible for both logic and creativity. The technical groups as the suppliers of innovation are major organizational assets. They create the organization's future, and optimum management of these technical operations is necessary if maximum effectiveness is to be achieved.

Situational management theory, modified and adapted to fit a particular situation, is a basis for the technical manager to develop the unique management hypothesis that is believed to fit the situation. The differences between repetitive, logical operations of *functional groups* and those of the nonrepetitive, innovative, short-term *project groups* seem to support a multiple input-output situational type of management theory. That situation is defined in three parts: the people, the structure, and the technology. Functions and projects may use similar quantitative tools, but there are major differences in the psychological (social) management situation.

Situational theory is both descriptive and prescriptive. The larger the groups of people, the more predictable this (and any) theory becomes. Because technical managers usually interact with relatively small groups of people and/or individuals, modifications of theoretical predictions are required.

Because you, as the manager, are the most independent variable, we dealt first with you and how your job of managing is the job of problem solver and decision maker. That job is based on the inferred general mental process called absorbing uncertainty and includes emotional processes, such as experience and intuition. Several prescriptions intended to assist in building your personal management hypotheses were discussed. The following chapter continues that discussion with analyses of the overall model of the organization: the general situation. It will cover major elements of the organization's people, structure, and technology. It will then provide a brief historical review of the beginnings of management theory.

8.0 "ONE-SENTENCE" SUGGESTIONS

1. Functions are "immortal," they go on "forever." Therefore, they should have clearly understood, predefined internal procedures that change very little with time.

2. Projects are relatively short lived. Their management processes have to be redefined each time they are started, and it is the project team that defines them.

3. Personal theories of management change whenever there is an appropriate change in either the internal or external environment.

4. Learn the ground rules of the organizational situation in order to build a viable personal theory. Who is *really* in charge, the formal or the informal leader? Pay attention to your "boss" (i.e., the "real" one, who may not necessarily be the formal one).

5. Don't let yourself be caught up in categorizing any group of people. In technical management, we deal with individuals.

6. Keep current with respect to management research to find out what others are doing. Maybe there is something that you can use. Don't reinvent the wheel.

and finally:
in learning "management," *"The virtuoso was once an amateur."* (Emerson)

and remember to keep your eye on your own goals even though when you're up to your armpits in alligators, it's tough to recall that your original idea was to drain the swamp.

SUGGESTED ANSWERS TO CASE QUESTIONS

1. Shelby's problems and Jon's complaints are merely "symptoms" of the underlying "disease," the disease of insufficient preparation for change. There was no training or guidelines to help people in setting up their teams to match the company needs, the organizational structure, and the people concerned. The teams themselves should set up their own internal organizational structures and processes, then have them matched against overall company policies. If there are any contradictions, either the team procedures or the company procedures should be immediately changed. Then there should be a series of meetings to familiarize everyone concerned with the new methodologies.

   (a) No "standard" management system will work. It always has to be modified to meet the needs of the particular situation. Something that may work in a small "model" form may not work the same when applied to an entire company because the people and the situations are different. In addition, to really have something change, one must change the position and relationships of people in the company. Tony didn't spell out new behavioral requirements. Thus, "symptoms" such as Shelby's problems and Jon's complaints are appearing.

2. Shelby must have a clear distinction made between the work she is to do for the team leaders and her boss's work in the electrical department.

   (a) Because each project is being organized as a separate task, there must be management rules established that both the team leader and the functional manager agree upon.

   (b) Conflict between functional managers and leaders of cross-organizational teams will occur without fail unless priorities are established at the beginning of each project. Those priorities may even have to be adjudicated by senior management, in some instances.

3. Jon has not become a manager. He is still doing his old job. As a team leader, he must be guided into learning to delegate and define what will be required from each member of his staff.

   (a) For example, the rule could be that
      ■ At the beginning of every project, an informal contract must be agreed to by both the staff member and the team leader. If there is a conflict between the present workload of the staff member and the needs of the

team leader, there should be an immediate meeting between the team leader and the functional manager to resolve the problem, or
- If a staff member is assigned to a project, he/she cannot be diverted from that task by the boss without approval of the team leader, or
- The functional manager for each group will represent that group in the initial meeting with the team leader. At that point, agreements will be reached regarding the services to be provided by the functional manager to the team leader, or
- (The answers themselves are not as important as the particular answer that you develop to fit your own situation. Remember that your situation is unique. The people, the customer, and the tasks are rarely standardized. All management systems have to be modified to fit.)

(b) They would be evaluated on a subjective scale according to the forecasted "value" of each project to B.D. Machinery. One possible scale could be subjective from 1 (most important) to 10 (not important at all). The evaluation scores or mechanisms should be developed by Tony since he is developing the changed systems. (Why?)

4. Any change in which there are interactions among people requires a change to the individual's responsibilities and organizational position. It then becomes a "new" situation and new behaviors are needed. Tony never really changed anything. Just telling people to work differently is not effective. Changing their jobs is.
(a) Do you feel that this answer will work in your company?
(b) Why?
5. Is this a familiar situation?
(a) What happened?—Was it resolved? How?

Do my suggested answers agree with yours? In either case, if so why do they? If not, what's the difference between our answers?

REFERENCES

Boettinger, H. "Is Management Really an Art?" **Harvard Business Review, On Human Relations.** Harper & Row, New York. 1979. pp. 195–206.

Chestnut, H. **Systems Engineering Tools.** John Wiley & Sons, New York. 1965.

Cleland, D. I., and Kocaoglu, D. F. **Engineering Management.** McGraw-Hill, New York. 1981.

Herman, S. M. **A Force Of One: Reclaiming Individual Power in a Time of Teams in Work Groups, and Other Crowds.** Jossey Bass Publishers, San Francisco, CA. 1994.

Kuhn, T. S. **The Structure of Scientific Revolutions** (2d ed.). Univ. of Chicago Press, Chicago, IL. 1970.

Sullivan, J. J. "Human Nature, Organizations, and Management Theory." **Academy of Management Journal.** 1986. 11(3): 534–549.

Wortman, L. A. **Effective Management for Engineers,** John Wiley & Sons, New York. 1981.

FURTHER READINGS

Barnard, Chester J. **The Functions of an Executive.** Harvard Univ. Press, Cambridge, MA. 1938.

Cummings, Thomas C: **Systems Theory for Organization Development.** Wiley, New York. 1980.

Cyert, Richard M., and March, J. G. **Behavioral Theory of the Firm.** Prentice-Hall, Englewood Cliffs, NJ. 1963.

Fiedler, Fred E. **A Theory of Leadership Effectiveness.** McGraw-Hill, New York. 1967.

Filley, Allan C., and House, Robert J. **Managerial Process and Organizational Behavior.** Scott Foresman, Glen View, IL. 1969.

Follett, Mary Parker. **Dynamic Administration.** The Collected Works of Mary Parker Follett, Metcalf, Henry C., and Urwick, Lionel (eds.). Harper, New York. 1942. pp. 185–199.

Kahneman, Daniel, and Tversky, Amos. "The Psychology of Preference." **Scientific American.** January 1982. 246(1): 160–173.

Pelz, Donald C. "Conditions for Innovation." Hill, Walter A., and Egan, Douglas (eds.). **Readings in Organization Theory,** Allyn & Bacon, Boston, MA. 1967.

Taylor, Donald W. "Age and Experience as Determinants of Managerial Information Processing and Decision Making Performance." **Academy of Management Journal** (March 1975); 74–81.

Weiner, Norbert. **The Human Use of Human Beings** (2d ed.). Doubleday Anchor Books, Garden City, NY. 1954.

CHAPTER 4

The Whole Organization Design: A Little History, Many Models

Case Study:

The Reluctant Organization

Cast

Paul Sliffer: Vice president, marketing, administration, and project management

Walter Medlock: President and general manager

Maude Finch: Controller and vice president, systems

Michael Moriarity: Engineering manager

The background: Maximum Motors Incorporated had a spotty history of cyclical profits. For the past several years, there had been a tremendous growth in its markets because of the increase in world demand for small powered equipment. However, Maximum never seemed to be able to deliver as fast or as well as its competition. Its reputation in the field for product quality and integrity was excellent but deliveries were almost always late. Therefore, any growth in sales occurred after competing companies had completely filled their order books, leaving the remainder to Maximum.

A project management system had been instituted a while ago to act as a buffer and expediter between the company and its customers, but deliveries were getting longer and there was more conflict within the organization than ever before. It appeared that the company's systems were not working. Therefore, Walter decided to call a staff meeting of his top people to try to come up with an answer that would work. The meeting started promptly at 10:00 A.M.

Walter: Good morning. You have all received comparisons of actual motor deliveries vs. contract delivery dates for this year and last. They seem to be getting further and further apart. What's going on? Every quotation is thoroughly reviewed by everyone in this room and approved before it is sent out. Each of you agrees that the work can be done on time, and yet it never is. Paul, you're our main contact with our customers. What's your opinion?

Paul: Well, as you pointed out, Walter, every contract is signed off before we finally submit it. The budgets are made up and there is another coordination meeting after the contracts are signed, so we seem to be starting off right. It's what happens next that is frustrating. About two or three months into the project, my project managers begin to fail. Engineering seems to take forever to answer our customers' questions, and after we get those answers, the cost estimates that are provided by accounting if there is a contractual change take too long to be of any use. We have to give the customers answers promptly. If we don't, they send their expediters in, and that often really interferes with our work. Our customers claim that we're not responsive to their needs. Then when they finally do agree to pay for changes, there are additional charges because of rework that our delays caused and that never should have happened in the first place. In fact, one of my best project managers just quit because of frustration.

Michael: Look, Paul, we are trying to work with your project managers, but they seem to have a million questions about every little detail and they insist on written answers to everything. Your people are the real causes of the delays. My engineers are trying to cooperate, but we only have so many hours in the day and if you involve them in answering questions, they can't concentrate on engineering, which is their real job.

Walter: Well, Paul?

Paul: Walter, Michael knows that our business is based on supplying specifically designed motors for specific applications. We don't sell from a catalog. Customers always have had questions, and a lot of our reputation has been based on producing to meet those customers' needs. The past few years have seen a tremendous rise in competition. Our customers are demanding more service and getting it from our competition, but the engineering department still treats customer inquiries as annoyances, rather than as opportunities to provide service. Time has moved on and it is no longer adequate to propose a design, negotiate a price, sign a contract, and deliver exactly as proposed about a year later, as we used to. Things change between contract signing and delivery of the hardware, and we have to adjust our organization to respond to those changes, or our problems are going to get a lot bigger.

Michael: Just a minute, Paul. My people just follow company procedure. They answer questions, and one of the reasons that we have maintained the reputation that we have is that every change is thoroughly documented and checked before being implemented. We've worked that way for years and we haven't had some of the failures that a couple of our competitors have had. We may be slow, but we're sure. Perhaps, Paul, some of your people can answer some of the obvious questions that have no major effect on safety or function themselves. That would save time.

Paul: We tried that, but your design engineers then refused to sign off on the changes that we accepted. For example, they insisted on running new load tests on some hydraulic motor piping changes that the project manager had approved, even though everybody knows that those pipes are overdesigned. It was just a waste of time, and then the accounting staff insisted on adding the extra costs of the tests to the contract instead of charging them to quality control or overhead.

Walter: Well, it looks like this attempted solution isn't working either. Does anybody have any suggestions? Maude, you're in charge of costing and general systems. What can we do to allow more flexibility?

Maude: We've tried to change our systems before with very little success because of our organizational inertia. Perhaps things are bad enough now that we can get something done. Maybe a study committee can handle it. Our auditors insist that all financial changes go through the company books, and they always want to know who made the decisions about a change of any scope. Another thing to consider is that our present engineering and accounting systems have kept us out of trouble. Remember the problems that the pump division had with the government several years ago because of inadequate engineering and cost data? Walter, you've been here about two years now and you know how difficult it is to change the way that Maximum Motors does things. We have familiar ways of doing things and it's tough to get any changes made around here.

Walter: OK. I think the study committee idea is great and you three are it! I suggest that you get started by picking your own committee chairman and deciding on an agenda fairly quickly. We'll get together next Thursday. This meeting is over.

QUESTIONS

1. What are the major reasons for Maximum Motors' problems?
   (a) How would you get to the bottom of the problems?
   (b) Do you think that Walter can effect a change in the organizational culture? How? Why do you think so?
2. (a) How would you write the agenda for the meeting if you were selected as the committee chairman?
   (b) What would the problems look like from each of the participants' viewpoint? How do you explain the differences?
3. Why do you think these three managers were selected to be on the committee? Would you have changed the committee's composition? Why? How?
4. How does certainty or uncertainty affect decision making here?
5. What examples from your own experience seem to apply to this situation? How were they resolved? Do you believe the results would have been the same if you had been an active participant, knowing what you know now? Why?

1.0 OVERVIEW AND INTRODUCTION

In Chapter 3, we outlined some initial ideas about some methods for building the foundation for personal theory and developing management hypotheses within the general ideas of situational theory. There were some basic steps that could be used in the development of your personal management micro theory within the situational macro framework. The first step was definition of management and started with some classical, but outmoded, definitions that did not differentiate among organizational management jobs. A new definition that described absorbing uncertainty, delegation and the development of an initial decision-making model was proposed. The definitions and processes were generally descriptive but could be prescriptive if modified by you to fit your perceptions of your organizational situation. In this chapter, we continue the descriptive route by defining the major parts of our initial model: people, structure, and technology; surrounded by leadership and repetitive decisions. Then we'll review some of the important historical bases for our management theories.

2.0 THE MANAGER: HANDLING CHANGE

We started the development of personal theory using the scientific method as a preliminary methodology. The last step in that methodology was to evaluate the variance between what you expected and what you got when you tested your hypothesis.

When there is a large variance between prediction and actuality, you might consider discarding the whole idea and starting all over again. The first alternative could be to discard the data as suspect ("That will never happen again" or "The data are no good because the test tubes were dirty" or "The boss won't like these answers so I'd better try again."). The other alternative is to take a hard look at the hypothesis and possibly the theory and try to determine what happened ("I guess I was wrong when I expected everybody to work harder because we gave them an across the board raise." or "Boss, when you make them show up at a standup meeting at 7:00 A.M. Monday morning, it doesn't seem to support your requirements for greater innovation in the design group. It just causes resentment.").

In either case, when the variance is very large (and you are the only person to determine what "very large" means), a logical next step is to restart the scientific method considering the new data you have ("As the data received doesn't support the predictions, I will revise my ideas. This new data has disturbed what I thought was happening.") Revising your theory/hypothesis is probably the most important ongoing management task that you can perform because it is these revisions and your consequent behavior on the job that enables you to learn through corrected practices. The modifications teach us how to handle different or expanded job responsibilities and act as guidelines in the future.

Changes occur as we modify our cognition and/or our consequent behaviors. It's an unstoppable thinking-behavior interaction. When our attitudes or cognitions about something change, our behaviors will also change. Conversely, when we change how we act towards others because of new data received, our thinking is also modified. There are both logical and emotional elements in this two-way, interlinked process. For example, it is very logical to read the management or technical literature (Why reinvent the wheel?), and to selectively add new information to our mental data banks. We also do that by watching and interacting with others during the constant on-the-job training that all managers experience.

When we know and can avoid the problems with on-the-job training, we use only those parts that we believe apply specifically to our situation. That is reasonable and logical, but logic is insufficient. As human beings, we often use other criteria when we are managing or problem solving and decide that we have enough data to stop searching for the very best solutions. It is almost an emotional decision at that time. We know that it is emotionally harder to keep searching for answers when we have found one that is satisfactory, even if we suspect that it might not be optimal; i.e., "satisficing," by Simon's description (Simon 1957, p. xxiv). So we stop with a "good enough" answer. This "good enough" answer is "satisficing" and that would be the end of it, if one were to take a strictly limited viewpoint of the responsibilities we have within our managerial jobs. If it were only a logical process, there would be a requirement to search for solutions and make only those decisions that are only in our area of responsibility (i.e., to make decisions under our organizationally limited levels of uncertainty). But we often decide (emotionally) to extend beyond our uncertainty level and temporarily take charge of other organizational responsibilities. As an example, "I see that this casting has a crack in it, so even though I have nothing to do with castings in my company, I'll bring it to the attention of the group who has the responsibility to fix this problem."

On the other hand, perhaps one could say that this extension beyond a strict interpretation of our level of uncertainty involves both logic and emotion. "Logically, if I don't do something about that casting crack, we may have a major failure and the company would lose revenue, which in turn might affect me. Therefore, even if it's not my responsibility, I'll take it on temporarily and make an emotional decision to do something about it." That logic-emotion combination can either limit (i.e., by "satisficing" too soon) or expand our management skills. To try to separate the two qualities won't produce distinct elements for analysis; these elements do not stand alone. That's one of the difficulties in managing. As human beings, we are total systems. The parts themselves are incomplete when examined separately, because the interactions among them are not defined.

The personal theory that you develop as a manager is a type of "gestalt" or whole theory that integrates all the parts into something else. It "starts from empirically observed general principles of phenomena . . . and deduces from them results of such a kind that they apply to every case that presents itself" (Helson 1973, p. 79).

It is suggested that the major rationale for management itself was the need for a human being as the only resource that can make decisions that always have missing and/or inaccurate data as a basis. Making these decisions must come before the usual management goals of achieving results, continuing growth, or anything else. As a manager, your decisions are expected to be nonrepetitive in order to minimize human effort and always involve the intent to modify other people's behaviors. Otherwise you are no better than a computer using machine-like algorithms that cannot take advantage of the positive emotional (nonrational?) elements that the unique you (as a gestalt) would use in solving unfamiliar problems.

3.0 AFTER ALL THAT—DO WE EVEN NEED A THEORY?

Even though we have to consider every situational challenge as requiring novel responses that may not be obviously based on any predetermined management theory, there is always something that can be used from others' experience in the field. The first step of the scientific method, that of definition, assumes that you have familiarity with the recommendations of others. Today's situation may not exactly match previous situations, but possibly there are elements of the past that can be used today. You should not be bound by any one set of historical concepts; just consider them for possible useful elements to assist you in your search for answers to current problems. It's always useful to have some point to start from. In other words,

> The challenge and response approach takes a pragmatic outlook on management. It does not attempt to build a consistent framework of thought or knowledge but stresses that management is a practice which should employ all ideas developed by sciences and arts to increase achievements in business performance. The manager is challenged by the situations in which he finds himself and must seek answers to his particular problems unrestricted by any single conceptual framework. This approach . . . does not itself seek to discover generalizations; its orientation is not toward the theoretical or scientific increase in general knowledge, but toward the answers to specific problems faced by managers (Massie 1965, p. 417)

An example of this type of challenge and response approach is typified by the following:

> One great mistake management makes is to look things up in a book. Management is a diagnostic discipline. One does not first read a book. One first looks at the situation. (Drucker 1981, p. 12)

There's rarely the time to start educating yourself about management when you have to make decisions on the job. Learn all the ideas and theories of others beforehand. Therefore, after discovering what others have done in the past, evaluate and deal with the present status of a new situation. Of course, if that situation is repetitive, deal with it by referring to the documented answers of the company's past. Complete pragmatism has its limits. Why repeat the mistakes of others when you can learn all about them beforehand? (We're all creative enough to make our own mistakes.) A completely creative approach might be justified if management were all art and little science. But even art is rarely totally dependent upon the artist's innate abilities; it also depends upon extensive, successful training and experience, as well as upon the acceptance of the artist's output by the social environment. However, let's not exclude the genius artist entirely. One *might* be born—but even Michelangelo spent time as an apprentice, learning basic stone-cutting techniques from others. Of course, he was a natural genius. But even he had to learn how to handle cutting tools and what the results would be when applied to the marble block.

In management, very few of us are "intuitively correct" managers. Perhaps it's not too presumptuous to consider learning how to handle past tools such as historical theories, ideas and concepts and understand them so we can then carve our own personal management theories from the raw blocks of the organizational situation. Learn from the past and create in the future.

4.0 HOW SHOULD WE MANAGE

4.1 The Historical Answer—Classical Theories

Prior to the 1800s, most technical organizations were very small. They were usually owned and managed by a proprietor, who set the standards, hired the employees, supervised the design/manufacturing process and sold the output. This setup was simple and straightforward, and it worked—for a limited market that was often situated locally. With the expansion of market demand, there was a proliferation of a new way to organize: the factory. Things had become a lot more complicated. It was impossible for the entrepreneur to handle everything and therefore a new job was created: that of the manager. The manager was not necessarily an owner but did have special skills in many areas. It was the beginning of the idea of a professional manager operating in a so-called universal model of the "perfect" organizational structure.

4.2 Universal Models: There Has To Be "One Best Way"

The universal models of management were developed by management pioneers such as Taylor, Fayol and others during the end of the nineteenth and the beginning of the twentieth centuries. These models were expected to apply to every organization. The question they tried to answer was: How should the "perfect" structure work? The models ranged in size from the broad, large-scale, rational, logical, and mechanistic recommendations of bureaucracy (Weber 1957) and those of administrative-classical theory (Taylor 1911), through the middle-scale human relations theory (Roethlisberger and Dickson 1939), and into relatively modern times by very specific individualistic open organization theories (Likert 1961). The common idea in these models is that there is one best way to organize, and the closer the structure is to this ideal of the best way, the more effective that structure will be.

Of course, there were differences in defining what that effectiveness was and how it was to be measured. Minor differences were allowed within the overall scheme, but these overall models were still supposedly universally applicable. Even though more modern structures now are available, some of the elements of classical theory might still be useful. For example, the emphasis on management itself as a central factor in the success or failure of the organization is still partially valid (but there have been questions about this in some recent research that are beyond the boundaries of this book). Other considerations such as changed market conditions and economics now modify this straightforward approach, but it is still partially useful.

That might account for the contemporary existence of organizational structures that are designed in accordance with one or more of these universal models and yet are slightly different from each other. It poses an interesting problem for the technical manager who wishes to select the most effective structure. These differences may not have been tested but may have simply grown from the assumptions of the people who originally proposed them when the company began and from the historical environment in which they were developed. We will start with one very important model in this supposedly universal classical theory group: bureaucracy.

4.3 Bureaucracy: The Professional's Model

Because of the growth of large organizations during the end of the nineteenth century and the inevitable separation of management from ownership, something had to be developed that would operate the company without the continual guidance of owner-entrepreneurs. Nepotism with its consequent social and familial ties had usually been able to guide newly appointed managers into following the structure developed by the owner-manager, but when the owner and the manager were different people, this no longer applied. The owner couldn't call a family meeting to discuss things when the hired manager wouldn't respond to the inputs from the owner's family.

A need quickly developed for well-trained, objective managers who could provide goods and services at low cost, develop markets and conduct servicing those markets at a profit (or as in the case of governmental agencies, at a minimum expense), and do so without checking with a nonresident owner. The solution to this problem was found in the development of the professional manager within the "ideal" classical structure. That was quite different from management in an extended family. That solution was put forth by Weber (1957). He described the duties of the managers and his ideal organizational structure in which they were to operate. He provided the theory behind the objectivity and logic that are the hallmark of the professional, even today. The design components have been summarized as contrasts between the professional and familial organizations in Table 4-1.

The organization staff was expected to carry out the decisions of top management by means of the following:

- A system of rules
- Impersonality of interpersonal relationships
- A well-defined hierarchy of authority
- A division of specialized labor
- A system of procedures that implemented the rules
- Employment and promotion based on technical competence (Shannon 1980).

Order and discipline were implicit. Information was passed upward and decisions were passed downward. Authority and responsibility increased as one climbed the organizational pyramid and were clearly defined for each position in that pyramid. Conflict and disagreement were dysfunctional and were obviously results of poor organizational design. In Weber's attempt to describe the ideal designs, he included requirements that vertical specialization (i.e., covering a vertical section through the organizational pyramid) was to be matched with more control, and horizontal specialization across various parts of the pyramid was to be matched with more coordination.

Table 4-1. Contrasts Between Organizations

Professional	Familial
1. Rules, policies, procedures	1. Personal direction by charismatic leader
2. Equal treatment based on performance	2. Rewards based on kinship and/or emotion
3. Division of labor and stress on expertise	3. Multiple job assignments, depending on need
4. Roles in a hierarchy	4. Roles based on personality
5. Career commitment	5. Temporary association
6. Separation of ownership and control	6. Consolidation of ownership and control

Source: Osborn et al. 1980, p. 277.

Although his work was translated from French only about the middle of this century (1949), Henri Fayol was another early writer of the late nineteenth and early twentieth centuries on management and organization. His ideas incorporated the professionalism of bureaucracy into classical management. He was also concerned with developing the best administrative structures for any organizational design. He suggested "ideal" design principles that depended on and reinforced the familial pyramidal structure that is part of many formal organization charts of today. These principles were:

- Division of labor
- Authority and responsibility
- Discipline
- Unity of direction
- Subordination of individual interest
- Adequate remuneration of personnel
- Centralization
- Scalar authority chain
- Order
- Equity
- Stability of tenure of personnel
- Initiative
- Esprit de corps.

With these principles in place, he predicted that an optimal organizational structure would always result. This design approach toward an ideal, the search for the "best way" to develop the organizational structure, was typical of these early attempts at defining management. At that time, it was thought that most problems could be solved through a logical, scientific approach. Markets were expanding and companies were becoming larger. The pressing need to coordinate the efforts of larger groups of employees was becoming a major management task, as the alternative was no coordination or, worse yet, coordination by the employees themselves. Employees were considered to be incompetent and irresponsible. Only managers were, by definition, competent and responsible. Managers of that day felt that output would invariably remain low unless there was strong, central direction with an organizational structure to implement that direction.

Also great scientific discoveries were being made and industry benefited. No one seemed to doubt that the positive benefits of science's objective methodology could equally be applied to organization design. There just had to be a scientific best way to develop and operate the organization. As a matter of history, many of the companies that were organized in this machine-like model did grow more rapidly than those that were not. The machine model was logical and predictable, and it worked. Markets were expanding and people always had some money to buy new products, if the price was right. However, these ideas provided little room for deviation in markets, in technologies, or (especially) in the behaviors of organizational participants. Efficiency in management was the central issue. If the company was managed in this "one best way," it would always succeed, or so they proposed. A central figure in this push to efficiency using logical and scientific methodologies was Frederick Winslow Taylor.

4.4 More Classical Theory: The Production Model

Taylor and his contemporaries developed their ideas during the end of the nineteenth century and published them in the beginning of this one (Fayol 1949; Taylor 1911). Before these pioneers developed their ideas, work was done in a nonconsistent way. Methods were rarely, if ever, standardized. However, much of the work on organizational designs produced by these two men (and some of their contemporaries) are still being used today. Their management models were intended to duplicate the efficiency of the physical processes of the manufacturing plant. The efficiency, logic, and objectivity of bureaucracy suited them well. For example, job descriptions were similar to the operation sheets provided by industrial engineers that outlined every step in the manufacturing process for employees to follow. There were similarly detailed prescriptions for managers to follow. Cooperation and efficiency were supposed to result automatically from science and rationality: inefficiency, only from poor management and inefficient, poorly trained workers. According to Taylor, waste in industry "arose from lack of expert engineering knowledge, failure of management to erect suitable standards, and the withholding of effort by the workers" (Krupp 1961, p. 16).

Time study measured physical activity, wages were related directly to measurable standards of productivity, and each job and organizational function was supposed to be explicitly defined. The essence of the scientific, mechanical organization was a result of integrating the work to be done by the organization with effective studies of plant layout, material flow, and workplace design (Taylor 1911, pp. 36–37). The management task was to follow the *one best way* to organize the work.

All organizational levels in this mechanical production model had their job descriptions, covering criteria for workers and managers alike. Management's tasks were defined as decision making, goal setting, and control. These tasks were centralized, with direction flowing down from the top through the structure and information about production flowing up. The formal organizational position was supposedly independent of the person in that position because the position itself contained the authority and the power. With unity of command (i.e., reporting to only one boss) and unity of direction (i.e., orders downward and information upward), the structure could function efficiently, with inefficiencies or disruptions of the model easily detected and eliminated.

In all fairness to Taylor and his associates, they were quite aware that this organizational structure was a very difficult thing to achieve. They placed most of the responsibility for an organization's failure to achieve the optimum with the managers, not the workers. They were eminently practical people who knew how to solve problems. As one of the early management analysts wrote,

> The type of management which regards the exact definition of every job and every function, in its relation to other jobs and functions, as of first importance, may sometimes appear excessively formalistic, but in its results it is justified by all practical experience. It is in fact a necessary condition of true efficiency in all forms of collective and organized human effort. (Mooney and Reilly 1931)

According to this model, the formal, directive, authoritarian structure leads to high performance, and this structure is universally applicable to all organizations. The traditional roles of the master-servant relationship implicit in the society of those times were embodied into the employer-employee relationship. This was truly grand theory in the best traditions of science, and it matched the social and cultural environment of the day. It was honored by its industrial environment and adopted without question.

Efficiency in organizations was supposedly increased when the wasted motion in poorly defined structural relationships was eliminated. The organization chart was expected to determine the authority and the responsibility of all participants. When I studied this many years ago, it seemed very logical and possible. It all fitted together so well. I liked it at that time, before I had sufficient experience working in different industries.

> In effect, power (authority) is exercised downward through a company's organization and becomes weaker (covers less area) as it travels down the line. . . . Theoretically, authority and accountability are coupled together in their initial travels downward. (Silverman 1967, pp. 78)

Although the logic was intellectually inviting at that time, my opinions about their overall applicability have changed. These concepts may apply to some of our organizational designs, but in my experience, in many technical organizations authority is dispersed with little concern for the formal organization chart (if indeed one even exists!) and it is not necessarily associated in any direct way with the accountability for results. There are subtle reasons for this that are not always apparent. For example, external uncontrollable events or even mismanagement by people operating in strict accordance with an organization chart can cause negative results. Distinguishing the causes between them is difficult, requiring expensive information systems, if it can be done at all.

Even at the time, several of Taylor's contemporaries understood that the classical model was not as adequate as it could be (Gulick 1937; Follett 1942). For example, Gulick injected the variable element of the specific function into the otherwise uniform and coldly efficient machine model. Accordingly, he felt that the structure cannot be uniform in all cases but depended on the functions or tasks to be accomplished.

> Students of administration have long sought a single principle of effective departmentalization just as alchemists have sought the philosopher's stone. But they have sought in vain. There is apparently no one most elective system of departmentalization . . . organizations must conform to the functions performed. (Gulick 1937, p. 41)

Of course, this doesn't help the manager designing the organization very much unless there is a recommendation of how these specific functions affect the design of the structure.

And although Follett (1942) shared the general idea of that time that it was possible to define general principles to guide the organizational design, she recognized that the situation was often the major determining factor. When the situation changed, the authority relationships might have to change and become more dispersed and less unified. She and Gulick disputed the assumptions of classical theory that authority was unitary, homogeneous, and explicitly defined by the organizational structure. If the situation defines the structure, universal designs cannot work, because situations are not "universal." They may in some ways resemble each other or be typical of some organizational "variable" such as technology, but the people who are part of the situation are not universal; they are unique and that makes the structure unique.

If one considers the tenets of the classical theory, both *the workers and the managers* were interchangeable. Human beings and the social groups were to be adapted to the plant structure, and this adaptation was defined by the organization chart. That formal organization chart defined the relationships coordinating and directing the pieces of the machine to perform most effectively. But, research by Roethlisberger and Dickson (1939) that was originally based on this theory resulted in findings that, at first glance, seemed to be almost in direct conflict with these major parts of classical theory.

4.5 Human Relations: The Social Group Model

This research was done in the Hawthorne Works of the Western Electric Company between 1927 and 1933. The original purpose was to test the effect on worker output of single changes in the physical environment. According to classical theory, workers were interchangeable and it would behoove management to determine the amount of work this "interchangeable" person could produce without being overworked. Overwork, naturally, would decrease production so it was a bad thing. The typical variables to be measured were conditions of work, fatigue, and monotony. By varying the lighting and rest breaks, for example, researchers expected the workers in the machine model of the organization to respond through changes in output. Output was expected to increase as conditions improved and decrease as they worsened. This was a linear and very classical structural hypothesis. It reflected the structure of the Hawthorne plant itself, which employed about 40,000 people, and most of the other large industries of the time.

The results did not support the hypothesis. Changes in the physical environment (e.g., rest periods or improved lighting) did not result in related changes in production output. In the first set of experiments, production output gradually increased, regardless of the experimenters' positive or negative manipulations of the work environment. The researchers were confused at first but then discovered that the workers were really responding to the attention paid to them by the researchers. This changed the idea that workers were like machines that responded linearly to work environment changes. The workers, as a social organization, had responded with more production even though the formal organization (the researchers) had, for example, made it more or less difficult to produce by decreasing or increasing the light where they worked.

> In place of a controlled experiment (there was the) . . . notion of a social situation which needed to be described and understood as a system of interdependent elements. This situation included not only the external events but the meanings which individuals assigned to them their attitudes toward them and their preoccupations about them. (Roethlisberger and Dickson 1939, pp. 183–184)

This was a change from the idea that workers and, by inference, managers were machine-like parts of the organization. Machines are not affected by the social meanings of their work. The machine-like model seemed to be less universal than the classical theorists proposed. However, there were other unexpected results of these experiments. The original interpretation that the workers were responding positively to the attention being paid to their ideas and to themselves as individuals by the experimenters was confounded by another experiment during the Hawthorne series. With another group of workers, production remained constant under the same varying experimental conditions that were imposed during the first experiment. In fact, the experimenters noted that whenever one of these workers attempted to exceed a group-imposed production level, one or more of the other workers warned that person to get back in line and decrease his output. Similar warnings to increase production to the group standard were given by the workers to any member whose output lagged below the group standard. The interpretation, in this case, was that the social organization had frustrated the goals of the formal (i.e., the experimenter's) organization. How could this conflicting result be correlated with the initial result of gradually increasing productions?

These seemingly divergent sets of production results support a conclusion that is at direct variance with classical theory. In *both cases,* the social organization, which isn't supposed to exist if classical structures are working well, actually controlled the results. The social organization was really in control. When the perceptions of the social group indicated closer goal congruence with the company's organization (e.g., "You're paying attention to my ideas. Therefore, your interests must be similar to mine."), production increased. When they indicated goal divergence (e.g., "You're just trying to make me increase production so that the work will be finished and I can be laid off."), production was fixed at a level that was just sufficient to satisfy the experimenters' expectations and continue the experiment. (These experiments were conducted in the middle of an economic depression when jobs were hard to get.) The social organization's control over the behavior of individual workers depended on the mutual dependence of the workers upon each other or, in other words, the group cohesiveness. When the workers felt that it was in their own self-interest (however they defined that state) to produce, they did. When they felt it was not, they fixed the level of production at the minimally acceptable level to stretch out the job.

Although the company organization included all the logical elements of control and harmony of classical theory, the social elements of work or the social organization reflected the more *emotional elements* in people. Because those emotional elements really controlled production, they now had to be considered by organizational theorists and managers. They were variables that had to be dealt with. Differences between goals of the company and the social organizations had occasionally proven to be destructive to increased production. The social organization controlled output without approvals of the formally constituted management structure. Management felt that this lack of predictability and control had to be avoided. The company organization was to be followed, and the social organization (which management recognized could not be ignored) modified to fit it. Cooperation and logic had to prevail if the best organizational way to manage was to be developed.

> The formal organization of an industrial plant has two purposes; it addresses itself to the economic purposes of the total enterprise, it concerns itself also with the securing of cooperative effort. The formal organization includes all the explicitly stated systems of control introduced by the company in order to achieve the economic purposes of the total enterprise and the effective contribution of the members of the organization to their ends. (Roethlisberger and Dickson 1939, p. 55)

When the social structure had goals that conflicted with those of the formal organization, such as when the work group rather than the group's manager controlled output, the company organization was expected to investigate the conflict and through communications and training secure a closer cooperation with stated organizational goals. Interviewing workers, exhibiting management solicitude, and modifying working conditions (as long as there was no interference with production) were the new tools to secure optimum output.

> Changing plant culture was a one-way stream. The pathology of organization was the disruptive or irrational changes that did not initiate with management behavior that deviated from the norm of plant structure and business goals. (Krupp 1961, p. 30)

Human relations theory, therefore, could be considered as a continuation of the classical "machine model" organizational structure but with a broader, more pervasive approach, because it dealt with the thinking and emotions of people. Although it was also a universal model, it provided for greater variability in behaviors and increased the allowance for human differences at work *just as long as production was maintained or increased*. It softened the complete authoritarianism of classical theory, improving cooperation and even, in some cases, increasing productivity. But management, not the worker, manipulated it and management still operated on the basis that unique human differences had to be subordinated to organizational goals. In effect, individual creativity could be a disaster because it upset predetermined, repetitive organizational decisions.

4.6 Classical Management vs. Human Relations Management

Human relations theory, with its attempts at manipulation of the social organizational structure, is actually a more flexible and limited, midrange theory than the proposed unlimited range of classical management. Although similar to it in recommending a universal approach (i.e., improving output by manipulating worker behaviors to match formal organizational needs), it does not lay down the rigid and exact specifics classical theory does. The methodology and the implementation are more flexible, depending to a great deal upon the emotional content.

Moreover, classical theory deals with larger organizational structures, such as the total organization, whereas human relations theory deals with the structure of smaller groups within that total organization. The theories' resemblance to each other comes from their common origin: the scientific ideas of the machine model. In both, nonpredictable human behaviors are considered to be dysfunctional, but the cure is different. Classical theory either ignores it or treats it as conflict that must be erased, and human relations theory acknowledges that it has a right to exist but attempts to modify it to fit the formal organizational goals.

Classical and human relations theories were expected to cover all organizational possibilities, but different organizations attempted to implement these seemingly universal approaches with varying degrees of success. In some cases, human relations ideas meant that people's views were listened to by management. This didn't always result in economic success; some companies that used it even failed. Conversely, other companies that didn't seem to use either of these theories succeeded, apparently in spite of the obviously lowered quality of the organizational structural design. There seemed to be missing theoretical areas in the design of organizations. One was that there was no overall concept. Both classical and human relations theories had no central integrative idea that allowed for modification to fit differences in various companies. It was only "one best way" or failure. But there are many ways. The following proposal covers one of them that seems most appropriate for technical organizations.

5.0 A MODIFIED CLASSICAL THEORY: THE PERSONALIZED APPROACH—LINKING PINS

A possible extension of both classical and human relations theory was proposed in a "linking pin" theory (Likert 1961). This theory suggested that basic work groups would be formed and given the responsibility to achieve certain production goals. Each group would have its manager also assigned as a "link" to the group that was next higher in the organizational hierarchy. That higher group, in turn, would have a manager who was assigned as a linking pin to the next higher group and so on. See Figure 4-1. The intent was to develop harmonious groups of people who would be able, through the "linking pins," to ensure communication throughout the organization. According to Likert, there were four stages in

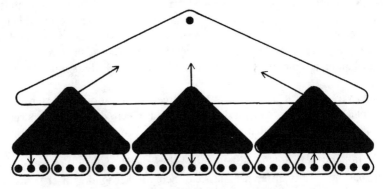

FIGURE 4-1: Linking Pin Organization

organization design. System 1 was the traditional structure in which power and authority were distributed according to the manager-subordinate relationship, similar to that of Taylor's organization. Systems 2 and 3 were intermediate in form, but System 4 was the ultimate in which there was extensive group participation in supervision and decision making. The research indicated that achieving System 4 was a lengthy, several year, in-depth training process. Even though this theory seems to be more humanistic and flexible, it has a simplified view of how people work and react in organizations and, as far as technical structures go, it doesn't provide for individual creativity. It has some advantages over classical theory because it provides for aspects of human motivation and communication, but it suffers from the central problem of advocating *one best way* to organize. Although some of the ideas in these frameworks, such as accountability and communications within organizations, are still important and useful today, these historical "universal" frameworks cannot be adopted entirely in today's rapidly changing social, economic, and political environments. Therefore, we will use those elements that are desirable and attempt to build a more flexible theoretical model that you can modify.

6.0 REDEFINING THE ORGANIZATION AS A FLEXIBLE TECHNICAL ENTITY

We start with a new and operational definition of the organization and its purpose, then move along to the general theory and model before concentrating on the specifics of our own bilateral (functions and projects) technical operations. The components of the model are defined briefly in this chapter but explored further in following chapters, as are the recommended methods for management: including both repetitive and nonrepetitive decision making. Then we will combine the specific situation and you, the technical manager, in order to look at applications of situational theory that are useful in building your personal operating hypotheses.

An operational definition is unusual in that it is determined by the operation itself. For example, the prime operational definition is that of "intelligence"— as the score that one receives on an intelligence test. Admittedly, it's a bit circular, but since it seems to fit best, we will use an operational definition to define the model, then describe how it is used by others, and finally offer some prescriptions for you to modify and possibly apply. That description and prescription design process also starts with the relatively independent variables first and then moves on to those that are supposed to be more dependent. The sequence of the definition and description of the model, therefore, is "people," "organizational structure," and "technology." (See Figure 2-1.)

For our purposes, the *organization* can be defined simply as "people who have fairly stable, regular sets of activities and interrelationships" that have expected outcomes (Litterer 1965, p. 6). Organizations exist because of the unreliability of the behavior of an individual human being and the consequent need for predictable, stable behaviors in people. Interpersonal cooperation and consequent group behavioral reliability are therefore the central reasons for any organization. Organizations support (and coerce) consistent and well-defined activities. However, organizational purposes often are neither consistent across various groups of individuals nor congruent with the purposes of the people in those interrelationships. Instead, they are more likely to be a complex arrangement of goals, activities, and relationships in which there may be competitions among objectives (Haberstroh 1965).

But organizations do provide a constancy of existence and purpose greater than that of any individual. Therefore, it seems likely that in addition to dealing with the unreliability of the individual, organizations also provide some support for the groups they contain that must be greater than the loss caused by intergroup conflicts. In other words, although it may not be optimal for each person, it does result in a greater return to the organizational participants than they could obtain by themselves. (This definition is a bit circular and not very exact, but it is still one of the better kinds of definitions that we have in the social sciences.)

Now that we have a definition, the next step is the design of the initial model of the technical organization. That model should be able to minimize conflict between the needs of the individuals and those stated by the organization. It should be usable as a basis for building the personal hypotheses of management within the organization, which can then be tested. Therefore, we will now proceed to develop a descriptive model of the generalized organizational situation and describe and prescribe some concepts to use as initial hypotheses in your specific organization.

7.0 PERFECT AND NOT SO PERFECT MODELS

There is a classically simple method of developing a model that was first proposed during the end of the last century (Weber 1949). This method was to develop a description of an *ideal type* of organizational situation, then measure actual discrepancies from that ideal in order to correct the situation that exists. It is similar to the procedure I have used throughout this book. (This is *really* an operational definition!)

> An ideal type is formed by the one sided accentuation of one or more points of view and by the synthesis of a great many diffuse, discrete, more or less present and occasionally absent concrete individual phenomena, which are arranged according to those one sidedly emphasized viewpoints into a unified analytical construct. (pp. 50ff) It is a matter here of constructing relationships (p. 69) which our imagination accepts as plausibly motivated and hence as "objectively possible" and which appear "adequate" from the normological standpoint. (Weber 1949, p. 92)

The model of the organizational situation that is proposed as the initial "idea" has three major components: **people, structure,** and **technology** and two types of managerial decisions: **repetitive** and **nonrepetitive.** That organizational model resides in a much larger situation called the **environment,** which was covered to some extent in the introductory chapter. The technical function within the larger organizational situation looks like a smaller model, but actually has the same shape as the overall model and also combines the three major components and two types of decisions. The technical organization is just smaller and has slightly different emphases on different components. The components and the decisions are held together with an organizational cement called **leadership** or **management.**

Both the importance of the particular component within the model and the relationships of the components to each other may change, thus temporarily changing the model's shape. The neat, elliptical shapes shown in Figure 2-1 may become large or small, but the geometry is not that critical. The ellipses may even change to another kind of geometric shape if you want to consider other components you think important. More important is the idea of having some kind of mental model. I expect this model or any other model you choose to eventually be changed into another one that may vary, depending upon your design ideas for both the components and their relationships to each other.

This model follows research that suggests how the organizations work and adjust their components to meet environmental changes. That adjustment (Cyert and March 1963) is proposed to be driven by a human-like desire to avoid uncertainty through a biased, overall, almost simple-minded search for a satisfactory short-term solution to organizational problems. Although internal conflict among participants is rife in this organizational model, because of differences in personal theories, that conflict is supposedly minimized through a sequential attention of the total organization to differing goals and the forming of shifting coalitions of participants to share any rewards that the organization gains as problems are solved.

The organization is supposed to adjust itself to avoiding uncertainty as a person does to avoiding pain. As an analogy, I reduce the pain in my head by taking an aspirin and ignoring everything else. The stomach pains that the aspirin may cause are temporarily overlooked until the headache subsides. Presumably, the stomach has formed a temporary coalition with the head not to make trouble till the head pain is reduced. It's just an analogy, but these mechanisms are proposed because the organization itself is considered to be person-like in its capacity to learn as a result of experience. There does not appear to be any major conflict between the descriptions in this research and parts of the descriptions of individual behaviors in attempts at self-optimizing. This description of the total organization treats uncertainty as something that should be avoided, like pain to a person. However, we know that real organizations do not make decisions and avoid uncertainty as the theory above suggests, but individuals can and do.

Our model, however, proposes that management is not about avoiding uncertainty in decision making but instead absorbing uncertainty in decision making as a function of the management position, i.e., the higher in the organizational structure, the more uncertainty in decision making that is absorbed by the manager. The organization and its people is only a stage upon which, or the situation in which, the manager acts.

8.0 THE MODEL-SYSTEM: CHANGES

This model of the technical organization with its three central components of people, structure, and technology is, by definition, a system. That definition of system includes an overall purpose(s) that is different from any components of the system. It also includes a description of the limits between the system itself and its environment. We have discussed how systems are partially similar to human beings in being gestalts, and it follows that any analysis of systems through disassembly or evaluation of separate parts does not provide answers about the overall purpose. It does provide, however, some very general indication of the functions of those various parts that might allow us to deduce the overall purpose. We'll never be sure of our answers, but at least we'll have more information than when we started and a firmer foundation for our consequent speculations about the system's intended and actual overall purposes (they may be different).

Therefore, keeping in mind the limitations of our analysis in order to understand, control, and modify this system, we'll make a useful simplified assumption that each of the three components, their interactions, and the system's environmental limits can be described and prescribed independently of each other. We'll also assume that our model is "ideal," that is, it is only an "ideal" starting point and it will be changed as we compare it to our real world. With these assumptions, we can start our analysis-understanding with a brief definition of the environmental limits of the model: the line at which the organization stops and the environment begins.

No organizational system is totally independent of its environment. Therefore, the limit is always permeable and slightly artificial. There are few rigidly defined limits to organizations, just as there are few to people. In both cases, there are always additional external factors that could justifiably be included. For example, assuming that a person is a system, if we were to ask that person to define himself or herself, the definition would probably depend upon the person's own peculiar viewpoint. The professional anatomist might limit that definition to the physical body lying on an operating table. An economist might extend that definition to the consuming and producing actions of the living person. A theologian would have some ideas about less tangible concepts. Finally, the person might include many elements involving social and political points of view.

Therefore, the system limits would be in the eye of the beholder. However, in order to have a starting point for our system, we must, understandably, have somewhat arbitrary limits. The limit is assumed to be the repetitive economic division between the company and vendors, customers and stockholders. But this artificial definition of the model "limits" is only part of the definition problem. Consider the definitions of the rest of the model:

(a) Limits (as discussed above),

(b) Components, and

(c) Interactions of the components with each other and with the environment as we have defined it.

These are discussed in following chapters. We will dissect our model for analysis even though we know the dissection process can give us only a limited glimpse of the model's interrelationships. Accordingly, we shall use the phrase "holding all other things fixed" implicitly, if not explicitly, from now on in our definitions and discussions, although that is really an impossibility.

As an example, let us consider the component "structure" in a small R&D company that has the social organizational practice of issuing purchasing requirements by telephone. We can define that structure as the repetitive human behaviors exhibited in the group and attempt to describe and predict possible changes when the company is acquired by a large conglomerate. Assuming that this R&D structure will change into a formalized, documented, requisition-based, three-quotation process and that a printed purchase order is now required, it is obvious that the "structure" has changed. We can count on equivalent impacts on the other two organizational components of people and technology in the model, even though the acquiring conglomerate seems to change only a few minor internal purchasing operating procedures.

However, we shall still implicitly assume "all other things are fixed" as the initial step. It simplifies the problems and prepares us to understand these easier ideas before we tackle the real complexities of our organizations and the world. The medical student studies the cadaver, knowing that it is not the human being that was there, but a part of the human being that must be understood if the student is to become a physician capable of helping live human beings. No matter how many dissections the medical student who becomes a physician participates in, he or she examines the patient who comes into the medical center as an individual. The individual is a gestalt, more than the sum of the parts. Similarly, you should understand the components of this organizational model so that you can understand the general way technical organizations function. Then you can compare your real, existing organization with this model and eventually redesign that organization to be what you intend it to be.

9.0 THE CHANGES AND TIME

Time is an uncontrollable condition that affects our understanding of the model. Some researchers suggest that changes over time can be forecast, controlled, and optimized if there is some sort of balance or constant equilibrium among the various system parts (including time) that cause change (Kotter 1978). But for such a balance to exist, there must be an implied state of optimal equilibrium from which change moves the system or model and toward which it should be moved. It's as if there were one best balanced goal to be achieved similar almost to a continuation of the "one best way" of classical theory. However, in any event it seems to be a very limited view of the situation for the following reasons:

1. The model we have selected may not be completely representative: Remember that our model is not the only one possible. We are never really certain that all the relevant variables have been included.

2. Multiple changes impact every organization, and it is impossible to react to them all: Management reactions to change may seem to be sequential and almost in a fire-fighting mode (Cyert and March 1963). If things get out of balance because of multiple changes (e.g., the chief engineer has just resigned, the law on patents has just changed, there is an organizational dispute about new vendors, and the data processing division has just suffered a disastrous fire that wiped out all the accounts receivable), which of these changes is to be handled first and how does equilibrium occur if one problem is fixed and another pops up? (When "Murphy" strikes, he rarely strikes in any single place or time, except, of course, if that place and time happen to be the most crucial.) The more apparent questions are what the organizational equilibrium is supposed to be and which of the multiple changes will affect the move toward equilibrium and which are to be ignored.

3. Equilibrium may not be a desired organizational state: In your position as technical manager, your major task may be to produce the innovative products and services that continually move the organization into a different relationship with its environment. In effect, your management task could be to maintain a relatively *constant state of disequilibrium.* Perhaps balance or equilibrium are not even desirable. You may even be aiming at a moving target. Additionally, equilibrium for one section of the company is not the same for all the others. Manufacturing managers may be pushing for long production runs to minimize costs while sales managers may be pushing for short runs to meet diverse customers' short run needs.

4. Time automatically changes equilibrium. The model in this book is explicitly defined as relatively static because it is easier to describe and the variables have to be limited in some fashion. That's the way books that are intended to teach are written, but time does affect the organization (and even the contents of books), causing new states of disequilibrium.

Therefore, the concept of change that introduces organizational disequilibrium as a state to be corrected by moving to an optimum state of equilibrium seems to be even less relevant than our less than perfect method of trying to understand the total system by analyzing its parts. That equilibrium concept implicitly assumes much more than "holding all other things equal"; it assumes a definable (and thereby fixed) optimum equilibrium for the whole system. There is no such thing. Our model, however, is intended to be analogous to the ones that Weber proposed; it is supposed to be as perfect as it can be, but redefinable. You may have different ideas. We hope so. That is what we are trying to foster.

10.0 REVIEW AND PRACTICAL TIPS

Using situational theory to analyze some of the data that are available about technical management, we divided this general body of knowledge into two major sections: the manager and the situation. The manager's main purpose is to be a decision maker, categorizing nonrepetitive problems according to the rules established by the organization (the formal system) or his/her own rules (the informal organization) and then solving them appropriately by absorbing uncertainty. Uncertainty is determined by the organizational position. It increases as the decision maker moves upward in the company. We then begin the foundation for personal management theory by reviewing some historical models in order to determine what we can use (i.e., professionalism) from them.

After reviewing major historical models that could be applied to technical organizations, we found them to be flawed because they all assumed a universal applicability; they were proposed to fit every situation—and they don't. The suggested model is multielliptical in its center (see Figure 2-1) and circular in its interaction with the external environment. It shows a relationship among three organizational components: people (the employees themselves), structure (how these employees interact on a repetitive basis; the total organizational design, both formal and informal), and technology (the techniques people use to change inputs into valued outputs). This "ideal" model's three ellipses are contained by nonrepetitive decisions made by the manager (called "leadership or management"). These nonrepetitive decisions are in turn surrounded by prior repetitive decisions that allow the internal elliptical components to interact with the environment. Using Weber's definitions, this model is proposed as a "perfect" model not because it is flawless but merely to visualize otherwise difficult relationships. The size and position of the various components will vary according to the specific model-image that the experimenter (i.e., you, the manager) develops to represent his/her perception of a specific organization.

Both this model and any other that you develop has to be defined operationally because you, as the manager, are part of the model and that means that complete objectivity is not possible. Therefore, by definition, every visualization or model of the organization has as many subjective parts as there are managers because each of us views the company situation differently. Taking a very broad view, even in science there cannot be complete objectivity because

> . . . according to quantum mechanics, there is no such thing as objectivity. We cannot eliminate ourselves from the picture. We are a part of nature, and when we study nature there is no way around the fact that nature is studying itself. Physics has become a branch of psychology, or perhaps the other way around. (Zukav 1979, p. 56)

It should be obvious that when we study our own situations, our acts of observation are affected by our past history and our expectations of the future. We see, in many cases, what we want to see and as it is impossible to be a completely "disinterested" (i.e., scientific) observer, it is possible that our observations themselves could change the situation. It's almost as if the physics of quantum mechanics applied here because we, as observers, are part of the situation. However, for the purposes of this chapter, let us assume minimal interaction between ourselves, as the managers, and the model being proposed here.

This minimal interaction assumes that we, as managers, affect the way the model operates through our nonrepetitive decision making. However, the model's components, people, structure, and technology, are intertwined with each other and really cannot be studied apart. At this point in our learning program, they must be defined in terms of each other, because there are few independent measures to use. For example, one of the central elements of the "people" part of our model is "motivation." Obviously, any study of motivation includes how the elements of the organization's structure and technology are interpreted by the "people."

However, when one understands that the model is a totality or gestalt, and the components are not really independent, we can assume that some parts of those components can be intellectually studied by themselves before assembling them again to consider the whole model. A partial analogy would be to attempt to define a human being in terms of the water and chemicals in the body when we know that these two components are interactive in the body. We can begin our understanding by defining how much water and chemicals there are, but we know that it is only a necessary beginning. That kind of understanding might help us to define the gestalt human being, but we know that it is far from a complete description.

The interaction between the total organizational model and the external environment is through repetitive decisions framed as information processes or systems that assist managers to communicate with that environment and with each other. Those processes or systems surround the organization, and they typically include most of the data processing, personnel, financial, and contractual relationships that organizational outsiders first encounter.

The repetitive decisions are like a circle around the central ellipses of the model; the nonrepetitive ones fill the interstices between the triangular organizational model and that circle of repetitive decisions. In my opinion, the major kind of nonrepetitive decision is that provided by managers, called leadership or management, which also resides in those interstices. Because nonrepetitive decisions are by definition unique, if they tend to become repeated, they should be automatically placed into the repetitive information process, often called the organization's standard operating procedures. Typical examples of procedures that fall into the repetitive area would be those applying to the company's selling of its products/services, purchasing raw materials, and hiring or dismissing company personnel. All of the repetitive decisions surrounding the organization are not intended to precisely limit it but rather to provide two major assets: an outer communication ring through which the organization transmits and receives data from the external environments, and guidelines for internal procedures.

A brief overview of this model follows in the next chapter.

11.0 "ONE-SENTENCE" SUGGESTIONS

1. Although historical models of management organizations cannot be used "universally," there are some parts that are still useful, such as "professionalism" and "bureaucracy."

2. Employees are obviously not interchangeable and not always irresponsible, and managers are not always responsible.

3. Organizational conflict is neither good nor bad—it can be either depending upon the outcome.

4. Organizations, like humans, have "memories" and "learn." That data becomes part of the "decision matrix" (see Figure 2-4).

5. Every management model changes with time or else it dies.

and finally:

Those who don't learn from history are bound to repeat it. (Santayana) and *He who reflects on other men's insight will come easily by what they labored hard for.* (Socrates) but after all that "An ounce of application is worth much more than a pound of abstraction.

SUGGESTED ANSWERS TO CASE QUESTIONS

1. The rate of change of the environment has greatly exceeded the ability of Maximum Motors to respond. The organizational model, including people, structures, and technologies, must be redesigned.

   (a) I would start the diagnosis by trying to find any commonalities. I would do a frequency distribution of the problems and try to determine:
   - Which people are causing them repeatedly? Are there particular groups that seem to be in the center of things?
   - Where in the organization are problems coming from? Are there particular relationships among groups that seem to be causing conflict?
   - Is the technology of processing inputs into outputs effective? Is it necessary to increase the response speed? If so, why and how (i.e., computers?)? This could be the initial process to select the right kind of technology.

   (b) It is difficult for any one person who has been in the organization for any period of time to change it alone. A new president who brings in a new supporting staff can sometimes do it quickly, but the on-site manager has a very difficult and lengthy diagnosis, design, and implementation program ahead. However by changing people's jobs and responsibilities, you can get changes installed. Familiarity and inertia often exist together.

2. The agenda would probably look like this:

   (a) ■ Problem definition
   - Recommended solution, including costs and benefits
   - Preliminary system design and timetable
   - Resources required (i.e., personnel, facilities, funds)
   - Tentative implementation program after the design is approved.

   (b) *Paul* would state that the problem is the technical group's insistence on taking excessive time to check each little detail. It's an unnecessary action and results in delays.

   Michael would say that engineering shouldn't be bothered with details that don't affect safety or function. That should be the responsibility of the Project managers.

   Maude would say that everything is O.K. as is and it's marketing responsibility to communicate better with customers.

   Questions: What would your answer be?

3. The problems seem to center in the technical groups managed by three people. If that is so, they are probably the experts in determining what is happening

and are best equipped to evaluate potential solutions. They are the ones to absorb uncertainty in their departments.

4. Walter seems to feel that the amount of uncertainty is too high to make a decision at this time, so he has delegated the problem to the experts for their opinions and recommendations. The groups answers may reduce the uncertainty, but eventually a decision will have to be made. At that time, unless it is shown that parts of the problems can be resolved under certainty and risk, a final decision under uncertainty will have to be made.

5. What examples from your own experience seem to apply to this situation? etc.

REFERENCES

Cyert, R. M., and March, J. G. **Behavioral Theory of the Firm,** Prentice-Hall, Englewood Cliffs, NJ. 1963.

Drucker, P. "An Informal Talk." **Boardroom Reports.** July 1981. 10(15).

Fayol, H.—**General and Industrial Management,** Storrs, C. (trans.). Pitman, New York. 1949.

Follett, M. P. **Dynamic Administration: The Collected Works of Mary Parker Follett,** Metcalf, H. C., and Urwick, U. (eds.), Harper, New York, pp. 185–199.

Gulick, L. "Notes on the Theory of Administration." **Papers on the Science of Administration.** Gulick, L., and Urwick, L. F. Institute of Public Administration, New York. 1937.

Haberstroh, C. J. "Organization Design and Systems Analysis." **Handbook Of Organizations,** March, J. G. (ed.). Rand McNally, Chicago, IL. 1965. pp. 1171–1211.

Helson, H. "Why Gestalt Psychologists Succeeded." **Historical Conceptions of Psychology.** Henle, M.; Jaynes, J.; and Sullivan, J. J. (eds.). Springer, New York. 1973. pp. 7–82.

Kotter, J. P. **Organizational Dynamics: Diagnosis and Intervention.** Addison-Wesley, Reading, MA. 1978.

Krupp, S. **Pattern in Organizational Analysis.** Holt, Rinehart and Winston, New York. 1961.

Litterer, J. A. **The Analysis of Organizations.** John Wiley & Sons, New York. 1965.

Likert, R. **New Patterns of Management.** McGraw-Hill, New York. 1961.

Massie, J. L. "Management Theory." **Handbook of Organizations,** March J. G. (ed.) Rand McNally, Chicago, IL. 1965, pp. 387–422.

Mooney, J. D., and Reilly, A. C. **Onward Industry.** Harper & Row, New York. 1931.

Osborn, R. N.; Hunt, J. G.; and Jauch, L. R. **Organization Theory: An Integrated Approach.** John Wiley & Sons, New York. 1980.

Roethlisberger, F. J., and Dickson, W. J. **Management and the Worker.** Harvard Univ. Press, Cambridge, MA. 1939.

Shannon, R. E. **Engineering Management,** John Wiley & Sons, New York. 1980.

Silverman, M. **The Technical Program Manager's Guide to Survival.** John Wiley & Sons, New York. 1967.

Simon, H. A. **Administrative Behavior** (2d. ed). Free Press, New York. 1957.

Taylor, F. W. **Principles of Scientific Management.** Harper & Bros., New York. 1911.

Weber, M. **The Methodology of Social Sciences,** Shils, E. A., and Finch, H. A. (trans.). Free Press, New York. 1949.

———. **The Theory of Social and Economic Organizations** (2d. ed.), Henderson, A. M., and Parson, T. (trans.). Free Press, Glencoe, IL. 1957.

Zukav, Cary, **The Dancing Wu Li Masters.** William Morrow, New York. 1979.

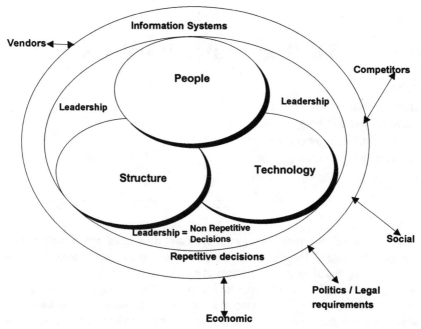

External
Environment

CHAPTER 5
The Model Components

Case Study:

The Unsolvable Problems

Cast

Bob Grossert: Chief estimator

Blanche White: Designer

Mark Kapital: Engineering manager

Richard Cells: Sales manager

Paul Upper: Company president

Paul was one of the most successful company presidents in the multinational Dieppe Corporation. The Dieppe Corporation was a producer of special-purpose manufacturing machinery. It had plants in many parts of the world and each one was called a "division." Paul Upper had been with the company throughout all of his professional life. He had joined the company in England as a junior drafter when he graduated from one of that country's finest technical colleges. Over the years, he had risen from that job to engineer, then to design manager, and general manager. Then about five years ago, he was promoted to be the president of the company's largest machine-building division in Europe.

When the Board of Directors needed someone to take over Power Ops, a large manufacturing company that they had acquired in the midwest section of the United States, Paul was the likely choice. Even though the United States would be new cultural environment for him, Paul felt that it wouldn't be a problem as he had visited the States many times on business trips. Power Ops was a manufacturer of special heavy-duty machinery for the construction industry. For the past few years, the profits at Power Ops had declined and Paul intended to change that situation. The implied promise of the Board, if Paul was successful, was a promotion to the Board itself.

Paul's transfer from Europe to the United States was relatively painless and, after he was settled in, he quickly began an analysis of Power Ops operations. The division was set up on a functional basis.

The sales department brought in customer's requests for quotations. The estimating department coordinated the gathering of the costs and recommended the selling price for each special machine. Management then reviewed that price and either revised it or approved it. The price was then sent it to the sales department for presentation to the potential customer. When the customer sent in a purchase order, the manufacturing department assumed general control of how the job would be done. It outlined a general delivery schedule for engineering/drafting, purchasing and testing. Then each of those departments did its assigned job. There was only a generalized overall budget set for each department with no specific allocations for the actual labor required to design, engineer, manufacture, assemble, and test each job. Purchasing worked from an inclusive budget for materials, but there was no direct feedback during the job progress of actual purchases vs. the budget that was sent to management. Because each department manager in Power Ops had been with the company for many years, each felt he was qualified to manage his part of the various jobs effectively.

Manufacturing operated by setting up general manufacturing process sheets for each machine component as the design came out of the drafting department. These process sheets guided the manufacturing of the machine parts. The machine operators completed time sheets indicating the time that they had spent in manufacturing the part and the time spent was compared to the standard time noted in the process sheets. The accounting department gathered all the labor costs together and assigned them to the various jobs.

At the end of each job, when the special machine was shipped to the customer, there was a "postmortem" and costs were compared against the selling price. That's when the profit (or loss) was determined. Usually, the profits were acceptable but lately they had begun to drop off drastically.

As far as Paul was concerned, this was not acceptable. There was no way to determine if the machine selling price was too low or if the company's engineering-manufacturing costs were too high. In addition, there was no way to correct a job while it was being worked on; costs were measured in total when the job "postmortem" occurred after the job was finished. Something had to be done. Paul called a meeting with Bob Grossert, Blanche White, Richard Cells and Mark Kapital.

Paul: We have a problem here. Our profits are decreasing almost on a monthly basis and the information we have doesn't tell us anything of value. We've got to do something. Any suggestions?

Richard: Paul, I can tell you this, the problem is not in low selling prices. We have many new competitors that have begun to take part of our markets. These competitors are all over the world and many times, we have lost major jobs to them. When we analyze their quotations against ours, we find that they have been able to underbid us because of lower engineering and manufacturing costs. In many cases, they are also able to deliver faster. We're not even sure that they are making any profits because their prices are so low. Maybe they're just trying to get market share before they raise their prices.

Within the last year, we have had to cut our selling prices way down in order to get the jobs that we have. I think we have to be able to produce better engineering and manufacturing processes if we are going to stay in business.

Paul: Richard, how can these new international competitors build and ship to many of our customers here in the United States at a lower price than we can? Are they smarter than we are?

Richard: I know some of their technical people very well because I run into them at various international conferences. Although they seem to be very capable, I don't believe that they are any better than our own people. I haven't been impressed. Their companies are able to sell at a lower price because they keep a tighter control on their internal operations than we do. They know where every item of labor and materials costs are and they watch it constantly. We don't know if we're making any money until the whole job is shipped and that doesn't give us any time to take corrective action. We can't raise our selling prices because of competition, so the only answer I can see is to decrease our costs.

In addition, they maintain close relationships with their customers and respond to every requirement immediately, and customers appreciate that. Sometimes our engineers take weeks to answer customer's inquiries.

Paul: Well, let's take this step by step. Mark, how come it takes so long to answer a customer's inquiry? That should have a high priority.

Mark: Paul, my design engineers are very busy people. They have to lay out each job and watch every part of it as it is processed. They have too many meetings with the design group, the drafting department, purchasing and manufacturing. It's a wonder that they get anything done at all. When they get an inquiry from a customer about a job that's in progress, they try to respond, but sometimes it takes awhile to get the necessary data. I just don't know what to do.

Paul: Blanche, what's your opinion? For example, why does the design group take so much time and why do they have to meet so much with the engineering staff?

Blanche: I'm not sure that the design group takes any more time than it needs to. In the last few years we have had to contend with problems that our worldwide competitors can ignore. We have all kinds of extra quality tests, safety requirements, governmental regulations, and special customer inspections that make our work more complex and take a lot of time. When we get a contract in, we're never sure that estimating has taken it all into consideration, but since we can't take a chance, we cover all the problems for each job. We do a thorough and professional job. And as far as meeting with engineers is concerned, if they would provide a complete specification sheet on each job, we wouldn't have to keep going back to them for more information. Anyhow, how does anybody know if we spend too much time? The quotation never shows that. Ask Bob Grossert why our estimates don't give us any guidance on the amount of time.

Bob: Our estimating processes have been successful in the past and that's why we haven't changed since the company began, so don't blame my group for our problems. We estimate the cost of materials and production labor because they're major costs. Then we add factors that the accounting department gives us for engineering, design and overheads. Finally, we suggest a profit. Management—

that's you, Paul—then looks over the quote. If necessary they adjust the profit and then the quotation goes out. We're not at fault here.

Paul: Well, I appreciate the inputs that all of you have provided, but it looks like we have no clear solutions here. Why don't we think about it and meet again in a week? I need answers.

The meeting ended. Paul sat there and began to wonder if he had taken on too much with this latest promotion. He had never had to deal with such internal dissension before and wondered if this was part of this novel culture. It hadn't been like this in Europe.

QUESTIONS

1. Are these symptoms familiar to you? What seems to be the underlying problem(s)?
2. If you were in charge of sales, what would be your suggestions?
   (a) What plan would you develop to implement those suggestions? Why?
   (b) How would you measure the results?
   (c) If there was a cost impact, what would it be? How would you measure it?
3. If you were in charge of engineering, what would you suggest?
   (a) How would you handle the problem of customer inquiries? Why?
   (b) Would your suggestions require the assistance of other departments?
       (1) Which ones and why?
       (2) How would you get them to cooperate?
4. If you were in charge of estimating, what would you suggest?
5. Is there an internal "customer" in the Power Ops company? If so, who is it? If not, why not?
6. Is Paul correct in wondering about differences in culture? In your experience, does this describe a typical management meeting in which you participated? What happened?

1.0 OVERVIEW AND INTRODUCTION

In the preceding chapters, we discussed the role of the manager and suggested several methods for making those nonrepetitive decisions that are the central responsibility of every management position. There were several hypotheses suggested and diagrams that also partially outlined the relationship between the organizational position and the amount of uncertainty that the manager was supposed to absorb in those nonrepetitive decisions. We also suggested that each manager will absorb uncertainty based on both the hierarchical position and the personal attributes of the individual in that position.

We then began to cover some ideas about the relationship of the manager to the organization itself by reviewing some historical antecedents and pointing out that some parts of these historical models might still be useful today. Now we start the evaluation of a model of the "system" of the manager in the situation. A classical definition of a "system" is a set of components that work cooperatively to achieve a defined purpose. As noted before, it is generally a very difficult task to understand systems, as a totality. The usual method to understand an unknown is to analyze-understand the parts of the unknown and then synthesize those parts to see if they work together again. That process is seriously flawed in trying to understand systems. We rarely can understand a total system by analyzing its parts. When separated from the system, the component may or may not provide a clue as to how the total system operates. For example, examining an individual's lungs doesn't provide us with any information as to how the total person lived, thought, ate, or worked. It only provides us (if we know what to look for) with the health of the lungs and possibly a clue as to what the lungs do. Carrying the analogy a bit further, after understanding all the person's parts and understanding them, it's impossible (at this stage in our medical knowledge) to put them all together again and get the original living person.

It is, however, rare in management for any of us to deal with the entire organizational model. Even the president of a company concentrates on improving only structural components of the company when there is some end result in mind. ("We will extend our sales force into the South American market by hiring and training more sales engineers in that area.") The implied end result is expected to be a change to the company sales component that will affect the company "system" by an increase in sales, and possibly more profit for the company. The company "system" will then be able to divert some of those extra profits to other components to achieve other changes other parts of the "system." And so on.

Thus, we will review the company-model components so that we can understand how they operate and then use this review to visualize how they affect or change each other. It is these effects and changes that are the "situation": that situation in which management takes place. This chapter generally describes the model component and provides the descriptive background. Following chapters deal with each component in detail and provide prescriptions.

2.0 MODEL COMPONENTS

The model has three elliptical center components. (The choice of an ellipse has no significance. It was intended only to show that they interact, similar to a Venn diagram. A Venn diagram demonstrates the relationships or overlapping of different areas of interest.) The major components in the center of the organization are people, structure and technology. These three interrelated components are maintained by the nonrepetitive, uncertainty-absorbing decisions of the manager-leader. That in turn is surrounded by the repetitive decisions that the organization uses to repetitively interact with the external environment. Typically, these repetitive decisions include hiring-firing of personnel, procurement-sales of products, and other "standard" methods of processing data between the company and its environment.

We start the discussion with the first component, *people.* Many years ago, a very experienced engineer gave me a important piece of advice. He said, "Always consider the most independent or difficult part of any design first. If you can solve that, the easier parts will fall into line."

Before that I had always thought it was easier to deal with the simpler parts of any problem first and then concentrate on the more difficult ones later. But he was right. If you can't handle the difficult areas, it's a waste of time to deal with the easy parts.

Following this advice, it's appropriate to define the most difficult part to understand, *people,* first for two supporting reasons:

1. This component of the model is the only one that can exist independently of this (or any other) management model. Even though the organizational "*people*" are affected by the company environment (and of course affect it), they can and do exist wholly outside of this model. The other two components, *structure* and *technology,* do not.

2. In a technical organization, *people* control both the quantity and quality of the work that is done. The *people* are "knowledge workers."

2.1 First—People

We are concerned here with the way *people* interact as part of our company model and the interaction that interests us is called "work." When one looks at history, it's intriguing that the idea of work itself has not always been fixed. It has changed from the hard, grinding, physical labor in both agriculture and hourly workshops of the past century into the less confined, creative mental tasks of the modern knowledge worker. However, the idea of work as being difficult goes back a bit further than modern history. It seems to have started in high places. For example,

> In the sweat of thy brow shall thou eat bread, till thou return unto the ground; for out of it wast thou taken: for dust thou art, and unto dust shall thou return. (King James Version, Old Testament, Genesis 3:19)

Now that is a grand, and almost overpowering, descriptive and prescriptive definition! It covers a lot of ground, as it ties work to both life and death in a very specific working hypothesis. In perhaps less majestic prose, the following author seems to agree that there are connections among work, life, and death that exist in some places even today. It's obviously not a description of the knowledge worker.

> The first time that I ever considered suicide in my life was in that furniture factory as I would stare at the clock and think to myself, "If I have to spend the rest of my life in this hellhole, I would rather end it." Well, of course, I didn't; and as I look back, it was not the worst place I worked. But I was young." (Schrank 1978, p. 9)

That's quite a change in definitions. Even though it would seem that physical work, as a repetitive behavior pattern, has a consistent, long, and important history, within recent times it seems to have become less important in technical organizations operating in most of the industrialized nations. To be more specific, there almost seems to be an inverse relationship between the amount of physical, "sweat of the brow" work and the degree of industrialization. Work in general and, most important to the readers of this book, technical work specifically is no longer primarily a physical activity. We don't physically operate factory machines or agricultural tools by hand. We now develop computer programs for control and the work is done by machinery. There may be knowledge workers who could say that this kind of work is just as hard, and that the "sweat of the brow" is now psychological, but I've taken a more literal interpretation. Besides, one person's mental "sweat of the brow" may be another's nectar and ambrosia.

Therefore, even though the (mental) work may still be very difficult, it is now controlled by the knowledge worker rather than the manager. Some managers may still believe that they can "control" how knowledge workers do their jobs. That's impossible. The work is done in the knowledge workers' heads. In the past, when physical output was important, there were observable changes in the raw materials that could be counted or noted by the manager. With mental work, it is the knowledge worker who manages the output and controls both quantity and quality.

Now with the rationale for *people* being the most independent component of the model, we may move on to define who these *people* are. They contribute (work) in some cooperative fashion to achieve some organizational purpose and then repetitively and predictably share in the results of that contribution. The word "repetitive and predictable" is included in order to limit the component to the employees of the organization, not stockholders or vendors.

We have defined the connection between the model and the people component as "work." Although both the quantity and quality of that work is controlled (in this example) by the knowledge worker, the end result is dependent upon two major attributes of the individual: *abilities* and *motivation*. There are other less important attributes (see Tiffin and McCormick 1965), but these two are assumed to be adequate for our purposes.

Abilities can be defined as a combination of innate human capacities and all the conditioning the person has experienced. Because we are concerned primarily with the technical organization, we shall assume that participants in that part of the company have already demonstrated their abilities through some type of academic achievement and experience. These abilities, therefore, are fairly well defined for each person, if you accept these terms. *Motivation* can be defined as a manager's interpretation of another person's behavior. It is inferred to be "that other's mental selection of a particular behavior from those perceived to be available by the person. In other words, motivation is what you, as the observer, interpret it to be."

2.2 Second—Structure, Fitting It All Together

The definition of the organizational structure is: those explicitly defined formal and implicitly understood informal human arrangements or interactions that are intended to regulate or control the behaviors of the organization's employees.

Formal arrangements typically could include job designs, reporting hierarchies, and directions for departmental goals. Informal interactions typically could include who one approaches to get something done without following a documented organization chart.

This informal structure also includes other less formal arrangements that are nevertheless an equal, if not occasionally stronger, regulation of behavior, such as social arrangements (relationships among people based on power, influence, affiliation, trust, mutual needs, culture, etc.). These might stem from the relevant norms and values shared by most employees. Culture is defined as shared behaviors and ideas that are of long standing and seem to exist throughout the organization. When the ideas are subject to more rapid change and apply to limited groups within the total organization, let us call it the social structure in order to differentiate it from the overall idea of organizational culture. As you can gather, there are no exact or sharp demarcations between culture and social structure.

By comparison with either culture or social structure, the formal structure is comparatively easy to understand and assimilate, especially when it is documented and in accordance with actual operating relationships. However, that is unlikely. It is rare in most companies that a current set of organization charts exist and even when such charts do exist, they are unlikely to reflect accurately and completely what is happening in that company. Moreover, the definitions of the culture and social structure are even more difficult to grasp than that of any formal structural design because there are fewer organizational guidelines available. For example, when starting with an attempt to understand the general organizational culture, you may find some broadly understandable commonalities across particular industries. Then there are unique cultural elements for particular companies within those industries.

Consider some of the cultural similarities in the original development phases of the aircraft manufacturing industry. There were significant differences in each aircraft company's organizational methods and procedures. Although these companies reflected the strong engineering personalities of their founders, their markets were different. The developer of helicopter flight, the Sikorsky organization, was modeled to respond to commercial needs and Curtiss Wright to military needs. But time affects organizational cultures. Today, although the Sikorsky Aircraft Corporation is still involved with the development of commercial helicopters, Curtiss Wright Corporation seems to be moving into an investment type of organization and away from its original aircraft culture. In both instances, the culture has changed but in the latter company, the cultural change has been great indeed.

When the organization's culture is congruent with the economic environment and its formal structures match that culture, it can achieve major strengths. For example, Southwest Airlines concentrates on customer service at minimal cost. Employees will substitute for each other in order to keep baggage moving and the planes in the air because the organizational culture (including the social structure) reflects an overall congruence between the intent of top management (i.e., the formal structure) and the interests of the company's environment.

When there is a separation between them, there are fewer risk takers in management. There is less desire of subordinates to absorb uncertainty and make decisions when new situations occur. When the organization culture supports non-decision makers, eventually there is less congruence with the changing economic environment which can lead to failure. It is a poor consolation when the whole company goes under because of an organization structure that penalizes managers who are willing to make appropriate and independent decisions.

2.3 Third—Technology: The Way We Do Things

One definition of an organization is a purposeful group of related parts or a system that uses materials, knowledge, and methods in order to transform various kinds of inputs into valued outputs. Therefore, a definition of the technology of the organization is the processes chosen for this transformation of inputs into valued outputs.

Several other definitions of technology assist in classification and consequent understanding, such as a variety of techniques intended to change some specific object and provide feedback for this change. The object itself determines the combination, selection, and order of application of the change method. As an example, there is the design and construction of a chemical processing plant where the terrain, the weather, and the interactions of the various labor unions on the site often determine how and if the project will be completed within budgetary and time constraints. As defined, technology can range from simple applications, such as calculating using pencils and paper, to the use of complex computer-based algorithms for the same purpose. Technology is not the end result or the product being created, it is the *processes* through which that creation occurs.

Those processes do not always result from technical improvements; they can be affected by the demands of political, social, or even environmental requirements. In recent years, these demands can come from unforeseen elements outside the organization such as consumer-oriented governmental agencies or other political bodies. Responses to these new types of demands have occasionally resulted in changes to both the formal and informal organizational structures. Examples include the chemical industry that now maintains a permanent group of technicians whose work is necessary to meet legal requirements by monitoring the organization's adherence to environmental concerns. This work does not add value to the products produced. Even some industries that may not be directly impacted now have structural groups that ensure that the technology used does not endanger customers or the environment. These structural groups often must provide recommendations to management for corrective changes to the company's technology in response to demands of many external groups besides those of customers and stockholders.

However, irrespective of the sources for changes in technology, there is one general technique that is used a great deal in modern companies: serial processing. Whether the end result is a management-directed improvement in the products being manufactured or change in the way the company interacts with the environment, action can occur only when sequential steps are completed successfully. In other words, action C can only take place when action B has been completed, and that action depends in turn on the completion of action A. This technique is used in all planning when the organizational management thinks it knows how to produce some desired outcome and the total task to achieve it is broken down into a series of consecutive steps. There is a planned flow and the performance of each step depends on the completion of the steps before it. Most manufacturing processes have this emphasis because there cannot be any ad hoc production. Serial processing is found in a wide range of technologies from automobile production to petroleum refining.

There are many technologies appropriate for different organizational situations or models and they are not necessarily mutually exclusive (for details, see Makridakis and Wheelwright 1978, p. 472), and selection of the appropriate technologies is part of absorbing management uncertainty. As noted before, those attempts begin when:

1. The problem rightly belongs to the particular manager.

2. No algorithm exists.

3. The manager has decided that this problem should not or could not be delegated upward, downward, or even sideways.

In the best of all worlds, the manager is empowered to make unilateral decisions, but most technical organizations consider the selection of the right technologies to be a major concern. Therefore, even though it would appear that this would be an individual's choice, its importance requires a broader (and perhaps more pragmatic) approach. When the situation itself presents no obvious clues to draw upon, technology selection usually requires input and support from other areas of the company. Accordingly, it might be wise to set up a task force with a limited life. It's interesting that this task force will use technology to reach a decision on selecting the company's technologies. The group first has to select the processes (i.e., the techniques as part of the management processes/ technology) that it will use to come to a recommendation (i.e., the output or product). Typically, task forces might select some form of nonstructured process such as brainstorming or Delphi forecasting before narrowing down the choices that can fit the company's capabilities.

(Briefly, brainstorming is a multiperson meeting in which everyone comes up with all their ideas which are recorded by the chairperson with no evaluation of practicality until much later. Delphi process is presenting a series of questions to several "experts" on a serial basis. The questions and the first "expert" answers are then provided to the second "expert." The totals are then provided to the third, and so forth until all the responses are received.)

Choosing the best technologies to fit the demands of the particular task force situation is almost as complex as determining the optimal motivational patterns for personnel or developing the best organizational structure for a company.

Fortunately, when the guidelines are few, the general prescription would be similar to that noted in the previous chapter: select the least complex technology and methods with which you have some familiarity before going serially on to another.

The manager's decisions and consequent behavior are the "glue" that holds the internal components of the model together. The interrelationships of these components, *people, structure,* and *technology* are greatly affected by the nonrepetitive "leadership" decisions and consequent actions of the manager.

2.4 Nonrepetitive (Management or Leadership) Decisions

Over the years, I have been asked if "leaders or managers" (assuming no difference for the moment) are really necessary because most technical organizations are supposedly staffed with competent, trained and motivated people. Even presupposing the prior sentence would be true, "managers" would still be required because of one central reality—we never have all the data at the right time. Therefore, there is reason to expect that there will always be a need for human managers to accept the responsibility to fill in the gaps using their unique mental processes and make decisions. Even with insufficient, inaccurate, or subjectively changed data, human beings can make decisions in strange and wonderful ways. People can do this—computers can't. People carry mental images of their situation and when parts of their observations are incomplete, they can mentally fill in the gaps. It is done so quickly that many of us don't even realize it when we do it. (Have you ever wondered how other people make important decisions, like where to live, what job to take, what school to go to, etc.?) Computers can extrapolate, make estimates, etc., but they do it based on preprogrammed definable rules. They always require complete data. Even when they calculate risk, the calculations that they produce must be programmed before an answer can be provided.

In every organizational situation, there will always be inadequately defined problems that require human management input. Computers can help by screening the new problem against any repetitive systems or past solutions that might fit. If it's been solved in the past, we may not want to do it again that way. Any remaining unsolved parts are then worked on a nonrepetitive mental process by managers who can and do make decisions under uncertainty, using the appropriate technology. In other words, it is the human managers who provide the missing parts or the unique answers intended to either keep the organization from injury or help it to grow whenever a new situation occurs. And one way that a manager-leader can begin the process of providing nonrepetitive solutions is to develop some type of mental model of how that manager envisions the situation. Quoting Warren Bennis,

What makes a leader. . . . Management of attention through a set of intentions or
a vision, not in a mystical or religious sense but in the sense of outcome, goal, or
direction. . . . Management of meaning; leaders make ideas tangible and real to
others, so that they can support them. For no matter how marvelous the vision,
the effective leader must use a metaphor, a work or a model to make that vision
clear to others. (Bennis 1989, p. 37)

The suggested model here emphasizes the areas between the outer circle of
the repetitive systems and the rest of the organization, noted as intersecting
ellipses. (See Figure 2-1.) Typically, many management nonrepetitive decisions
in technical organizations involve conflict resolution among individuals, design
tradeoffs, and all the forecasting-planning activities directed toward the possible
solution of future problems or grasping of future opportunities. The results of
those activities are often represented financially by the operating budgets. By
definition, leadership (i.e., nonrepetitive decision making) always comes before
the repetitive decisions in the company's standards; otherwise, how could those
standards have originated?

2.5 Repetitive Decisions (Standards)

Although the multi ellipses of the organizational model are maintained and
aligned by the nonrepetitive decisions of the manager-leader that are intended
to solve unique problems, when some of those decisions appear to be repetitive,
they can become part of the company's culture. When documented, they are the
formal standards of management. Conversely they may not be documented but
instead become part of the informal "the way we do things around here." If
understood by the people in the company, when repetitive decisions or the culture
involve organizational externals, those decisions may be simultaneously a buffer
between the organization and the environment and an information transfer mechanism. As a external buffer, those decisions separate "us" from "them." As a
transfer mechanism, they guide typical repetitive processes between the organization and the environment and even occasionally within the organization itself.
Typical external interactions are selling products/services, buying raw materials,
paying debts and hiring/discharging employees. Typical internal interactions are
salary classifications, data controls, and individual promotion policies. The model
places internal decisions into a "decision matrix" (see Figure 2-4).

Of course, initially these repetitive decisions or interactions resulted from
nonrepetitive management solutions to problems. But when the solutions were
expected to be necessary again, it became part of the culture either through
documentation or "standard practice." Accessing repetitive decisions or solutions
usually begins with some standardized inquiry or input. Have we ever done

anything like this before? The information transfer process (or repetitive management information system) is supposed to provide some standardized response or output to this question. If that standardized response is inadequate or even inappropriate, for whatever reason, that inquiry is then directed to a responsible manager for a new response or solution. That always happens when the inquiry is both nonrepetitive and important. (This is fairly familiar ground by now.)

2.5.1 Typical Repetitive Decisions (Closed and Open Loop)

Following this, the repetitive decisions are defined as both the information flow and the channels through which the organizational model corrects itself and interacts with the environment. Most systems that are developed to solve repetitive problems are intended to be closed loop rather than open loop systems. Closed loop designs occur when some type of feedback from the output automatically affects or controls the input. In this design, minimal independent intervention is required to see that the action taken either has occurred or is appropriate. Management has set the standard and determined the tolerance from that standard that the action may have. The control of the feedback loop that is built into the system corrects the situation.

The classic example of a closed loop system is the heating thermostat that turns the furnace on when the temperature drops below a predetermined point and (through feedback of the temperature rise affecting the heat-sensing elements of the thermostat) shuts it off when the amount of heat is satisfactory. One of the prime problems in designing any closed loop system is minimizing system "hunting" when a variance occurs. A system variance is the difference between an expected measurement of some kind and the actual measurement. The system "hunts" when the thermostat temperature drops and the heat is first turned on (because the room is cooling down and the temperature goes through the turnon point on the way down). Then when the heat is turned off (as the heat builds up and the temperature on the upswing goes through turnoff), it continues to rise, then again begins to fall. The variance in both cases causes the system to continually "hunt" for the right answer within predetermined variance limits. (Most thermostats today eliminate this problem by turning off and on according to the rate of heat change, not only the actual temperature.)

In another kind of system, known as open loop, the results of the comparison against the standard are stored somewhere, and this implied stored standard and other controls affect how the variance is used. For example, when reports showing deviations from budgets are delivered to managers and they do not take action on those deviations, an open loop system may be in use. Reports to management are similar to the change in the room's temperature as in the thermostat example above. The manager is not like the thermostat control in the closed loop system reacting according to predetermined rules, because she/he has the choice of taking some kind of decision or not, or even of letting the whole system fail.

2.5.2 Dynamic or Concurrent Information Systems

When feedback data are being collected during a process and management evaluation is proceeding in accordance with predetermined standards, the controls are dynamic or concurrent. All closed loop systems have these controls. An example is the use of statistical quality controls in manufacturing, where measurements of the deviations in process are automatically fed back to the equipment itself in accordance with predetermined variance limits. That feedback automatically corrects any excessive deviation.

These repetitive types of subsystems provide data to every part of the organizational structure that has any contact either with the environment or with another department in that structure. Personnel, accounting, production control, purchasing and field service are examples of typical departments with many repetitive decisions incorporated in their dynamic information systems. Those subsystems should have very short time periods between the error or feedback signal and some kind of planned reaction. If the rate of change in the company's environment is increasing, it should mean an equivalent movement toward the design of faster responding, closed loop systems. One important support for implementing dynamic information systems is the use of computers in technical organizations. When appropriately programmed, that kind of equipment is well suited to the task of providing immediate and dynamic feedback.

2.5.3 Other Considerations

However, all change does not come from interactions between management decisions and the environment outside the organization. Much of it can be an internal response to some new demand from the outside, and in some cases this response can be even more difficult and intractable. For example, major product failures (i.e., environmental economic input) almost always result in some kind of immediate change within the organization. It may be a change in people ("Who's responsible for this latest fiasco?"), the structure ("Let's increase inspection. Maybe we can pick up the problems here rather than out in the field."), or the technology ("Let's form an ad hoc group to develop processing solutions to this unique problem."). In this event, the information system cannot be the primary tool for solving these problems. It is intended to handle only problems or changes that have been handled before and responses appropriately programmed. It is, however, the first place the problems land.

Internal changes also occur because of a wide variety of internal positive or negative factors. Typically, an internal positive reason could be the recruitment of a particularly brilliant, intuitive manager who sensitively anticipates problems or reacts to the very small stimuli that are recognized as potentially important. Similarly, negative factors could range from outright incompetence to minor oversights. The repetitive information processes of the model do not apply to these examples. They require the nonrepetitive problem solving called management.

No repetitive information process design can service the needs of all equally and too many information processes are designed to suit the most influential internal and external demands of the moment. Consider the rule for computer usage set up at the data processing center that establishes tax reporting ahead of equally vital engineering change notice reporting in the processing queue. In that situation changes in financial reporting will definitely take precedence over changes in documentation of standard valves for the company's product lists. The financial people are obviously more important to this organization than the technical people. Situations, however, are never static and a rapid drop in product acceptance in the market may result in this organization's engineering people achieving a higher priority than the financial people, at least until the crisis is passed. Therefore, adaptations should always be made for supposedly minor information processes, but if these adaptations (or responses to change) are not attended to constantly, the system will invariably begin to suffer major failures after several years. When maintenance and modification of existing information processes have a low priority, trouble in the organization grows.

2.5.4 Designing Information Processes—Inadequately

Besides inflexibility and lack of updating, another source of poor design of information processes is an implicit design assumption that it is somehow always possible for positive feedback to result in positive achievements. When profits are reported to be up, there is often additional pressure to increase these profits even more for the coming year. The process is intended to reward more these supposedly positive achievements and punish negative ones continuously, regardless of the reasons behind these performances.

In a strictly theoretical sense, process stability requires negative as well as positive feedback. Of course, positive feedback itself is not inherently bad. It simply should not be the total basis for all decision making and should not be designed as an open loop process. For example, we can now consider the "marvelous" thinking processes of some managers when profits begin to go up and therefore provide positive feedback with respect to the company's financial picture. Without some analysis this positive feedback could lead to increased pressure for more profits and increased production by possibly lowering budgets for expenditures such as maintenance. The problems then begin when maintenance costs begin to drop and another positive variance occurs between the budgeted number and the actual cost. With no management forethought, this might be considered as a great achievement instead of an indication of impending disaster.

When the results of decreased maintenance cause the eventual catastrophic failure, and they eventually will (remember, both "Murphy" and Mother Nature rarely take a day off), maintenance is begun again, production is temporarily de-emphasized, and the increased profit cycle is reversed for a short while. So, only positive feedback can cause trouble if no investigation of the causes is made.

Attempting to develop a method of repetitive decision making and designing information processes to assist in making decisions require the inclusion of both negative and positive feedback using closed loop concepts. Easy to say but difficult to do.

3.0 REVIEW AND PRACTICAL SURVIVAL TIPS

This brief overview of the elements of the suggested organizational model should be modified to fit your perception of your organizational situation. This model includes people, structure and technology as internal elements that are held together by nonrepetitive decisions of the managers. It is the manager who provides the missing and/or accurate information that is invariably lacking in the problems that the company faces. That is the necessary human input. Those nonrepetitive solutions to problems are decisions that are included into the company's culture when it seems that the problem will reoccur. That culture defines past decisions into a current standard set of expectations that can either be documented or understood by company participants. When documented it is called "standard operating procedures." When undocumented it still guides and controls as "the way we do things here."

The "way we do things here" as well as "standard operating procedures" often become cast in steel unless there is a definite sense of change. Organizations do "learn" (Cyert and March 1963), but that "learning" should not occur only as a reaction to a disaster. Some companies have internal auditors whose major task is to periodically examine existing processes to determine if those processes should be canceled, changed, or continued. With senior management approval, this is easy to do. Just stop sending those reports and see who complains. Another way would be to request that every recipient of a report provide, at least yearly, a justification for the report or a response to a memo from the auditors that says, in effect, "Show cause why this report should not be stopped."

In the following chapters, we will discuss the model's components in more detail in order to assist you in selecting those parts that you consider useful in your situation.

4.0 "ONE-SENTENCE" SUGGESTIONS

1. An important repetitive decision is to be sure that the responsible manager defines the hiring process for his/her group and that human resources (or the personnel department) defines the processes for firing in the company.

2. Develop a specified "life" for every repetitive report and process. Have interested people then justify, on a periodic basis, why those reports and processes should continue.

3. Develop action-plans for variances in reporting. For example, how large must a variance be to initiate corrective action? Have that plan applicable differently but justifiably to different groups. It cannot be applied universally to the whole company.

and finally:

 A good thing to remember and a better thing to do, is to work with the construction gang and not the wrecking crew. (Anon.) besides designing your own systems is easy because nobody really knows better than you do what you are really doing.

SUGGESTED ANSWERS TO CASE QUESTIONS

1. Being undercut on selling prices and having internal costs too high is a never-ending problem for every management. These are symptoms of an underlying lack of response to the environment. The central problems seem to be slowness in responding to increased worldwide competition and an inability to take an unbiased evaluation of the present company operations. Because a company has been successful in the past does not always mean that it will be successful in the future. Repetitive solutions that helped to transmit data between the company and its environment have to be reevaluated on a regularly scheduled basis.

2. Establish targets for new selling prices that would be competitive. Provide immediate feedback to management on all competitor's prices and product designs.

   (a) Advise engineering, purchasing, and manufacturing of these changes and request immediate feedback on internal, nonrepetitive decisions that would reduce internal costs. Without the cooperation of internal groups, management cannot exist and the company would fail. Set up a group or individual in the sales department to be responsible for logging in all requests, scheduling them and notifying customers.

   (b) After providing a fixed time scale to get a response, require a comparison of prior costs vs. expected costs and a forecast from the internal departments as to the methods they intend to use to achieve those cost savings.

   (c) Every change should be measured by the cost effect and this data should be reported by the departments involved. A supporting feedback should be documented by the accounting department.

3. There are many alternatives but several obvious ones could be:

   ■ Beginning a program of design standardization,

- ■ Requesting quality assurance to decrease the number of vendors to include only those would agree to cost reductions in turn for greater purchasing from them,
- ■ Setting up a small internal design team to review several past designs in an attempt to simplify designs and decrease costs on common parts of the product line. Then institute a team bonus plan that rewards cooperation among departments.

4. Customer inquiries would be handled by the lead engineer or designer on each job. Before any job was begun, that lead engineer or designer would be directed to contact the customer and become as familiar with the customer's requirements and contact personnel as possible. The customer is the reason the company exists.

5. Yes, the "internal customer" is the next department to whom you deliver your work. For example, the "internal customer" for engineering may be the manufacturing group. Therefore, the manufacturing group would either approve or send the work back to be redone if they do not accept it.

6. Yes, there are major differences in the social environments of groups. For example, sales is entirely different from engineering. That's not "good" or "bad"—it just is that way.

REFERENCES

Bennis, W. "Why Leaders Can't Lead." **Training & Development Journal** (April 1989): 35–39.

Cyert, R. M., and March, J. G. **Behavioral Theory of the Firm.** Prentice-Hall, Englewood Cliffs, NJ. 1963.

Makridakis, S., and Wheelwright, S. C. **Forecasting Methods and Applications.** John Wiley & Sons, New York. 1978.

Schrank, R. **Ten Thousand Working Days,** MIT Press, Cambridge, MA. 1978.

Tiffin, J., and McCormick, E. J. **Industrial Psychology** (5th ed.). Prentice-Hall, Englewood Cliffs, NJ. 1965.

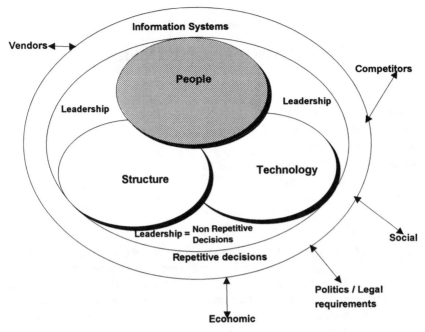

External
Environment

CHAPTER 6

People—Basic Building Blocks Concerning Motivation

Case Study:

The Disinterested Designer

Cast
George Wilcox: Chief engineer
Marge Gustav: Vice president
Millard Myer: Designer
Phil Josephs: Engineering design checker
Peter Maxon: Section chief

———————

It was late Wednesday afternoon when Marge Gustav phoned George Wilcox on the shop floor and asked him to stop by in the morning. George wondered what the call was all about and tried to go over all the open items since he had last talked to Marge. His budgets were on schedule, he had completed the appraisals of his section chiefs, and most of his "hot" design projects seemed to be coming along well. What could she want?

When he got to his office the next morning, he found Peter Maxon and Phil Josephs waiting for him.

———————

George: Well, you guys are here bright and early today. What's up?

Peter: George, I've been trying to keep this problem from landing on your desk but I think that it's gotten to be unsolvable and I want to do something drastic to get rid of it. It's that new designer, Millard Myer; he's driving me up a wall. For the past few weeks, ever since he was transferred in here, he has been coming in late anywhere from half an hour to two hours. When he does show up, he sits at his board all day long and half the night too, so I'm told. I know that when I leave here, sometimes after 6:00 P.M., he's still here. His time sheets are never in on time, and the only way they do come in is when I send my secretary down to get them. He seems to know what he is doing, though. Every time I talk to him the answers are right on the button. I've tried to get him to straighten up on his administration, but after a few days he just slips back into those sloppy habits of his.

George: Wasn't he the guy you said would be great for those advanced development projects? You remember the ones that needed the combination of physical chemistry and hydraulic controls?

Peter: Yes, he's the one, but I didn't think that he would be this much trouble.

George: Well, Phil, I guess that you're here about the same thing. What's your position?

Phil: The guy is a gold-plated A No. 1 genius! It took him about half an hour, using the back of an old envelope to write on, to calculate the residual loadings on that model "E" widget that was giving us so much trouble. He is a different kind of guy, though, never has lunch with any of the boys; in fact, he never even goes to lunch. Just sits there at his board, or at the little desk behind it, and stares up at the ceiling. Sometimes he works at the computer. Of course, I don't know what he's doing but I guess that he does.

George: Pete, have you talked to him? Told him what you expect? I'm going to ask an obvious question, but have you showed him how to complete those forms? They are a bit complicated, you know. Those accountants don't seem to care how much time it takes, just as long as they get their data.

Peter: Like I said, I've talked myself hoarse. He's very pleasant and always says that he'll get right to it and he does, but after a few days it starts all over again.

George: Okay, fellows, give me your suggestions in an informal memo because I'm late for a meeting with Marge. Talk to you later.

Ten minutes later in Marge's office

George: Hi, Marge, sorry that I'm a few minutes late, but some of my boys grabbed me when I first came in, and I just got away.

Marge: That's okay, George. What I want to talk to you about is not something that can be fixed right away, so a few minutes delay is no problem.

The Board of Directors was complaining at the last meeting that we never seem to have any revolutionary products that sweep the market anymore. I guess they still remember the time that we captured a majority of the market for nuclear control valves several years ago with our new TFE models. They seem to see a more gradual approach to our product improvements now. Our competition has been biting into our markets and we have had to reduce our prices to remain competitive. With our overheads, that has a direct effect on the bottom line. Do we have any answer for them: anything new and better that's on our drawing boards? If not, do you have any suggestions? They want to see some results within the next year. How about giving me your analysis informally by the end of next week?

George: Okay, Marge, be glad to. We're working on some improvements that should help the bottom line as soon as development is completed. I'll have a memo drawn up for you.

Later that day, as he began to lay out the memo, George realized that some of those new products would be coming from Peter's group. He decided to walk down the hall and check things out with Pete. On the way, he passed Millard's cubicle, and on a hunch, he stopped. Millard was deeply engrossed in something on the computer screen and was slowly pecking in some instructions with one finger. Seeing George, he stopped.

Millard: Hi, Mr. Wilcox, how's it going?

George: Fine, Millard, fine. I didn't want to interrupt you because you seemed so involved in what you're doing. (George recognized the symbols as a second order differential that applied to turbulent flow in valves.)

Millard: Well, I guess that I do get involved. It's usually pretty noisy around here, and when things calm down a bit, I can lose myself in these equations and then they come out right. I think that I almost have the answer to that cavitation problem. If I can get it, we can decrease the wall thickness by about 25% and cut the weight down. That might even get into that new flight development program that I've been reading about in the papers. Wouldn't that be great? Do you think that the company would let me have an interconnect to our computer, so that I could work on it at night at home?

George: Well, I don't know, Millard. Why not take it up with Peter?

On the way back to his office (it was too late to talk to Peter, as he'd gone for the day), George thought over his conversation with Millard. When he got to his office, he found the informal memo from Peter. It recommended that Millard be discharged.

QUESTIONS

1. What do you think the main problem is in the engineering department?
2. Should there be any changes in the recruiting and/or reporting systems?
3. What differences do you see between the culture of the company and that of the engineering department? How do Millard and Peter fit in?
4. Describe the motivation ideas that might be intrinsic here.

(a) How would they apply to all the people in this case study?
(b) How would extrinsic motivators be described?
5. What should George do about Peter's recommendation?
6. What should George's response to Marge be? What should he say about future developments?

1.0 OVERVIEW AND INTRODUCTION

In Chapter 5, we reviewed general descriptions of the components of the organizational model. Those major internal components, people, structure, and technology, are intended to be used as a starting point in the development of your own specific management model. They are part of a typical organizational model that you, as the manager, might want to modify to fit your own situation. Because the component *people* is the most important one in the organizational model, it is described first. *People* are the knowledge workers that produce the goods and services that support the organization, and, when they are managers, they provide the missing or incomplete data necessary to make the nonrepetitive, creative decisions that solve unique problems and guide the organization's future.

This chapter is intended to provide the beginning ideas about that most important component, *people*. These ideas include suggestions about recruiting people, acclimating new recruits to the company, and then motivating them to contribute:

In Chapter 7, we continue by prescribing how to provide feedback and rewards as *people* contribute. In later chapters, we deal with other components of our organizational model, including *structure* and *technology* before moving on to nonrepetitive decisionmaking (i.e., leadership) and repetitive decisions (i.e., information systems).

2.0 PEOPLE: AS THE MOST INDEPENDENT COMPONENT

As noted before, *people* is the only component in the central area of our simplified organizational model (*people, structure,* and *technology*) that can exist separately outside of that model. People can do things that are independent of the work situation. Although it might be argued that the second component, *structure,* might have some slight independence if work-related social groups continue after working hours, that component is itself dependent upon *people* for existence. Even though the work-related discussions during the engineering department's weekly bowling tournament about the latest design problems could meet the definition of structure (i.e., repetitive work-related behaviors), they are not completely predictable and repetitive interactions (one definition of structure). Therefore, *structure* is not independent and doesn't exist consistently outside the company. Finally, the third component, *technology* (or the methods by which organizational inputs are converted to outputs), is the one with the least independence of the three as it exists totally within the organization. *Technology* includes all the ways that the organization converts inputs into desired outputs and this might blend computer operations, machine tool controls, and even pencils/paper design methods. Because each company is unique, it follows that technology used is also unique and it therefore presents another reason for defining it as the least independent of the three central components.

However, the independence of *people* (as far as the organization is concerned) makes this component least susceptible to management directed change. Those of us who have been subjected to repeated changes in organizational structures at work and, of course, the unending intended but not always successful improvement in the way we do things, know how important the component *people* is. When we observe people's on-the-job behavior and then try to understand what caused that behavior (i.e., the underlying thought processes), we never have complete information because we never know if the observed behaviors are typical of all the behaviors the individual can provide. What about the behaviors that we, as managers, do not see? We can only draw conclusions on the basis of our interpretations of the partial behavioral data we do see. People have both past working histories and lives off the job, and we are not aware of how these affect their behavior in the organizational situation. Therefore, the description of this independent component, *people,* is necessarily limited as are the applicable research findings. We are dealing with limited data here but that's what management always does.

We begin by describing the processes that are supposed to affect the motivation of the knowledge worker including initial effects such as recruiting and acclimatizing the new recruit before dealing with more directly applicable concepts about the motivation to work. But first, let's classify or generally define motivation ideas.

3.0 CLASSIFYING MOTIVATION

There is much to review in findings about human motivation. We'll start to classify by generalities before proceeding to specifics. The closer we get to the individual, the more variability there is and possibly the more careful we have to be in applying any theory to a specific person. It's not easy. If we didn't follow this classification path, we might be tracking through a featureless body of general information, with no landmarks as guides to the specific people we are dealing with. The framework that I use divides all the models of motivation reviewed into two major areas: those that are supposed to apply universally and those that are supposed to apply specifically in the situation.

Typical examples of the division into the universal models are general theories such as those about people's physical needs (typically the need for food, safety, affection) and cultural pressures (social needs as a job title, e.g. "What do you do?"). The specific models are commonly dependent on the definitive social situation. Admittedly, this classification (like so many others in the social sciences) is a bit artificial as an argument could probably be made for placing one or more of the models in either division and you could undoubtedly improve the classification by rearranging the models in those divisions. However, the purpose of this framework is merely to assist in relating the various motivation concepts to your own framework or theory. When you develop that framework, you may want to use some of these concepts. If they work for you, so much the better. After classifying and relating these models, we'll develop some very general prescriptions that can be applied in technical organization. Now, on to the beginning: recruiting.

4.0 RECRUITING AND MOTIVATION

Some recruiters have not completely thought through the screening process and use faulty selection criteria as, for example, academic grades. There is a less than meaningful relationship between the candidate's academic grades and later management success according to the following researchers:

> Grades . . . were found to be weak predictors for . . . on-the-job performance. Participants were reassessed after 8 and 20 years with the overall indication being that although college experiences can be a meaningful determinant of managerial performance, one needs to relate particular university experiences to the specific criteria emphasized in different jobs. (Gattiker 1990, p. 107)

and

> It seems reasonably clear that the grading system (at all levels, including the graduate one) rewards the conforming plodder and penalizes the imaginative student who is most likely to make a significant contribution to engineering or science. (Shannon 1980, p. 118)

Therefore, in an interview of a neophyte engineer-scientist, the individual's academic grades would be a poor predictor of future performance on the job. Some companies may rely on another weak screening criterion: psychological testing.

Although psychological test data might be useful to detect major personality problems, they are not very useful as indicators of a future successful behavior that is desirable in technical operations. In a classic research text that challenges the assumption of relevance between this kind of test data and future organizational success, the author states:

> Global traits and states are excessively crude, gross units to encompass adequately the extraordinary complexity and subtlety of the discriminations that people constantly make (p. 301) and . . . With the possible exception of intelligence, highly generalized behavioral consistencies have not been demonstrated, and the concept of personality traits as broad response predispositions is thus untenable. (p. 146), (Mischel 1968)

Although personality research findings are very limited and not specifically helpful to the recruiter-interviewer, one might consider generally that

> People who do things better . . . are characterized by precision, reliability and efficiency. . . . They seek problem solutions by using tried and true methods. They . . . would be expected to be risk avoiders. . . .
> . . . People who do things differently are called innovators and are characterized as tangential thinkers who may approach tasks from unexpected angles. . . . Innovators can help work groups break with the past . . . would be risk takers. (Duchon et al. 1992, p. 114)

In response to the above quotation, the questions might be: How does one measure for these ephemeral qualities of precision, reliability, efficiency, innovation, etc.? The first step is to establish the criteria for selection.

4.1 Setting Criteria For Selection

Every recruitment or selection process must begin with the recognition of some predetermined, unmet organizational requirement. Why will we recruit and what is expected of the new recruit? To assist in defining those requirements specifically, these are some important questions that might be appropriate.

1. *What* do you want the person to accomplish?

2. *How* will it be measured and tied specifically to the accomplishment?

3. *When* do you expect it to happen? and finally, if you are responsible for the person's salary,

4. *How much* will it be worth to the company *at that future time?*

In too many cases, these requirements are either not defined or are poorly described. While developing the above questions, the recruiter should consider that it is not technical competence that is desired as much as social competence. Technical competency is relatively easy to acquire if one has the basic scientific background. Social competency or the capacity to "fit in" is more difficult. There is another view on technical competency.

It has been observed that for information workers, engineers and scientists, the pace of technological development gives the new graduate a half-life as short as five years . . . this means that within five to seven years, one half of the technical specialist's knowledge is no longer current. This has far reaching implications for . . . staffing . . . in addition to the most obvious needs for continuing development efforts (Coombs and Rosse 1992, p. 96)

Technical competency can be upgraded and assuming that the person being recruited has the basic scientific knowledge, upgrading should be a normal part of every corporate training effort. In my experience in many technical companies, it is the social competency, or the person's potential ability to mesh with the ongoing organizational situation, that is a major consideration if the recruit is to survive. The first question noted before: What do you want the person to accomplish? must be modified by: With whom and in what situation do you believe that accomplishment should take place? When all the answers to the questions noted before have been formulated by the recruiter, it is appropriate to take the next step—developing a screening process. The process of defining the organizational environment is often overlooked in the recruiting process. It's often, "We need a "turbo-encabulator" engineer with ten years of experience in semiturbulent flow of partially compressible noxious gases," instead of "We need someone who has a strong basic technical knowledge with the mental flexibility to quickly learn the state-of-the-art in our company and be able to work in a team environment with our people." Each company is different and, in my opinion, technical knowledge is much easier to acquire than social competency. Learning about semiturbulent flow is easier than learning how to get along with the other team members.

Even objective technical testing of candidates can be a trying and difficult process. Most tests have been shown not to be either valid or reliable. A test is valid when it predicts what it is supposed to predict. For example, the results of a person taking a dictation test should be predictable of that person's ability to perform well in similar situations. Reliability means that the test result would reappear if the test would be administered again. In other words, if the test results measure the same thing consistently.

Therefore, although theoretical questions that would present typical, existing problems that the new employee might be expected to face in the future would be a possible, although inexact indicator of the interviewee's *present* thinking methods, it is still up to the interviewer to determine how much credence to place upon the interviewee's responses. In any event, however, the questions should refer to both present and expected future organizational situations. It seems to be the best we can do.

4.2 Screening Processes: Interactive Interviewing

Consequently, after the recruiter develops the optimum interview questions and the position has been adequately publicized both within and outside the company, it is the interview that is usually the major screening device. When the recruiter-interviewer is satisfied that the general defining questions noted before (which are specific to the organization) have been developed, it is appropriate to create more specific questions that would apply to the particular job opening. Some researchers suggest that there are detectable differences in behavior and responses to these types of interview questions. The questions generally begin with, "If this is a situation that could have occurred in this job, how would you (the interviewee) handle it?" Another type of question would be to ask the interviewee to, "Tell me about how you designed the etc., etc., or accomplished the etc. etc." (referring to a part of the interviewee's resume).

A personal note: don't ask the interviewee for his/her strengths and weaknesses. It would be a complete neophyte who would *not* respond (as I have done so innumerable times), "My strength is that I work hard and long at the job. My weakness is that I really don't want to stop working at any task that I have." Really now!

An interview of a more experienced person would certainly be based on past experience and some familiarity with the situation being presented right now. As noted before, the interviewing process is always open to serious questions of reliability. During a one-on-one interview, there are the problems of the interviewer's predispositions, the interview environment and the interviewee's discomfort. These factors may negatively affect the interview. There is also the obvious factor that when several people are to be interviewed, the results of any one interview may be lost over time. For example, what were the first person's answers last month and how do they compare with the present person's answers today?

4.3 Comparing Interviewee Scores

By requiring each interviewer to ask very similar questions of all interviewees and then rating each of the interviewee's answers subjectively with an ordinal score of, say, from 1 (minimum) to 10 (maximum), it should be possible to gain *some* minimal semblance of reliability—at least for that interviewer across many interviewees. Then compare total scores for each interviewee after the interview.

(For example, "The third interviewee has a higher score than all the others. I guess we'll make her an offer.")

As the position being filled becomes more important to the company or higher in the structure, it might be desirable to have several interviewers talk to the potential recruit separately. A string series of interviews can increase reliability (i.e., the coincidence of evaluations across several interviews) when several people independently conduct an interview. If the subjective ordinal scores are very different among interviewers, the reliability would be low. ("Why do you have a score of 30? I have a score of 75 for that interviewee. What do you perceive that I did not or vice versa?") Conversely, when the scores are similar, the interviewers would all seem to agree on the interview results for a particular recruit.

For maximum reliability used when the position being filled is either critical or is at a high organizational level, the "mass" interview is suggested. This is one in which there are several interviewers in the process simultaneously. ("Do you mind if three of us interview you? We want to be sure that you are reviewed by several of us because this is a very important job in our company.") Then, when one of the interviewers asks a question, the other two or three takes notes and "score" their perceptions of the interviewee's answers. By requiring each of the interviewers to ask different questions while the others write down their own scores, a greater inter-rater, same-time scoring should occur. Inter rater means that the two raters are comparing scores of the same response, received at the same time, from the interviewee. If the subjective scores of the different observers don't match fairly closely, it indicates that the observers perceived different things in the interviewee's response. If the scores match closely, there is a greater indication of inter-scorer reliability.

4.4 Acclimatizing and Socialization

After an offer is made and accepted, the recruit joins the organization. In too many instances, the recruit is introduced around and then expected to produce without any extensive acclimatization process to familiarize him/her with the social environment. That can be a disaster. In some instances, if the recruit comes from another part of the organization, the acclimatization process can sometimes proceed informally and very quickly. There are mutual friends, similar administrative procedures and relaxed discussions about the new situation. That is usually sufficient. However, if the recruit is new to the company, it is wise to consider a formal process to speed integration and acceptance. As a minimum, that might include providing an employee manual that outlines organizational goals and philosophies, standard operating methods, compensation and fringe benefit packages, performance appraisal procedures, career development techniques and any other information designed to inform the recruit as to what the company is about and what it expects.

Another process would be to assign a coworker as a "mentor" for the first few months. Someone who is familiar with the situation can be a great help to the rookie. In very complex situations, even a mentor may be insufficient to assist in acquiring new habits, skills and values. In that event, formal training designed specifically to answer many questions that others have asked is often helpful in "learning the ropes." Sometimes companies provide retraining of employees when there are major organizational changes. In that case, the situations of the "new recruit" and the "old timer" are very similar because both have to learn what the new organization is all about. Learning the ropes is the first step. The next is developing the motivation to contribute.

5.0 MOTIVATION: SHOULD I CONTRIBUTE?

In technical (and nontechnical) organizations managing has been defined as an inferred mental process of making nonrepetitive decisions involving others by "absorbing uncertainty," with uncertainty being a function of the particular job and person in that job.

It is implicitly assumed in this process is that the decision maker has both the intent and the ability to make those decisions. With that assumption valid, the process would be followed and decisions would be made every time. That doesn't always happen. One reason many costly, well-intentioned management systems fail is that even though the managers get the information on time, there is no assurance that they will do anything with it! Surely it is obvious that managers are not always willing to make decisions. We next consider the willingness to make decisions, that is, individual motivation.

We start with a relatively safe and low-level assumption: that all behaviors are the outward results of some internal mental processes of the people component and that those internal mental processes are also at the core of decision making. It then becomes important to learn all we can about those behaviors to increase our understanding of the mental processes behind them. We're concerned with how, where, and why those processes came into existence. More specifically, we want to define, understand, and (if possible) prescribe how to modify the on-the-job behavioral subset positively, in order to provide the organization with better management through more effective decision making.

6.0 PRODUCTIVITY: PEOPLE BASED

The organization is usually expected to produce something of value, and irrespective of whether that value is in goods or services, one measure of how well they are produced is called *productivity*. One general definition of productivity is the end relationship between input and output (not how inputs are changed into outputs—that's technology). That relationship has two very general limits. The first limit occurs when we say that productivity increases when the input is fixed and the output increases, and the second limit happens when the output is fixed and the input decreases. In more general terms, when there is a minimal input for a maximal output, we have optimum productivity. We will not be concerned here with either of those two outer limits, but generally with optimizing the human mental input to increase the physical output without changing the economic input at all.

For our purposes, in order to better understand let us backtrack a bit into economics and define the total input as some function of the product of capital invested and human effort (capital × human effort). If we concentrate momentarily on the physical capital investment part of this equation, we find that in most industries, during the first part of this century, this capital investment per worker increased almost every year, with a resulting equivalent or greater increase in production output. This output increased because of capital spending on the methods of production (those methods being technology) and in internal operations (i.e., more effective and flexible management structures).

However, investment in another kind of capital (training the people) was not very great. And even when minimal funds were spent on training, it was almost impossible to connect those kinds of capital costs directly with increased productivity. Measuring the changes in output of improved machines and systems is a lot easier than measuring the output of people, especially when that output is intermingled with the efforts of many people such as in a team project. Therefore, most economic justifications of capital spending were based in installing faster and better machinery, in less expensive purchased materials, and in reduction in the measurable, direct labor content in the products. The worker was considered to be an adjunct of the machine; thus, decreasing labor hours was the same as increasing machine productivity by decreasing direct costs.

Within more recent times, there has been a change in capital spending because modern technical operations are quite different from those of earlier days. The knowledge worker, and not the production machine, now determines the amount of the input for a proposed output. Investment return on capital, therefore, is closely tied to the more or less independent efforts of the technical worker, who is definitely not a machine adjunct.

Reviewing just a bit, productivity was defined as a function of capital and human effort and that human effort became more critical with the increase in the number of knowledge workers in the technical organization.

$$\text{productivity} = [\text{some function}, f_1] \times [(\text{capital}) \times (\text{human effort})]$$

In recent times, the expenditure of capital has changed from cost reduction to becoming a requirement for survival. Updating obsolete equipment is necessary even though it may not result in an increase in productivity because competition never ceases. However, even under the best of circumstances, it is not easy to measure productivity improvement from capital investment, because the situation after the investment is never quite the same as predicted. Materials may change, vendor shipments improve, or other interacting events happen to make measurement fiendishly complex. It becomes even more complicated when we try to measure the effect of the human mental, not physical, effort. Human mental inputs including ideas such as commitment and inspiration are surely as important as the estimates of the materials by the cost accountant. Differences in those human inputs do exist and do affect productivity, but accounting cannot measure that. However, using some very generalized guidelines, it should be possible for you to come up with an overall definition of measurements that, although not exact, are adequate for management purposes. These measurements are in units that are all encompassing including those we generally use in scientific or technical areas.

6.1 Measuring Productivity or Motivation: Setting Ordinal Scales

All measurements are in nominal, ordinal, interval, or ratio systems. A *nominal* measurement is one that differentiates only between two or more categories. (For example, "the red checkers are mine and the black are yours.") An *ordinal* measurement is one that defines "more or less" than something else, but the amount of that "more or less" is *intended* to be linear though not actually so. (For example, "I like Remington's Paintings of the Old West better than I like Warhol's paintings of soup cans, and although I can't give you any objective measure of that liking process, I'll trade you three Warhols for one Remington.") That ordinal measurement may not be exact across many measurements but it's the best that we have when *people* assign values. In this example, we can even value nonobjective things such as our appreciation of a particular piece of art. It's possible to define our opinion of good and bad paintings and even measure that "goodness" using different measurements that are either nominal or ordinal. "I like black shoes more than brown ones (nominal) and I'd like to exchange two pairs of these new brown shoes I received as presents for a single pair of black ones." (ordinal). Management often must use ordinal measurement because it's useful for comparisons of human effort. "Charlie is one of the best hydraulic systems designers in this company. I think that he's able to do twice the job of anybody else." There are also *interval* and *ratio* measurements.

Interval measurements are linear but have no absolute zero. For example, 60 °F is not twice as hot as 30 °F, because there is no absolute zero on the Fahrenheit temperature scale. It is just 30 °F hotter. *Ratio* numbers are linear and *do* have an absolute zero, such as is found on the Kelvin temperature scale. Thus, 60 K is really twice as hot as 30 K, as well as being 30 K hotter.

Ordinal measurements are similar to those in decision making under conditions of risk, where the decision maker has to estimate the value of the possible solution and the probability of its occurrence subjectively. Therefore, as we have neither an absolute zero nor any way to define objective linearity between two points in our measurement system, we'll obtain the best measurement of productivity using a subjective *ordinal* measurement system to determine the amount of human effort. But we, as managers, have to define what the optimum is on, say, a scale from 1 to 10. For example, a "10" may be absolutely the best and a "1" may be the rationale for warning someone that they must improve. As managers, we are concerned with optimizing human effort, and because the individual's motivation is important as the prime determinant of human productivity in technical organizations, we will try to measure that motivation using ordinal numbers in an evaluation scale.

A useful definition of the human effort variable in productivity includes both the mental and the physical inputs of people and is a function of their abilities and motivation. As noted before,

$$\text{productivity} = [\text{some function, } f_1] \times [(\text{capital}) \times (\text{human effort})]$$
$$\text{and}$$
$$\text{human effort} = (\text{some function, } f_2) \times (\text{abilities} \times \text{motivation})$$

Abilities are defined as the combination of both the inherited characteristics and the training and conditioning of the particular person. Productivity can therefore be defined now through the substitution of the formula for human effort into the prior formula:

$$\text{productivity} = f_1 \times [\text{capital x } (f_2 \times \text{abilities} \times \text{motivation})]$$

For example, this formula tell us that the productivity or the music produced by a concert violinst is dependent upon the violin being played multiplied by the abilities of the violinist and the motivation to play. Abilities include an innate, inherited eye-hand coordination and years of corrected practice. The effort the violinist exerts in making music results from using those abilities and from a directed mental process called motivation. That directed mental process can only be *inferred* from observation of the violinist's behavior. The assumed selection of this particular goal-directed behavior is made from the total repertoire of possible behaviors perceived to be available to that violinist. Obviously, motivation is a major key to human effort, which is in turn the key to productivity. In technical organizations, productivity depends directly on people because people move the organization.

This rather lengthy development has resulted in an intuitively apparent idea.

Human motivation determines the technical organization's productivity as capital must be spent anyway to maintain the health of the company, i.e., "We need that computer software upgrade to remain competitive even though the old software still works very well." We measure human output (i.e., driven by inferred motivation) only through some kind of ordinal system (i.e., by determining whether it is better or worse than some management predetermined standard).

7.0 MOTIVATION AND BEHAVIOR

Motivation includes the idea of choice. In our example of what managers do, one choice is either to make or not to make a decision. Choices are vital in determining organizational productivity in the technical organization. We assumed that those mental choices/decisions can be inferred from what we, as managers, see and interpret the behaviors of other people when those people are at work. There are quite a few uncontrollable variables in these behaviors that we can only guess at without even considering if we interpreted what we saw correctly. For example, we know that behaviors are affected by time, off-the-job experiences, and the job itself. Some basic research has found that four facets of the job have important psychological effects. They are:

1. Organizational characteristics (position in the hierarchy, ownership, bureaucratization)

2. Degree of self-direction and occupational commitment (closeness of supervision, complexity, and work routinization)

3. Amount of job pressure

4. Level of uncertainty (Kohn and Schooler 1973).

But the data also show that with education and other factors of occupational experience controlled, the degree of self-direction has the strongest influence. We will now do some theorizing of our own. If self-direction is similar to independence, and if occupational commitment is related to motivation to work, we have some research justification for a preliminary assumption that technical people seem to be motivated to work when they believe that they are independent and self-directed. There are a lot of "ifs" but they seem to be reasonable, at least with respect to technical operations in which the processes of production are controlled by the person.

The effects of the present job can be added to the effects of the work that occurred earlier in the person's professional career. One experimenter (Kaufman 1974) investigated the relationship of the work challenge that a varied group of eighty-five engineers had at the beginning of their careers to their consequent work performance, professional competence, and contributions. For those who possessed high technical ability, early challenge reinforced professional contributions and competence later in their careers. This type of research seems to say that it is possible to have long-term effects from early work experiences, at least for technical personnel with ability. These two examples of research findings indicate some of the complexity of the connection between observed behaviors and our interpretations or our inferences of the other person's motivation. They are applicable to our attempts to use research findings that apply to motivation in technical organizations, but first we must explore some basics that could be foundations to ideas about the motivation to work.

7.1 Why Work? History

The motivation to work, and even work itself, has had a spotty history at best. During the early Greek and Roman eras, work was something to be done by slaves, farmers, or the lower fringes of society. It was not accepted as necessary for free men and was not considered to be the best way to live. Politics and philosophy were important; work was not. Work still has a bad name to some extent. Consider,

> This book, being about work, is, by its very nature, about violence to the spirit as well as to the body. It is about ulcers, as well as accidents, about shouting matches as well as fist fights, about nervous breakdowns as well as kicking the dog around. It is, above all (or beneath all), about daily humiliations. To survive the day is triumph enough for the walking wounded among the great many of us. (Terkel 1972, p. xi)

But, this attitude is not unanimous in many technical organizations. It can be contrasted to a spontaneous comment I once heard during a design meeting in a high tech company, "This place is so much fun, and I even get paid to come here! If this were a play, I might consider paying admission to join it." Differences in attitudes and approaches to work spring from many places. Freedom to control one's own work and early work experiences (see above), social or cultural norms, and other factors are responsible for that inferred totality called "human motivation." Even interpretations of our observations of behavior may not be valid.

> Any reader must be aware of the assumptions undergirding observation. One person's burdensome work may be another's releasing joy, for the human spirit works in strange ways between euphoria and martyrdom. In many respects, of course, the problem of work . . . is really the human task of responding creatively to existence in the cosmos. It is the problem of living. (Fairfield 1974, p. xiii)

I have noticed in various books about motivation theory some explicitly negative prejudices of the researchers when they were describing their inference about human motivations connected to difficult or physically demanding jobs. I felt better when I discovered the following quotation.

> There is a peculiar arrogance in those who discourse on the brutalizations of work simply because they cannot imagine themselves performing the job. Certainly workers often feel abstracted out, reduced sometimes to dreary robotic functions. But almost everyone commands endlessly subtle systems of adaptation; people can make work all their own and even cherish it against all academic expectations. (Morrow 1981)

Considering these quotations could throw some light on the potential motivational differences of others such as that between the person who is the head of the design standards department who seems to delight in repetitiously fitting every screw and valve into its proper specification sheet and the advanced product designer who seems to achieve equal delight when using novel screws or valves that are not even on those sheets. They both enjoy, but differently.

Now having been warned about the possible historical and social biases of everyone (including you and me) both for and against work, we are prepared to deal with some research on motivation theory and findings. But, as I said before, what you see in either the theory I will describe or the work behavior you observe may only be what you interpret. And it can be measured only subjectively, with ordinal measurements.

8.0 DESCRIPTION: MOTIVATION THEORIES

In Table 6-1 there are two broad vertical columns used for classifying universal and situational theories, and then a horizontal classification into rows within those columns of applicable motivation theory by *culture, group,* and *individual*. Hence, these theories are categorized first in general terms and then in more specific areas. The closer we get to the individual, the more variable (and more difficult) it becomes to define exactly what that person's inferred motivation is.

Table 6-1. Motivation Theories

	Universal	*Situational*
Cultural	Religion-culture	Society-class
Work Group	Human relations	Participation-morale
Individual	Psychological	Social-psychological

We will describe the theories classified in the Universal column first and then the Situational column.

The Cultural row across both vertical classifications include theories that apply to major themes of life, such as religion, general social culture, and social class. The Working Groups row applies to relatively stable peer groups and their effect on individuals. This row includes themes such as human relations theory based in the Hawthorne studies, morale, and participation. The Individual row applies to the psychological characteristics of the person, and the person-social group interactions. That is the most important row for us because it is the individual knowledge worker who is the supplier of technical productivity. The vertical column classification, therefore, moves from the general (i.e., culture) through groups into individuals. And the horizontal classification moves from universal theories to more specific situational theories. This classification scheme gives us the ability to relate one motivation theory to others and possibly pick and choose from them for our own special theories.

For example, considering the two broad vertical columns of universal and situational theories, a theory that states that there is an inherent "need for achievement" (McClelland 1961) in managers would probably be universally applicable and relatively independent of any situations or contingencies. On the other hand, the suggestion that an unmotivated person at work would respond quickly to the social pressures from people who are considered to be technically superior (Schachter 1959) would probably fall into a social-psychological, situation-dependent category. However, before going further, let me repeat that none of the research we shall classify suggests that motivation is completely independent of either the person or the situation: they are interactive. After all, we do not live in social vacuums and we all interact with each other and our interpretation of the situation to some extent.

This classification is useful because it seems to allow different ideas to be understood more easily. The theories do not always compete; in many cases, they supplement each other.

8.1 Cultural Inferences

In the first horizontal row called *cultural,* the person's motivation is assumed either to be *universally* applicable or to depend upon some contingency or interaction with the work-related *situation.* It's a matter of scale. If the theories to be explored assume that the motivation to work is universal in that culture, the motivation should be relatively independent of the organizational work situation, as it is an almost constant requirement of that social environment. On the other hand, if the source of motivation in that culture is assumed to be within some identifiable class within a society, it's classified in the second column as based in the situation and the motivation will not be universal. Only those who interact with that class in society are influenced by it. In the latter case, motivation is defined as society-class. In both columns, however unfortunate, there is always some overlap.

The Universals

The classification of universal motivation (religion-culture) to work as part of a culture is easily illustrated if one looks at the historical record that "millions upon millions of humans have worked at some of the most onerous tasks imaginable to perpetuate their society's values, both tangible and intangible" (Fairfield 1974, p. xiv). Religion has often resulted in the expenditure of tremendous effort, with some promised afterlife the only reward to those involved. According to recent archaeological interpretations, the Pyramids were built by Egyptian farmers during the annual flood of the Nile River, when farming was impossible, as a way to share in the eternal life of the Pharaoh in the Land of the Dead.

In more recent times, the Protestant work ethic has become part of western working culture. The concept itself has existed for a long time, but it was very well described in seventeenth century Protestantism, which proposed that the route to salvation and escape from eternal damnation in some afterlife was a laborious one in this life involving continuous mental or physical labor. Time spent in leisure and enjoyment was wasted and justified moral condemnation (Weber 1930). This work ethic was a major motivational push during the westward drive across the North American continent during the last half of the nineteenth century. It became an implicit part of early management theory (Taylor 1911) and, although it may seem to have lost much of its force in relatively recent times, it still appears occasionally in the motivational culture of some western countries.

Another of the broad bases for a universal motivation to work is that inspired by patriotic considerations. Of course, that could be a subset of culture. For example, in a country at risk from an external source or in a country at war, there are reports of factory workers who labor almost unceasingly and of members of the armed services who seemingly give their lives willingly because of responses to the apparently universal motivational forces that they feel. Similar but, of course, less emotional forces can be detected when an entire industrial organization happens to be caught in a temporary economic emergency (though not to the extent of requiring one's life, I hope). Overtime may be volunteered for extended periods of time or other personal sacrifices offered.

In these examples, a nation at war and a company in economic trouble, the motivational time span for individuals to participate is much shorter than the length of their working life, which is affected by general cultural and religious forces. Unless there is some additional payoff over the regular rewards that they receive, this motivational pressure quickly subsides. With no such payoff to the individual (either tangible or intangible), demotivation begins within a relatively short time.

The Situationals

General cultural forces may also affect situations. Consider observations of behavior directed at joining a different social class. The relationship between behavior and social class culture that leads to these expensive habits is beautifully described as follows:

> The quasi peaceable gentleman of leisure, then, not only consumes of the staff of life beyond the minimum required for sustenance and physical efficiency, but his consumption also undergoes a specialization as regards the quality of the goods consumed. He consumes freely and of the best, in food, drink, narcotics, shelter, services, ornaments, apparel, weapons and accouterments, amusements, amulets, and idols or divinities since the consumption of the more excellent goods is an evidence of wealth, it becomes honorific; and conversely, the failure to consume in due quantity and quality becomes a mark of inferiority and demerit. (Veblen, 1935. pp. 73–74)

Conspicuous consumption behaviors are alive and well today. And this is only one of many cultural determinants of human behavior.

In the more limited areas of the working organization, company cultures exert pressures on behavior just as those larger cultures do on life outside of work. For example, a fairly large company had promised to provide almost permanent employment for those participants who cooperated with both the written and unwritten company policies, defined here as the company culture. That promise was never completely and explicitly documented, but it was similar to other company cultures, in that everyone in the company just "knew" it existed. In this case, personnel policies were designed to reward human conformity and employment longevity, using undocumented social controls not usually considered to be company concerns.

Everyone in the company for any length of time knew what the policies were. These nondocumented policies directed the expected hours of unpaid overtime work to be contributed by technical professionals, suggested the dress code, and even recommended educational institutions for employees' children. It was a patriarchal or familial type of organizational culture; one that generally promises more than it can ever deliver.

As it turned out, that semifamilial culture was a major factor in the company's eventual economic failure. When the market moved away from its major product line, there was no one in the company who would point out to the patriarchal company president that the old products were no longer the best in the trade. Sales and profits dropped alarmingly, and when the president became ill, there was no successor to take charge. Then the company was bought by a large conglomerate and the extensive employee benefit plans (both explicit and implicit) were scrapped quickly in order to "bring this division into line with all our other divisions." New top management was brought in "to shake things up a bit," but the shaking up didn't work well. The ingrained company culture required a patient and lengthy gradual wearing away from the familial culture into a more objective one, but the conglomerate was not structured for patience. The conglomerate's culture was for the here and now, with profits that had to increase every financial quarter. Therefore, it attempted to change the company's culture radically by introducing new policies and planning methods. These were quietly sabotaged by the employees. That new process had violated their cultural expectations. Even though the people realized that the company itself had to change, the change was too fast for them. With more time and training, it might have succeeded. The actions of the conglomerate, of course, were not entirely wrong either; it had to move quickly because of the loss of the company's markets. Things continued to go downhill, and the company was finally dissolved by the conglomerate after several years of poor performance.

8.2 Summary of Cultural Motivations

Cultural pressures affect people's motivations in both the larger environments of society and the limited ones of the company. Those pressures have deep roots and must be considered in developing your personal theory about motivation in technical operations. Any hypotheses that include modifying company culture rarely succeed without the cooperation of both the formal (i.e., management) and the informal structures of your knowledge worker. There cannot be rapid violations of existing organizational mores without inevitable consequent damage to productivity.

However, mores can be changed over an extended period of time. One division general manager applied three years of staff training, repeated functional problem-solving meetings, and personal corrective action to his management group before he was able to move that group from the closed style of management that he had found when he arrived in the division to the more open style of problem confrontation and team cooperation needed to provide new products and innovation. Therefore, a hypothesis requiring modifying major parts of the company culture is probably not within your capability as a technical manager. We discuss it here only as general background material so you'll know what to do when you become the general manager. We now move to a slightly smaller arena and describe some effects of working groups within the company on the person's motivation.

8.3 Working Groups

The effect of the small group on the individual's motivation was carefully reported in the research done during the Hawthorne experiments, which were first started in 1927. We described some of the results of these experiments in Chapter 4, in terms of the effects on the formal organization. Now, we investigate the effects on individual motivation. To briefly review, at that time, the Hawthorne Works was Western Electric's largest manufacturing plant. Western Electric itself was part of the Bell Telephone system. The experiments continued for about five years, and their purpose was to test the effects of physical changes in the worker's environment on the work output. Those changes typically included temperature, humidity, and number of rest breaks during the day. The experimental environment of a small working group was selected in order to obtain better control over the data.

> In a small group it would be possible to keep certain variables roughly constant; experimental conditions would be imposed with less chance of having them disrupted by experimental routines. It would also be easier to observe and record the changes which took place both without and within the individual. And lastly, in a small group there was a possibility of establishing a feeling of mutual confidence between investigators and operators so that the reactions of the operators would not be distorted by general mistrust. (Roethlisberger and Dickson 1939, pp. 19–20)

Initially, the output of a small experimental group of five workers in a relay assembly room was measured. The experimenters found that changes in the physical environment, rest periods, improved lighting, and so forth produced no directly related change in output. There was, however, a general and persistent tendency of output to rise during all the experiments. This happened irrespective of the increase or decrease in the physical variables. Increased lighting increased production, but so did decreased lighting.

The explanation for this peculiar result was quite novel. As noted in Chapter 4, it was proposed that the group had achieved a shared work morale (informal support) that was perceived by the workers to be reinforced by the management structure (the researchers as the formal organization). They increased their productivity. The informal and the formal organizational structure were mutually reinforcing. The researchers reported that:

> "In human organizations we find a number of individuals working together toward a common end: the collective purpose of the total organization" (Roethlisberger and Dickson, 1989, p. 533).

The research direction was quite obvious. Strengthen the work group, build internal cohesion, and ensure harmony between the plant management and the group goals. However, there was a problem in implementing these ideas. Work groups and plant management goals do not always coincide. This was illustrated by another experiment completed in the Hawthorne studies. In the bank wiring room, the output did not have the same general and persistent tendency to rise during the experiments that had occurred in the relay assembly room. It remained relatively fixed throughout the experiments, even though all the variables that had been tried before were used again. The researchers even reported that whenever one worker in the wiring room experiments exceeded the production standard established by group norms, he was disciplined by the group through physical horseplay and discussions in order to reduce his output. The group's norms (informal organization), which were intended to "protect" them from their perceptions of an oppressive management (formal organization), were controlling output even though it was economically advantageous for the worker to exceed those norms. Payment was based on individual incentives, but the worker was apparently more concerned with group pressures than with personal income.

The Hawthorne experiments suggest that optimum productivity occurs when the pressures exerted by the informal and the formal working groups *on the individual worker coincide*. During the time of these experiments, there was relatively little management concern with the individual, because it was assumed that he or she was modifiable by the group, which in turn was controlled by the formal organization. Management concern for that individual was limited to determining what social pressures could best be utilized to promote group harmony. In other words, individual differences, which were supposed to result in discord, were minimized, because it was assumed that the group controlled the person and management could control the group.

In effect, the company culture was supposed to be integrated into the group's culture. Individual differences were discouraged if they didn't match the norms of the company culture.

Further research directed at geographical differences, however, reported consistent differences in individual social norms between workers in smaller communities and those in larger, industrialized cities. This made the task of making the individual conform through the group to company culture very difficult. If workers were really that different, no uniform company policy would work for all. Blood and Hulin (1967) found that the human relations concepts of matching group and company goals are more easily accepted by workers in smaller communities than in larger ones. This meant that the findings of the Hawthorne experiments were less universal than originally predicted, because the company now had to account for the individual's cultural and geographical background *before* he/she joined.

The Hawthorne results were not as universally applicable as had originally been claimed, but the company-to-group-to-worker motivation sequence was not entirely eliminated. It was just limited in that it had to make room for the person's antecedents. This is a classic example of original data being reasonable but incomplete. The theory, then, is also incomplete, and there are probably other data yet unexplored that will help to provide more complete answers.

8.3.1 Effects of Groups

The Hawthorne studies found that group norms might not coincide with those of management. Other data showed that individual norms might not coincide with group norms. The answer seemed to be that this happened because management had not been able to show the workers that coinciding goals were best for everyone and the group. Management's formal goals were supposed to prevail, of course, and it was management's problem to change the motivation or attitudes of the workers to match those goals.

Sometimes the employee is even more concerned about meeting group norms than about accepting some individual kind of payment from management as a result of his or her own behaviors. The group norms become an end in themselves and are as highly valued as any kind of compensation. That compensation may take several forms, such as exchanging social benefits with each other (Homans 1958; Schrank 1978) and attempting to gain control over one's work (Schachter 1959). For example,

> I have found that it is terribly important to feel part of a community in the workplace. There is something that the work itself can never provide us with . . . for instance, . . . the work as it was organized did not permit me enough schmoozing time, time to wander around the plant, visit and talk to people in other departments and not be stuck in one spot doing the same thing. We need to see workplaces as communities where people go to carry out tasks for which they are paid . . . what is most neglected . . . is the nature of the human relationships the rituals such as greetings on arrival, coffee breaks, lunch time, smoke breaks, teasing, in jokes, and endless talk about almost everything. (Schrank, 1978, p. 78)

These needs (as part of the individual's self-interest) are proposed to be universal and not dependent upon any situation. That is the social support on the job that comes when a person accepts the group's behavioral norms in return for the social interchanges he or she values. It is suggested that the social support exchange and the need to compensate for lack of control over the work are universal motivational factors associated with groups. Although this seems to apply less to our types of knowledge workers, it may be partially true for them as well.

Some research indicates that the worker needs the informal group to help in a defensive reaction against apparent loss of control over the work. When the worker joins others who are in the same predicament, he or she receives some psychological support against feelings of inadequacy. "Misery loves company" is a truism, according to this research, because the group is also miserable (Schachter 1959). This could explain some of the person's dissatisfaction when a work system excessively simplifies and rationalizes the flow of work to be done. If the worker's efforts are totally defined, he or she is totally controlled by the system, rather than the other way around. It is only when he or she becomes associated with an informal supportive group that the worker senses any capacity to affect the job, as the instructions of the formal management group are only confining.

8.3.2 Summary of Working Groups

The person's working group is an important link between that person and the overall organization. The amount of control that the group exerts seems to be linked to many variables, such as group behavioral pressures (Hawthorne experiments), internal needs to socialize (Schrank 1978; Schachter 1959), and the rate of environmental change (Lawrence and Lorsch 1967). While groups are important, they provide only some of the reasons individuals exhibit a whole range of unique characteristics. The next section takes up some of these other characteristics, dealing with the person aside from his working group.

8.4 The Individual

If we could control for the effects on motivation of the overall organizational culture and the working group, we would be able to describe the person independently of any external influences. Of course, that is not possible in the real world. However, it is intellectually possible and useful because we can then use a simplified classification scheme or model to untangle the many inferences about the internal motivations of people. That's one of the advantages of using models; we can *mentally* hold the rest of a model fixed while we examine its parts.

For example, one of the oldest models of individual motivation, *hedonism* can be traced as far back as the ancient Greek philosophers. Hedonism means that people are supposed to think about selecting alternatives that lead toward maximizing their pleasure and minimizing their pain. William James (1890/1950) agreed. He wrote, "But as present pleasures are tremendous reinforcers and present pains tremendous inhibitors of whatever action leads to them, so the thoughts of pleasures and pains take rank amongst the thoughts which have most impulsive and inhibitive power." (James 1950, p. 550)

The problem with this view of motivation is that there is nothing to test. Any form of behavior can be explained as either pleasurable or painful after the fact. It's almost impossible to predict in advance. Of course, the pragmatist would say that the particular individual's assessment of pleasure or pain could be based on her or his past behavior, and that past behavior would therefore be a good indication of future behavior. Although this might be arguable, we have made no progress, because we might not have an adequate sample of past behavior.

What does hedonism really mean? How are pleasure and pain connected with our immediate concerns here, those of work and life? Sigmund Freud's general answer to this last question was that effective work was one of the three positive solutions to the human dilemma of life (Fancher 1973, p. 227). (The other two, by the way, were creative play and enduring love. Not a bad combination).

These views of work (and the implied motivation to spend time at it) that place it as one of the central thought processes of people was supported by some classic research on the motivations of the American worker (Morse and Weiss, 1955). About 80% of the respondents reported that they would continue working even if they had enough money to live comfortably. Perhaps even more interesting was a finding that the percentage of people who would continue working was positively related to the amount of training required by their particular occupations. These respondents therefore considered work to be important and they seemed generally to be satisfied with their jobs, *especially those that were highly trained*. Other research (Katzell 1979) reported that overall satisfaction levels remained high and basically constant for over twenty years. But is this the whole picture? Some critics of this research say that the interviewees really were not satisfied but reported satisfaction in order to justify their work.

An interesting and subtle point might be made here:

> When up to 90% of the workers are reported to be satisfied with their work, the behaviorists say that workers do not really know what satisfaction is and they will lead them to a superior kind. That sounds oddly like the proselytizing of a missionary. (Fein 1974, p. 76)

Considering both sides of the question could bring up some problems. Do the interviewees really know? And are the behaviorist-missionaries right? Research should always be "tested" mentally against your own experiences and situation.

Let me give you an illustration. In one company, all the technical managers complained bitterly to me about the administrative restraints on their ability to make decisions. However, on the whole, all the managers reported, in confidence, that they were well satisfied with their jobs and the organization, and their behavior on the job seemed to agree with those reports.

Only much later, did a casual remark lead me to realize that most of the workers accepted the administrative problems rather than having to relocate away from the "sunshine location" of the plant. It seemed that the reported complaints about administrative controls were not as important to these people as the unreported satisfaction with the company's location. Therefore, can we really accept research without some in depth personal interpretation?

8.4.1 Need Hierarchy

The need hierarchy motivation theory attempts to deal with this problem of interpretation, but it does not do it well. (Maslow 1971, p. 35–58). It suggests that there is a pyramid or hierarchy into which five classes of human needs are arranged. Those needs at the bottom of the pyramid take precedence over those at the next level up and are the focus of the person's attention until they are satisfied. At that point, the second higher class of needs becomes most important. When these are satisfied, the next class of needs begins to dominate and so on until the top or fifth class of needs is reached. A lower or satisfied need is no longer an important motivator and, although more than one class of need may be acting on a person at a given time, the dominant need is the lowest one in the hierarchy. The most powerful and first needs are the elementary biological needs, which provide for life itself: food, drink, shelter, and so forth. Where life is primitive, these needs are never completely satisfied and they are the preoccupation of almost every waking hour.

When these very basic needs are satisfied, the next class of needs becomes important: that of security. If the lower class of needs is satisfied, these needs are no longer as powerful motivators as the class now occupying the person's attention, but if the hunt for security begins to clash with reawakened and unsatisfied needs for food or clothing, security again takes second position. The five-step hierarchy is sequential:

1. Physiological needs: The basic requirements of self-preservation that can be described by needs for food, water, and shelter.

2. Security needs: Both physical and economic, although only the economic requirements concern us here. People want to be assured that they will be able to have an acceptable living standard that will be maintained in the future. Examples of this class of needs are pressures for maintenance of income, jobs, and general employment security.

3. Social needs: The need people have to be accepted as part of a social group and to be able to influence other group members.

4. Esteem needs: The need to feel important in society. Status, importance, prestige, and recognition are the motivators here.

5. Self-actualization: This is supposed to be the pinnacle of the need pyramid, where people want to feel that they are trying to reach their full potential and they have jobs they like that are well suited to their skills and abilities. Job and individual importance, responsibility, advancement, growth opportunity, and challenge are included here. Although Maslow called this last need self-fulfillment in earlier works, in his later work (1965) he renamed it self-actualization. That seemed to be a more appropriate name for the needs behind people in this class of the hierarchy because these needs are attached to striving for goals rather than obtaining them.

Figure 6-1 shows Maslow's need hierarchy pyramid.

The idea of the need hierarchy is intuitively appealing because it promises all kinds of explanations for previously unexplained motivational drives. For example, few technical workers in Western cultures are driven by physiological needs. Therefore, security needs could explain the fear and despair of people who are out of work for extended periods of time, notwithstanding the size of their savings or the income that they might receive from unemployment compensation. It could also explain the greater success of unionization attempts during and immediately after a depression.

The social needs of people could explain the desire of workers to move around and socialize with others (Schrank 1978) and their desire to form cohesive informal work groups or continue social relationships formed during work after working hours. Esteem needs could explain the comments on conspicuous consumption (Veblen 1935). Finally, self-actualization could explain almost everything, as it is all inclusive.

And that is the problem with this theory; it is so all encompassing it is hard to use it for specific situations. There's that interpretation problem again. The idea that as needs are satisfied, other needs rise to take their place. This tells management that there is no single solution to the problems of motivation. As soon as one company benefit is distributed or a raise in salary is achieved, that benefit or salary change is no longer a motivator.

Consider the manager attempting to use this theory with his technical personnel. How can he determine which level of need (other than physiological) is

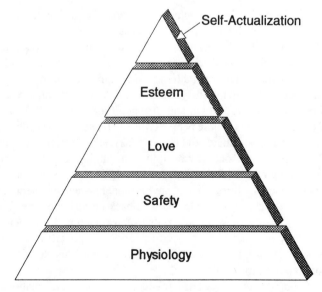

FIGURE 6-1: Maslow's Pyramid of Human Need Classes

affecting his people? And then there is the never-ending management task of evaluation of the value of the work to be done, and the determination of reward that is tied to some need. Finally, unless the recipient interprets the reward the same way the manager did and agrees that it meets his or her needs, the manager's actions may have no effect. The theory is simplistically appealing and has had little support from quantitative research. The one quantitative test (Hall and Nougaim 1968) did not support it. The slight qualitative support for the theory reported: "Workers in higher occupational prestige categories, and those who have more education, place greater emphasis on challenge, autonomy, and other intrinsic rewards" (Mortimer 1978, p. 5). That's hardly a ringing endorsement.

Hall and Nougaim (1968) tried to test Maslow's theory by evaluating the needs of managers working in a division of a large organization over an extended period of several years. Their hypothesis was that as the managers were promoted, their needs would change in accordance with Maslow's predictions. It didn't happen. The researchers reported that the managers apparently adjusted their occupational values, expectations, and aspirations over time so that they became more congruent with the rewards and satisfactions that they perceived actually existed in their work. The progression was not to satisfy a need and then work on another one higher up in the hierarchy as it arises; it was to adjust their needs to fit the opportunities available to them. That's quite a difference. It's no longer a universal theory.

8.4.2 Need for Achievement, Power, and Affiliation

Needs can be interpreted in hierarchies other than the five noted above. One research effort (McClelland 1961, 1962) indicated that most people have three related needs that affect the way they work. According to this concept, most people have one of these needs as a relatively constant and dominant force in their motivational and behavioral patterns. Although these needs may sound as if they came from a horoscope, they did not. There is substantial research that supports those descriptions (Litwin and Stringer 1968, pp. 14–24).

1. Need for achievement: These people typically make good executives, particularly in challenging or difficult industries. They enjoy activity and like work in which they can take personal responsibility for finding solutions to problems. The idea of winning by chance does not produce the same satisfaction as winning by their own personal efforts. They prefer moderate achievement goals and calculated risks. They also want concrete feedback on how well they are doing.

2. Need for power: These people will usually attempt to influence others directly, and they are seen by others as forceful, outspoken, and demanding. There is little concern for warm, affiliative relationships and they tend toward having strong authoritarian values.

3. Need for affiliation: As warm, friendly, companionate relationships are primary goals, these people often take supervisory jobs in which maintaining good relationships is more important than making decisions. This need does not seem to be important for many, but it is a fair assumption that building good working relationships with both superiors and subordinates is required to obtain other, broader kinds of personal satisfaction.

The research did not propose that any individual has only a single one of these needs, only that one of them is usually dominant. It is quite possible for a person to have all three needs. An example is the case of one organization's president who had a very high *need for power* combined with a moderate *need for achievement*. In that company's dynamic economic environment, he created a thriving and successful business. In another firm, the president had an equally high *need for power* but not as much *need for achievement*. That president dominated every meeting, personally made every decision, and finally became the limiting factor in the growth of the business. It was physically impossible for him to handle all the details of a growing business the same way he had when it was much smaller.

This need theory differs from Maslow's in that it has methods to measure the presence of the three needs. This method uses a derivation of the Thematic Apperception Test (TAT) developed by Murray (1938). The person being measured is shown a series of pictures, usually of people in ambiguous social and work situations, and is asked to make up a story for each picture in the series. The stories are recorded and then analyzed for evidence of the different kinds of imagery associated with the various motives. This test is projective because the person "projects" into the story her or his own thoughts, feelings, and attitudes. The stories are a sample of the kinds of things the person thinks about and include her or his dominant needs. Numerical scores of the imagery noted are matched to various scales, and the nominal amounts of the three needs (achievement, power, and affiliation) can be determined by a qualified experimenter.

8.4.3 Summary: Universal Motivation Factors

Reviewing Table 6-1 again, we find that the theories covered so far such as "culture" (i.e., the Protestant ethic), "working groups" (i.e., fit between people and their departments within the company) and "individual" (i.e., need theories) are all expected to be universal descriptions of everyone's motivations to work. Because they are all inclusive, they are almost useless when we have to deal with specific individuals. They explain why all people work, but that is their main problem as well as their attraction. Too general a description provides insufficient specific guidance for us to apply that description to our own situation. That step requires more tools than these theories can offer at this time. Therefore, we'll now move on to ideas that more specifically relate the person to the situation.

8.5 The Situation

Some results attempt to describe internal states of mind, and these also often depend on the interpretations of the researcher and/or the subject. These interpretations (of both researchers and subjects) might be understood better by using the next set of ideas about motivation. These ideas deal with perspectives of personal success at work (called competence). White's theory (White 1959) could be an introduction into this very personal area.

Research seems to show that individual motivation(s) can be modified when the individual in the group perceives a change in the personal situation that affects his or her self-interest. If the person now feels that it is not to his or her benefit to adhere as closely to group behavioral norms, there is less universality in those group-person motivational effects.

A major research area indicates how the situation can effect a lack of coincidence between the person's motivation or behavioral norms and those of the group. *Participation* comes into play. Participation refers to the opportunity a person has to share in group decision making and in supposedly helping to change the situation.

This research goes beyond the relatively simpler universal ideas. It now follows the reverse route, where the person is supposed to be motivated to change his or her own attitudes to attain and keep the group's support in gaining control over his or her job. In this case, the situation is held constant and depends on the tasks of groups within the company. These tasks are rarely the same for each group (the engineering group surely has different tasks and a different situation from those of the purchasing group) and the attitudes or motivation of participants depend on how well they participate in these different situations. Here we are no longer talking about groups of workers, which the Hawthorne experiments tried to do, nor are we concerned about all individuals satisfying personal social needs or controlling their jobs. We are concerned now with a particular situation's effect on attitudes and motivation of the individual through the intervening variable of the group.

8.5.1 What Is Participation?

Many times when problems arise that seem to come from some sort of human motivation, one of the first suggestions is to get everybody involved or motivated through participation. It's proposed as a simple solution to what is usually a very complex problem, and sometimes simple solutions just don't work as expected. The research shows that this technique does not always produce an unequivocally positive answer. It has mixed results.

In perhaps the best-known experiment in participative management (Coch and French 1948) the researchers used participative management to increase worker acceptance of the introduction of new production techniques in a manufacturing organization. Four groups were involved:

1. A control group, Group A, was simply given the new techniques and ordered to comply.

2. An experimental group, Group B, elected two members to confer with management and assist in working out the details of the new techniques.

3. Two other groups, Groups C and D, participated completely in decisions regarding the change.

The largest increases in productivity were in the participative groups, Groups C and D. The control group, Group A, stayed below prechange levels, and a high proportion of them quit in the first months of the change. The representation group, Group B, was closer to the full participation groups than the control group. The answer was obvious; participation increases productivity.

But the story does not end here. When the experiment was replicated in other situations, the preference for participation that appeared to be common to the American workers in the prior experiment was not repeated by Norwegian workers (French et al. 1961), by South American workers (Whyte 1959), or by German workers (Weiss 1956). These latter results may reflect cultural differences, but it is also possible that the reluctance of workers to accept participation in order to increase productivity is a result of their perception of the situation. It could depend on many factors, such as:

—Is this really participation or is it a management manipulation to get me to agree with decisions that have already been made?
—Do I have the authority to implement any decisions made during participation or do they have to be approved farther up the line?
—Does management really know what it is doing? If so why is it asking me? Its job is to plan and mine is to do the work. Why should I do its job too?
—Will I be blamed if the proposed change doesn't work out as expected?

The real question that management rarely sees in its drive for worker participation is: What's in this particular situation for the worker who participates? Many of these activities bring no rewards to the participants except psychic (or intrinsic) rewards. Many researchers (and some managements) believe these are very important, but the participants don't. It is quite possible that at best the worker does not share this need, and at worst feels that any increases in productivity become a detriment to future employment. Why cooperate in working yourself out of a job, especially if the situation promises no personally valued return?

A researcher describes the reactions of a worker named Toy Wynn as a typically motivated employee after she had participated in suggesting an improvement:

And what did the originator of an improvement receive for her efforts? Not any money, as one might expect, except through the company's employee profit sharing plan as savings in operation costs affected profit margins, a long term consequence.

No, the reward was intangible. Toy Wynn had the stimulation of looking at her job as something that she could improve for her own benefit and for the company's. She had the fun of shepherding a suggestion through team discussions and before the supervisor. And she had the satisfaction of being recognized by the supervisor and her peers for having come up with a good idea. (Wass. 1967)

In all fairness to the company, it is probably true that this achievement could be part of that worker's annual evaluation, but that kind of evaluation is not directly related in time or results to her contribution. A major problem in this kind of program is that there is an implicit assumption of a coincidence of goals between the person and the organization when that organization embarks upon any kind of participation program. The group (or in the case above, her peers) is assumed to have nothing to do with her attitudes. It is ignored. There is just as little validity in the idea of a coincidence of individual and organizational goals as there is in the idea that there is no direct connection between the worker's and the group's self-interest. Both ideas are invalid. Why should people accept goals when they often lead to consequences that are unrelated to their effort (i.e., higher company profits at the end of the year) or possibly to negative consequences ("We did such a great job of job simplification that we no longer need as many of you!").

I have often wondered about this selection of participation as a first answer to apparent productivity problems, because many managers do not act as if they really believe in participation at all. Haire et al. (1963) reported on a study about participation covering 3,600 managers in fourteen countries, including the United States, West Germany, and Japan. They found that in all the countries, managers had a relatively low opinion of the capabilities of the average person, yet when they were questioned about their own attitudes, they stated that they believed that participative, group-centered methods of leadership were more effective than traditional directive methods. How can one have participative leadership methods if you believe that the people being led don't have the abilities to participate?

One interpretation could be that those managers accept participation *intellectually* as a future method of managing, but right now, they will not or cannot follow up that intellectual acceptance with appropriate behavior in the particular situation. The participation these managers usually expected could be similar to the universal, human relations motivations assumed in the Hawthorne experiments, which attempted to manipulate the motivations of the working group (and the worker within it). Possibly participation is used to obtain that harmony between the group and the formal organization. Because managers may assume that the workers do not have very high individual capabilities, the manager must provide the guidance and goals that everyone is supposed to follow. Participation in this case becomes manipulation. This is an extension of the Hawthorne findings.

Satisfaction for the individual, cohesion within the working group, fusion of the work group to supervision, and collective purpose were believed to be mutually consistent as well as necessary conditions for plant survival. But satisfaction concerns individuals, cohesion is related to groups, and morale to collective plant purposes. These three possible levels of meaning were not distinguished; rather they were merged. (Krupp 1961, p. 25)

Therefore participation, like beauty, is unique to the individual and obviously determined by the individual's interests. When the person considers that participation will lead either to no direct personal reward (however that person measures it), an eventual loss of job, or a substitution of negatively valued company goals for his or her own, it would be stupidly self-destructive to participate. And people are not "stupid" when it comes to self-interest. This is particularly applicable to the people in technical departments. They have the physical freedom to move around the plant, to talk to others, and in general to learn what the past history of the results of participation has been. They can easily predict the possible future course of further participation.

On the other hand, there is research that indicates a predisposition of technical workers to accept some positive results from participation. Those data show that "job involvement was more strongly correlated positively with participation in decision making in the subgroup of the most highly educated participants" (Siegel & Ruh 1973).

What these findings may be saying is that people with higher educations are more likely to feel a closer coincidence between personal and organizational goals when they intellectually define those goals as close to their own.

That seems to definitely apply to technical groups. Either there is a positive mental alliance of goals that the technically qualified person assumes when joining the subgroup or the subgroup is modified to fit the goals of the individual because of that person's high value to that subgroup. The reasons are not as important as the result: higher goal coincidence between the subgroup and the individual's goals. This would suggest that attempts to install participative management in technical organizations will generally fail unless attention is paid first to achieving some agreement between the common goals of the person and those of the technical group (or of the person's unique situation). That usually means:

1. A personally valued reward can be distributed upon the achievement of specified common goals.

2. Achievement of those goals is controlled by the person.

3. Cooperation of or participation in the person's technical subgroup is necessary.

We can conclude that participation to increase people's motivations is not a simple technique. Its effectiveness depends, to a great extent, on how it is treated by the organizational participants. Those participants are all different from each other, but within those differences we have already found one common characteristic that affects their thinking: the level of education. The higher the education level, the closer the alliance with the professional subgroup.

8.5.2 Social Psychological Fit

But other important criteria distinguish participants (and their managers) from each other within the same company. Those criteria can define the *amount of fit* between the person and the situation in the subgroup. Some very interesting recent research deals with this fit of behaviors of people in different departments or subgroups within an organization (Lawrence and Lorsch 1967) with that subgroup's situation. The researchers defined the organizational economic environment as an independent variable that is seen from the viewpoint of members of a functional group as they look outward. They identified three main functional groups—marketing, technical, and scientific—which generally correspond to the sales, production, and research and development (R&D) functions of most firms. See Figure 6-2. Among other variables, the researchers investigated the ways of thinking and behavioral patterns of these three functional groups.

First, we have investigated the differences among managers in different functional jobs in their *orientation toward particular goals*. To what extent are managers in sales units concerned with different objectives (i.e., sales volume) from those of their counterparts in production (i.e., low manufacturing costs)? Second, we have been interested in differences in the *time orientation* of managers in different parts of the organization. Might production executives not be more pressed by immediate problems than design engineers, who deal with long-range problems? Third, we have been concerned with differences in the way various managers . . . typically deal with their colleagues, that is, with the *interpersonal relations*. Are managers in one part of the organization more likely to be preoccupied with getting the job

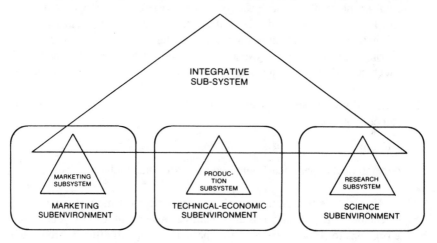

FIGURE 6-2: Differentiation and Integration

done when they deal with others, while those in another unit pay more attention to maintaining relationships with their peers? (Lawrence and Lorsch 1967, pp. 9–10) [Italics mine]

They were attempting to define "the difference in cognitive and emotional orientation among managers in different functional departments" (Lawrence and Lorsch 1967, p. 11). They found (as many of us did when we changed jobs and worked for different companies) that for two companies in two different industries, ". . . these two organizations were quite different places in which to work" (p. 155) and that the managers themselves were also quite different; i.e., "the managers in the two organizations had somewhat different personality needs" (p. 155). They also found that the managers *within* functional groups thought and acted more like each other than like managers in other functional groups *in the same company*. The differences among groups in the same company were called *differentiation*.

The greater the intergroup differences within an organization on the three attributes noted above (goals, time span, and interpersonal relations), the greater the degree of differentiation. Organizations that exist in a fast-changing, R&D-oriented environment (e.g., a plastics manufacturer) need high differentiation to be successful. The managers in each of the three departments were very different from each other because of the need to respond to rapid change and innovation. Organizations in a more stable, production-oriented environment (e.g., a manufacturer of cardboard containers) required low differentiation for success. The managers in the same three kinds of departments resembled each other greatly in the need to respond to little change within their own departments.

But in all cases, irrespective of the amount of intergroup attitude differences, the amount of intragroup differences seemed to be minimal. Thus, production engineers, whose job functions were definitive, would all be concerned mainly about getting the job done and less about maintaining social relationships, no matter what company they were in; sales personnel would all be concerned about maintaining good social relations with customers; and research scientists would fall somewhere between the two.

I have found confirmation of these results while consulting for a diverse cross-section of technical and manufacturing firms. This confirmation, of course, is subjective and very informal, since I didn't use the researchers' rigorous measurements and tests. Generally, when I found the same cognitive and behavioral patterns among managers who work *in the same functional area,* there was a higher probability of group success. In these research experiments, each group of managers (i.e., production, sales, and R&D) differed from the other two, but when the organization was in an economic environment of fast change (such as plastics), the more successful companies had group managers with greater differences (differentiation) *among the groups* and greater similarities *within groups* than companies that were not as successful. Conversely, when the organization was in an environment of slow change (e.g., cardboard containers), the more successful companies had functional groups of managers whose differences among groups were much smaller. They had less differentiation.

In other words, the amount of difference among groups was related to the rate of change in the environment; fast change equaled major differences among groups (high differentiation) and slow change equaled minor differences among groups (low differentiation). We have two intertwined ideas here:

1. Managers in functional areas really do think alike *within* their functions. That's not so surprising and is the probable reason that people in the R&D engineering group, for example, prefer to spend their time with each other. They think alike.

2. Managers in one functional area really do not think like managers in another functional area. Continuing the example above, "Those people in the R&D group are not being uncooperative with those of us in production or sales. They just *don't think the same way* we do!"

These research findings are explored further, where we can describe the management (the *structure* component) tools used to integrate these differentiated groups into one organization. At this point, your theory should be concerned with determining if you detect a similarity of thinking of functional managers within your own group, and if the thinking, behavior, and motivation among different functional groups of managers in your company is differentiated. (If you're really interested, you can find measurement tools for this differentiation in Lawrence and Lorsch.)

The amount of differentiation (or differences among functional groups) is positively related to the rate of environment change as perceived by participants. The participants, therefore, fit into their functional groups in terms of their own particular cognition and behavior. This seems to support the idea that goals are primarily individual and then oriented towards the group, rather than being oriented to overall company goals. That makes sense to me.

How can goals be constant for the whole organization, have different thinking patterns for different internal groups, and still be meaningful in terms of an individual's behavior? Increasing company profits, for example, may mean one thing to people in the design department, another to manufacturing personnel and obviously another to accountants in the finance department. Goals, therefore, should be defined within an overall framework but applied differentially to the self first, the group second, and then, in some general sense, to the whole company. But how does this happen, and how does the individual fit the group norms?

8.5.3 The Fit Between the Person and the Group

There has to be a close relationship between a person's thinking and behavioral patterns and those of the functional group in which he or she works. Highly creative, nonlinear thinkers and doers probably wouldn't succeed in a tightly structured, production-oriented group of people. They undoubtedly would work best in a loosely organized environment that would tolerate unusual ideas, methods, and, possibly, behaviors. Conversely, a more structured group would appreciate the work of a completely logical, careful knowledge worker who adheres to a relatively predetermined set of acceptable, organizationally defined behaviors. In order to have some sort of organizational success, this type of close fit between the person's thinking and behaviors and those of the group is important, as is the expected differences among various organizational groups. Fast situational change in the company's external economic environment demands greater differences among functional groups—higher differentiation. Slower situational change requires fewer differences among functional groups. In both cases, there are intergroup differences, but the amount varies and these differences are defined by the perceptions of the organizational participants.

And because in both examples there is always a greater similarity of thinking within groups or functional areas, it seems reasonable to believe that there is a greater concern by the individual for the goals of his/her own department than with those of the whole organization. As the rate of environmental change and the differentiation among groups increases, it seems reasonable that the goals of functional groups will also agree less with an overall set of organizational goals.

For example, applying these findings simplistically to human motivation, in a production-oriented company, the functional group manager should be more of a team player as part of the total company team. On the other hand, functional group managers in a successful innovation-oriented company would probably consider the company goals as secondary to their own departmental needs. In other words, if this research applies, those companies in slow-changing environments would have a more consistent overall internal culture than those in fast-changing environments. Those in fast-changing environments probably would be more inclined to consider their departmental culture above that of the company.

8.5.4 Proposed Useful Motivation

Unless we have a hermit working alone in some secluded monastery without the benefit of other human interactions, we have to consider the relationship between the person and some situation. Those ideas are the ones suggested by equity concepts (Adams 1963), two factor theory (Herzberg et al 1959), infantile treatment by the organization (Argyris 1957), and linking pins and supervision (Likert 1959). I believe that all these theories apply to the relationship between the technically qualified individual and the situation. And even though all of them have been expanded by other, more recent researchers, those more recent research findings don't seem to add any major breakthroughs to the original work that we cover now.

Equity theory was selected because it deals with a person's one-on-one valuation of the exchange of services for reward in a job. The two factor theory was selected because of the seeming applicability to technical personnel, as many of the subjects were engineers.

Infantile treatment was selected because (unfortunately) there are still companies in which top management seems to feel that it has all the knowledge needed to run the company. These feelings are obviously false in modern technical companies where design decisions can make or break the company; thus, it's important to be aware of this theory when dealing with one's superiors. In these companies, the knowledge worker is supposedly treated almost as a child, which results in conflict as an endemic part of the organization. Finally, linking pins and supervision theory was selected because the successful behaviors of the supervisors or leader is interpreted by the person as vital to the exchange between that person and the overall organization.

White suggests that people want to understand and manipulate their physical and social environment, and they interpret the results of that understanding and manipulation according to their perceptions of their own success. They like to make things happen rather than wait passively for them to occur. Both Freud and White agreed upon the importance of creative work to the individual's personality. They regarded how people develop. Freud considered the individual's personality to be pretty well molded at a relatively young age; White's idea of the person's sense of competence was one of ongoing development. White's theory seems to match most of my experience working in technical groups, because I have observed that sense of competence recover strongly after a success even when there were some initial failures. The sense of competence, however, does not keep on growing forever, because we are physically limited beings. It is a kind of self-fulfilling prophecy. The individual rarely achieves more than what he or she expects to. (Failure, of course, is also a self-fulfilling prophecy in this theory.)

On the other hand, we know that achievement does not always occur just because the person thinks that it is possible. Physical, psychological, economic and social limitations always exist. There is only one explicit limitation in this motivational theory; you cannot achieve more than you think you can. This sense of competence can be a major factor where innovation and creativity are required. Successful innovation begins when you think that it can be done. Then, if you are right and there is a success, the sense of competence becomes stronger. In turn, that sense supports further personal growth and change. (See Gellerman 1963, pp. 111–114 for further descriptions.)

Turning now to other findings, it has often been observed that so-called equal rewards and achievements are not accepted equally by all organizational participants. Even if it were possible at a given time to reward everyone equally and have them value the reward equally, its worth to people would change with time and circumstances. Therefore, we will now deal with the changes in people's values.

8.5.4.1 Equity theory. Equity theory is based on relative rewards and deprivations. The person provides the inputs, which typically include particulars such as education, experience, training, seniority, skill, creativity, sexual differences, ethnic background, social status, intelligence, and the efforts expended on the job. The company (i.e., the situation) then provides the outcomes or rewards to the person for those inputs. These outcomes typically include pay, fringe benefits, *intrinsic* rewards, seniority, and *extrinsic* status symbols such as a larger office, possibly pay also, or a change in title.

According to this theory, a person perceives that inputs related to outcomes are not fair or equitable whenever his or her perceived job inputs and/or outcomes stand psychologically in an obverse relation to what he perceives are the inputs and/or outcomes of others (Adams 1963, p. 424). If in the person's opinion, the inputs exceed those of another, but his or her outcomes are the same or less, inequity exists for that employee. Similarly, if his or her inputs are less than another's, with the same or greater outcomes, inequity also exists. There is a type of mathematical formula:

$$\text{Equity} = \text{Person}\left(\frac{\text{input}}{\text{output}}\right) = \text{other}\left(\frac{\text{input}}{\text{output}}\right)$$

if either subjective calculation, completed mentally by the "person," is not equal, the result is Inequality.

This perceived inequality involves tension, and the persons are assumed to be motivated to reduce this tension through psychological and/or actual changes in the inputs, the outcomes, and/or the selection of the comparison "other."

Although the theory doesn't provide any general guidelines between the amount of tension resulting from overreward and that resulting from underreward, there have been findings that might be useful intuitively. The overreward tension is often easily reduced by the person when the subjective value that he or she places on the input is increased. It's easier to come to the conclusion that you're worth more than that you're worth less. The underreward can cause decreased inputs or even searching for another job where you're paid more. This theory provides an explanation of the way in which people develop standards against which they judge the fairness and equity of the rewards received on the job, as compared to others.

When the person feels that unfairness and/or inequity occurs, the increased internal tension is supposedly reduced in two general ways: either the cognitive framework (what you think about the reward) or the actual physical conditions are changed (I'm leaving this job). Changing the cognitive framework means rethinking the value of the input and/or the output, or even changing the mental relationship to the "other," who is used as a standard. One way that it can be done is described in another idea involving cognitive dissonance (Festinger 1957). Accordingly, an alternative when the person stays on the job is to find something about the job outputs, other than money, to which one can attach a higher value.

For example, members of the armed services who are often paid less than equivalent managers in commercial organizations may regard themselves as "defenders of freedom" and thereby attribute additional social value to their work. That changes their cognitive framework and may decrease feelings of inequity. Another alternative could be to place a higher value on the work of those equivalent industrial managers, thereby concluding that, for example, a member of the armed services is not worth as much as those managers. And, finally, a third alternative could be to pick a different working group or associate as a comparison. One could also use any combination of these alternatives. Although those changes would definitely modify the cognitive framework, taking physical action to change the situation could also change feelings of dissonance. Such action might be decreasing the quality or the quantity of work done, asking for a raise, or, in some cases, changing jobs either within or outside the company.

An interesting and rather surprising prediction of equity theory is that people do not seem to attempt to maximum their outcomes (such as money or status) but instead try to obtain only an equitable or fair amount. With less than the "fair amount," there is a feeling that an injustice has been done and with too much, potential guilt about receiving it. This theory might be particularly applicable in describing motivations of people where everyone is doing essentially the same type of work. These conditions will occur less in the future in technical organizations but presently they might exist in administration or in engineering standards, where it is easier for one person to compare herself or himself against another. However, if the technical group is involved in creative design or innovation, this motivation theory is unlikely to be central. It is much more difficult to compare one's inputs and outputs against another's in that situation (although the human mind is capable of many unusual things), besides which many of the people that are drawn to this kind of work (creative and innovative) seem to be less concerned about comparing themselves to others than about satisfying some inner drive. Therefore, it seems reasonable that equity theory might apply best to people in production-oriented tasks. For creative or innovative people, some other theory of motivation might be more suitable; possibly the two factor theory.

8.5.4.2 Two factor theory. This research was based on an investigation into the reported causes of job satisfaction and dissatisfaction of engineers and accountants. The researchers (Herzberg et al. 1959) asked these two groups to tell them about the times when they felt very good and very bad about their jobs. The reports about the "good" times frequently concerned the content of the job. Achievement, responsibility, recognition, and the work itself were most frequently mentioned. Conversely, the "bad" times were concerned with the job context or the situation, not the job itself. Frequently mentioned in this latter category were supervision, company policy, working conditions, and salary.

These findings seem to mean that the job itself tends to produce satisfaction (and motivation). However, unfavorable job contexts such as poor supervision and bad working conditions tend to produce dissatisfaction, although good supervision and good working conditions do not produce satisfaction. These findings could be confusing until we realize that they indicate a nonlinear effect of both job context and job content on job satisfaction. In other words, improving the job context or surroundings will increase the person's job satisfaction only to the point at which that person becomes almost neutral about the job. At that point, the job itself must produce any positive feelings of job satisfaction and related motivation.

There were several problems based in the assumptions in the data-gathering process behind this theory. For example, there was the assumption that the people who were questioned had the ability (and the motivation?) to report the conditions of their jobs that satisfied and those that did not satisfy them accurately. They might not have reported objectively (if that were even possible), because it also seems reasonable to suggest that the people questioned might attribute successes and accomplishments to their own efforts (i.e., as a result of their own motivation on the job or the job itself) and dissatisfaction or inadequacies not to themselves or their work but to obstacles represented by company policy or supervision (i.e., the working conditions or the job context). These assumptions were the subject of some criticism in the psychological press. See Figure 6-3.

This theory assumes that people are mature, because they are supposed to be able to report their needs and independently respond to both the job context and the job content. It also could possibly connect with Maslow's ideas about a hierarchy of needs (see Figure 6-3). If the lower needs of physiology and safety (as defined by Maslow) are satisfied, they will no longer motivate. If the person is assumed (in the framework of the two factor theory) to have those needs satisfied (part of the hygiene needs for most technical personnel), the next motivating factors should be social. Most technical personnel have the freedom to socialize on the job if they wish to do so. Therefore, the two factor theory could have been tapping either esteem or self-actualization needs, which is reasonable, as these needs would be related to the job content and not the situation or job context. Eventually, we have come to the not-so-surprising conclusion that it is the job itself and not the context that is central for creative and innovative people.

8.5.4.3 Infantile treatment (maturity and supervision).
In contrast to the two factor theory, Argyris (1957) proposes a framework to correct the situation when the person is treated as if he or she were a child. In this theory, the management part of the job context and not the job itself is central.

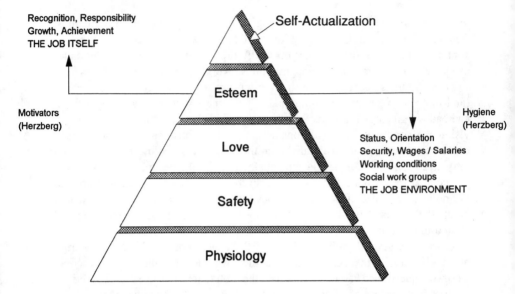

FIGURE 6-3: Maslow and Herzberg: A possible combination

Argyris proposes that most traditional industrial organizations treat people as if they were irresponsible children and not as the mature adults that they really are. He suggests that the seven changes that normally occur as an individual matures are completely ignored by organizations. These changes are:

1. . . . he moves from a state of passivity as a child to one of increased activity as an adult.
2. . . . gradually outgrows his total dependence upon others and develops a capacity to shift for himself.
3. . . . has a limited repertoire of behaviors, but as he grows up, he can respond to a situation in different ways.
4. . . . does maintain his interest for a very long period of time.
5. . . . time perspective is very short (but increases with maturity).
6. . . . develops from being everybody else's subordinate to being an equal or even a superior.
7. . . . does not have much of a "self" to have attitudes about, as a child [but apparently will when he or she becomes an adult]. (Argyris 1957, p. 73)

When these changes are ignored, people are coerced to be dependent, subordinate, and submissive.

There are three major sets of variables which cause the dependence, subordination, etc. The formal organization structure is the first variable. . . . Directive leadership is the second, and managerial controls (budgets, incentive systems, quality control, motion and time studies) is the third. The degree of dependence, subordination, etc., that these three variables cause tends to increase as one goes down the chain of command and as the organization takes on the characteristics of mass production. (Argyris 1959, p. 119)

According to this theory, the person's reaction is appropriately immature, involving informal activities such as apathy, gold bricking, and/or rate setting. However, if the organization changes its treatment of people, the following predispositions (Argyris' word) or mature feelings and attitudes of persons should appear:

1. The need for togetherness in relation to other employees.

2. Viewing wages as guaranteeing a fair standard of living and a secure job.

3. Noninvolvement about upward mobility in the company.

4. Control over one's immediate work environment, including the need to be left alone.

If these predispositions are encouraged by the organization there will be "the combination of the informal employee culture and the psychological work contract" (Argyris 1959, p. 148), the employee will accept change, and (implicitly) productivity and organizational growth will improve. This connects to the Hawthorne experiments which attempted to show that the best organization was one in which the person, the working group, and the total organization had the same goal: increased productivity. Of course, in that case, the formal organization defined what productivity was; in this case, the individual supposedly defines it. This theory seems to be more applicable to technical operations and could be placed in groups in which either production or creativity is stressed.

Although I have observed examples of organizational treatment of technical personnel as immature, as time goes on these examples will diminish because the demographics and economics are against it. With more of our production now dependent upon the output of the innovative "human capital" of our industry, when the company treats this "human capital" as childlike, there is little creativity. Individual innovation requires freedom and does not evolve in a constrictive, limiting management situation. How can anyone *demand* creativity?

There is a major difference between the viewpoint of those researchers who assume the inner motivations of people at work and those who ask the people themselves to report their motivations. The two factor theory asks people and seems to assume that the respondent's internal mental state depends on her or his work. On the other hand, other theories of Argyris and Likert (whose research we will evaluate next) state that everyone has inner motivations or basic requirements for independence and opportunity and these requirements are not being satisfied in the work situation at all.

That position is partially supported by research that investigates the centrality of work to the person's life interest. We have already noted the results of Morse and Weiss, who found that most people would continue to work even if they had no economic motives, and recent demographic data show that there are indications that work itself is becoming more important even when there is a diminished economic need. In all occupational categories and especially in technical areas, people seek "both intrinsic and extrinsic satisfactions from their jobs" (Mortimer 1978, p. 16). Intrinsic means things like self-satisfaction and achievement. Extrinsic means pay, position, etc. This seems to give more support for the two factor theory and less for the inner motivation theories, which are not as job dependent.

The definition of what is intrinsic (besides the obvious definition of being within the person's mind) and what is extrinsic (what he or she physically receives from work) could determine which theory applies best. Definition and measurement seem to be the major sticking points in all these theories and obviously the complexity of the human being has not been completely described so far. Supervision does play some role in all the theories covered. That is the concern of Likert's person situation interactive theory. As in other theories, there are some interesting definitions of sources of motivation that he claims everyone has in the work situation.

8.5.4.4 Linking pins and supervision. Likert (1959) is concerned with reinforcing major motivational forces that he says everyone at work has. These forces include economic motives, ego motives (such as status, recognition, approval, and acceptance), security motives, and a desire for new experiences (i.e., creativity or curiosity). His position is that "an individual's reaction to any situation is always a function, not of the absolute character of the interaction, but of his perception of it." In other words, it is how the person sees things that counts, not objective reality (if that can ever be defined). Consequently, ". . . an individual will always interpret an interaction between himself and the organization in terms of his own background and culture, his experience and his expectations" (Likert 1959, p. 191).

If the person is to satisfy his internal motivational forces, the organizational situation should be viewed as supportive and one that contributes to the individual's sense of personal worth. Part of that support is assumed to be a satisfactory level of economic reward, but so is the decision process by which the levels of economic reward are established, according to Likert. See Figure 6-4.

In other words, people have intrinsic needs; but the satisfaction of those needs is related to the person's perception of organizational support (the interaction between person and situation). And that support is shown primarily by having enough income and influence in the process of deciding what that income and influence should be. That influence occurs mainly through the person's direct

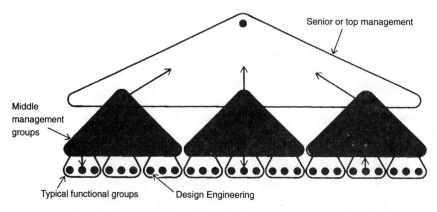

Senior or top management

Middle management groups

Typical functional groups

Design Engineering

FIGURE 6-4: Likert's Linking Pins

supervisor. This concept is the basis for a "bottom-up" directed management structure.

The supervisor is a "linking pin" (Likert 1959, p. 203) between the person and the next higher level of management in the organization. Therefore, a person is supposed to be able to influence his or her supervisor, who in turn influences his or her boss, and so on. When this situation applies, the major inhibiting factors against productivity, such as punitive budgets, become unnecessary. Productivity increases, because the linking pins ensure that the great majority of the people have been able to influence the relevant generalized goals of the entire organization. This tends to produce greater cooperation with the organization in attempting to achieve these goals. Gellerman sums it up:

> The real business of management is to assure the profitable use of its assets; therefore productivity is the goal, and control is merely one of the possible means to achieve it. But the effect of control concepts is too often an inhibition, rather than a stimulation of productivity. . . . Sustained high productivity eluded the man who was preoccupied with it and fell lightly into the lap of the man who was chiefly concerned with creating an atmosphere. It was almost as though the most sophisticated way to serve management was to ignore it and serve the employees themselves instead. . . . At the root of this paradox is a simple idea (so simple as to seem almost self-evident) which actually has some rather revolutionary implications. That is the responsibility for production is inherently the province of workers, not supervisors. The proper function of supervisors is to provide information, materials, and organization that workers need to do their jobs and otherwise stay out of their way. (Gellerman 1963, p. 45–46)

9.0 CAVEATS

There seems to be some common problems with most of the descriptions that we have been reviewing; these are listed as follows.

1. Standard internal motivations: Whether you assume that all people have the same basic internal motives either by themselves or interacting with the work situation (Argyris, Freud, Likert, White) or a succession of motives (Maslow, Herzberg), there really are no tools to determine what those motives are or how to measure them, even when you ask people directly.

2. Changing of motivations: The relationships of various motivations within people (McClelland) or how they are changed with circumstances (Adams) is also subject to question. Needs for achievement, power, and affiliation have, of course, repeatedly been found in people and the Thematic Apperception Test that is used is perhaps one of the best tests in terms of validity and reliability that psychologists have. The test results, however, are subject to fairly casual and transient kinds of unintentional manipulations by experimenters, most of the validation studies were done on college students, and there is some uncertainty as to whether they are as valid for females as for males (Sechrest 1968, p. 529–628).

3. Reporting: People do not perceive situations the same way, nor can they be completely objective in their reporting of feelings, interpretations, reasons, and so on.

4. Research descriptions: As I have pointed out before, researchers are just as human as the subjects on whom they report. They (and I) have implicit (and sometimes explicit) positions from which data are recorded and analyzed, theories formed, and hypotheses tested. These positions sometimes include a moralistic tone that should be considered.

> Humanists' goals and behaviorists' objectives appear similar. Both accept Maslow's self actualization concepts as the preferred route to self fulfillment. But by what divine right does one group assume that its values are superior to others and should be accepted as normal? Both the selection of goals and attitudes toward work are uniquely personal. The judges of human values have no moral right to press their normative concepts on others as preferable. (Fein 1974, p. 72)

However, in summarizing the data (and in all fairness to the researchers noted above), there do seem to be repetitive and important themes that appear in most of the theories on motivation. My interpretation of these themes is summarized next.

1. Work is important to people, and that importance generally seems to be related to the amount of training, personal investment, and organizational position a particular person has.
2. Economic security and satisfactory compensation levels are prerequisites for most people before they can begin to devote their full efforts to the job. There

is an important motivational interaction between the person and the situation, and that interaction is interpreted differently according to the perceptions of the person concerned. The perception in turn is modified more importantly by work experiences rather than inherited characteristics of the person.

3. People function best in groups in which there is a similarity in attitude among the members of the group. On the other hand, different groups within the same organization do think differently from each other and usually do not have the same overall goals as the organization does.

5. There is no one single way to describe motivation. Organizations can move toward getting more effort from technical people by developing structures that are more open. However, there are important considerations of the needs of the cultural, social, economic, and technological environments that affect the amount of openness.

6. Managers can attempt to gain motivation from employees through the manipulation of mutually agreed upon contributions and rewards associated with those contributions.

All the descriptions that we have looked at so far seem to indicate that motivational concepts that are wholly based either on situational or personality variables are not likely to be of much help in building our own motivational theories. They attempt to explain a complex problem with oversimplified formulas. The complexities of the subjects being studied—human beings—tend to minimize the use of simple theories. They rarely can explain complex phenomena. They can be used, however, as building blocks for more complex theories and as guidelines for the construction of specific hypotheses to be used in specific situations.

Then, of course, we may use more than one theory to account for differences in individuals and in situations. When we do that, we are no longer even dealing with the relatively simple person-situation interaction. We are moving past description and using these theoretical building blocks to predict and possibly to change and control others.

10.0 EXPECTANCY THEORY: PRESCRIPTION FOR MOTIVATION

People in technical organizations are supposed to make decisions. There is always an element of risk in decision making. Whenever an: "individual chooses between alternatives which involve uncertain outcomes, it seems clear that his behavior is affected not only by his preferences among these outcomes but also by the degree to which he believes these outcomes to be probable" (Vroom 1964, p. 17).

In very general terms, the decisions and the motivations behind those decisions depend upon some function of the person's subjective total value of the probable results of that decision multiplied by the total expectancy that the results will actually occur. The subjective total value is the sum of all the expected values of that decision. The total expectancy of occurrence is the sum of the potential achievement of those values. In the following equation, M = motivation, A = sum of values, and B = sum of expectancy of achievement.

$$M = f \times (A \times B)$$

In Expectancy theory, motivation is a function of values multiplied by the expectancy of achievement.

Using a simplified example, let's examine all the alternatives for a decision to complete or not to complete a machine design today. On a Friday afternoon, these alternatives are limited by whether the designer will work overtime this weekend or let it go until Monday. The selection of the alternative chosen could be as follows:

Summing all these subjective values (to the designer) for working overtime:

a. Overtime pay (positive)

b. Compliment from the boss (positive)

c. Increased status and recognition in the drafting department (positive)

d. Loss of free time (negative)

e. Complaints from the family (negative

f. Others, both negative and positive.

As in all motivation theories, implicit assumptions abound here. One is that it is possible for people to define and add both negatively and positively valued outcomes to determine a sum of values. That's reasonable. People do that all the time, although the cognitive mechanism by which they make value judgments is not really known. In this theory, when the mental sum of the values is positive, motivation is increased; when negative, motivation is decreased.

Summing all the potential expectancy of achievement (again to the designer):

a. The design will be completed (positive)

b. The design will not be completed (negative)

c. The design will be partially completed (either way)

d. Others, both positive and negative.

In addition to being able to add expectancies of achievement cognitively, there is another characteristic that applies only to expectancy. The range of expectancy can vary, in my opinion, only from +1 to zero to 1. In other words, +1 means that, if the decision is made, the person absolutely expects the achievement to occur. A zero means he or she expects that nothing will happen. That occurs when you multiply any value by zero. Multiplying a positive value by a positive expectancy should result in some positive number that determines the person's motivation. On the other hand, nothing will occur when multiplying by zero, and demotivation should occur when the expectancy is negative or less than zero.

Because the person's values and expectancies are based on that person's perceptions of the situation, the idea of "expectancy" as a part of motivation neatly includes the whole range of potential causes. The historical context of an individual's expectancy is quite important, because it affects how he or she now values *future* outcomes.

As an example, the management of a large industrial manufacturing company attempted to develop a motivational plan based on this theory. They used a modified type of management by objective to establish measurable goals and the suitable rewards for achieving those goals. But very few of the technical personnel were willing to estimate the expectancy of achievement (of the contingent reward for reaching the predetermined goal) to be greater than zero, because in the past, reaching a goal at the company was not rewarded. Because of the time that had elapsed between goal setting and goal achievement, there were often changes between the setting of those goals and their accomplishment. The variances between the two invariably became excessive and therefore the company decided that there were few rewards. The people, on the other hand, felt that management never paid off on its promises. Their perceptions were very different from those of top management. Top management felt that changes had made the original goals no longer applicable.

The changes of goals during the time period between setting objectives and measuring progress toward them have to be considered if the final variances are to be considered. There are those changes due to the modifications by top management and those due to the employee concerned. The emphasis must be on today's happenings and the reasons behind them. Implementing these expectancy concepts takes time. In my experience, it takes about three years of consistent effort before they are accepted. The administrative management follow-up involves restructuring goals and objectives as required, but also *paying for partial achievements* if the restructuring caused by management greatly modifies the original objectives. Of course, if the restructuring is caused by the employee, there may not be any payment at all, because the objectives were not met. I say "may not" because sometimes restructuring is required because of unforeseen technical problems and not personal motivational or unforseen technical failures.

10.1 The Design—Expectancy

The closer the direct connection between some organizationally valued goal and the person's expectancy that she or he can achieve it, the higher the probability that she or he will subjectively assign to it some value between zero (it will never happen) and $+1$ (a sure thing). The goal must be both definable and measurable, so that interpretation of the degree of goal achievement is minimized. Optimally, the goal should also be highly valued by the person. "A better attitude toward the company," for example, is not exactly a measurable goal unless there is an agreed-upon attitude measuring instrument (whatever that may be—I've never seen one). But completing a specific design within a specific time period, given specific resources, is measurable.

The probability score assigned in the mind of the person doing the work is, by definition, subjective and therefore impossible to determine objectively. But subjective measurements do work e.g. a scale from 1 to 10, you can be sure that the employee mentally calculates both the probability of achievement and the payoff.

10.2 Values

The other part of the evaluation, the values, are limited by the rewards available to and deliverable by the organization because industrial organizations have legal, economic, ethical, and other limitations. Some highly valued rewards may be impossible to deliver. For example, no matter how much the motivational value of the painting of Mona Lisa by Leonardo da Vinci, it's not likely that the painting will be a reward in any motivational scheme. Some potential rewards just cannot come from the job, and it is always helpful to get a clear understanding of this. (On the other hand, considering it strictly from a theoretical viewpoint, if the problems and risks in removing the painting from the Louvre were less than its value to the person, a strictly logical approach would be to attempt to remove it—an example of a completely valid but almost certainly unachievable goal.)

Therefore, the motivational system design includes negotiating a measurable, mutually agreed-upon, and personally valued goal; supplying the resources necessary to achieve it; and then paying off for achievement. In simpler terms:

Set up goals.
Determine the rewards, and measure achievement.
Deliver!

The last item is difficult if you, as the manager, do not have control over such things as salary changes, office assignments, time off, or any of the other rewards that could be valued and are within the ability of the organization to provide. On the other hand, it is possible that public approval of a job well done could be a valued reward, and that is certainly within your ability to provide, as a manager. If you do have control over extrinsic rewards, the next pitfall to avoid is the obvious one of goal change during the achievement process. If that is inevitable, a partial reward must be delivered and a new goal established.

Table 6-2 shows a summary of motivation categories and researchers.

11.0 BURNOUT—WHAT IT IS

There are instances where people, over extensive time periods, who perceive themselves as being under extreme pressure at work begin to exhibit behavioral characteristics that can be interpreted as "burnout." According to one researcher in the field, burnout is

> . . . characterized by a lack of energy and a feeling that one's emotional resources are used up. . . . A common symptom is dread at the prospect of returning to work for another day. Another component, depersonalization or dehumanization, is marked by the treatment of clients as objects rather than people. Workers may display a detached and an emotional callousness, and they may be cynical toward coworkers, clients, and the organization. . . . Individuals experience a decline in feelings of job competence and successful achievement in their work or interactions with people. (Cordes and Dougherty 1993, p. 624)

Table 6-2. Motivation Theory and Researchers

	Universal	Situational
Cultural	Weber, Veblen	Coch and French
Work Group	Roethlisberger and Dickson	Lawrence and Lorsch
Individual	McClelland, Maslow	Adams, Herzberg, Argyris

When there is a perception that the person is not making any progress or even losing ground, it is possible that there are several interacting bases for this perception.

1. *The manager:* There has not been an explicit requirement noted for success in the job.

2. *The individual:* There has not been adequate training received to support competency in the job.

3. *The situation:* There have not been adequate resources requested by or delivered to the individual that would support meeting those success criteria.

Individuals cope defensively by limiting their involvement and by providing an emotional buffer between themselves and others, as excessive chronic work demands drain their emotional resources. Technical operations occasionally place demands on the knowledge workers that are perceived by them as excessive and the outcome can be "burnout." However, this can be avoided using the motivational process noted before. Just the act of cooperative goal setting would quickly diminish the possibility of an overload. A contingency or expectancy approach in which the individual agrees to the task that must be done and understands the processes of measurement with consequent rewards would diminish any probability of this problem.

12.0 REVIEW AND PRACTICAL TIPS

The theories covered so far are primarily descriptive and provide few specific guidelines that deal with people as workers. Contingency theory is relatively new and differs from the preceding theories in its specific recommendations for appropriate management behaviors in specific situations. I suggest that the difference between the other theories and contingency is that situational theory describes and contingency theory prescribes.

Situational theories usually are meant to show what happens as the result of the person-situation interaction. Contingency theories suggest what to do when a certain situation arises. Of course, because no two situations are *exactly* alike, the prescriptions always have to be slightly modified, but that's what managers are paid to do: make modifications/decisions by absorbing uncertainty. Expectancy theory is an important, useful subset of the overall contingency theoretical framework. Using minor parts of all the other theories that we have discussed as bases for analysis of the person(s) and the situations, we have come to a very limited (and therefore more useful) personal theory. We can then develop motivational hypotheses to test in your own organizational situation.

This brief review of major ideas about motivation that I believe are most applicable to technical operations is intended only to provide you with basic building blocks for your own unique operating theory. Obviously, not all of these theories and consequent suggestions will apply; there is tremendous variability when we deal with individuals. It's fairly straightforward to develop procedures for a large group; i.e., "All engineering employees will use only the corporate design standards when selecting hydraulic valves." It's a different problem when you have to improve the work output of a specific person. What motivates an individual? Therefore we'll use expectancy theory because any motivation hypotheses must take into account our perceptions of the values that the individual would expect to receive and his/her belief that those values will actually accrue.

13.0 "ONE-SENTENCE" SUGGESTIONS

1. Setting goals for groups of people is easier than dealing with individuals.

2. Recruit for social competence first and technical competence second. When technical people work in teams, it is vital that all cooperate. Technical competencies can be easily learned.

3. Do not hire unless you have a plan to integrate the recruit into the work group.

4. As part of the plan for recruitment, define what you expect in six months—a year—or whatever, and most importantly, what it will be worth to you at that time—in terms of salary, position, etc.

Finally:
One is only motivated to achieve a goal when it is possible to define it in tangible terms. (attitudes are not measurable—physical results are) (Anon.) and, of course, there's an answer to every problem; the real difficulty is finding it.

SUGGESTED ANSWERS TO CASE QUESTIONS

1. The engineering department has not adjusted to the changing economic environment. Both the company and the engineering group have stopped being innovative possibly because there wasn't any need to be at this point in the company's growth cycle. The engineers seem to be quite able to support a production-

oriented environment, but are not "different" enough from production to begin to produce new products.

2. Is it necessary for everyone to be treated alike? What does that mean? I have been in organizations where administrators were required to fill in forms, even to the extent of sitting down with designers once a week to find out what they were doing. As far as recruiting is concerned, my feelings are that creative people are the best evaluators of other creative people, and I would suggest that Millard be one of those selected to interview potential new employees. Reporting systems should fit the situation. If daily time sheets are inappropriate, perhaps weekly progress review meetings with resulting minutes providing the reporting would be sufficient. Systems should suit the people. When the people have to suit the system, there is little creativity.

3. Peter seems to fit the existing culture of the engineering department, which in turn fits the company culture very well. Millard fits neither. It is ordinarily difficult to change culture quickly, but the present situation George faces calls for drastic action, such as forming a separate and special new products group, physically away from Peter and the existing group.

4. Intrinsic motivators for creative people like Millard could be a quiet place to work and no requirement for filling in time sheets.

    (a) They would be less useful for more active people like Peter.

    (b) Extrinsic motivators could be adequate salaries and an interconnect to the computer from the home.

5. George should discuss his own evaluations of the situation with Peter. If I were George, I would suggest that Peter come back with other alternatives to solve the problem of Millard, as Peter sees it.

6. Although it might appear that Millard is onto something, I would recommend that George follow very conservative accounting concepts when responding to upper management requests; i.e., report a loss as soon as it occurs, but don't report your profits until you're sure of them.

    He should respond by telling Marge about several potential possibilities for improvement and what would be required to achieve them: for example, different recruiting and training practices, changing the company policies on computer interconnects, implementing a different "cafeteria" motivation program. He can then subjectively assign risk and value factors to each of the recommendations for future developments.

Those are my suggestions. Do they agree with yours? There is another question you might want to work out for yourself. How would you develop a motivation program for these people? Select one person and work out a hypothesis. Could you use these ideas in your company? Why?

REFERENCES

Adams, J. S. "Toward an Understanding of Inequity." **Journal of Abnormal Social Psychology.** 1963. 67:422–436.

Argyris, C. **Personality And Organization.** Harper & Row, New York. 1957.

———. "Human Behavior in Organizations." **Modern Organization Theory,** Haire, Mason (ed.). John Wiley & Sons. New York. 1959. pp. 115–154.

Blood, M. R., and Hulin, C. R. "Alienation, Environmental Characteristics, and Worker Responses." **Journal of Applied Psychology.** 1967. 51:284–290.

Coch, L., and French, J. R. P., Jr. "Overcoming Resistance to Change." **Human Relations.** 1948. 1:512–532.

Coombs, G., Jr., and Rosse, J. G. "Recruiting And Hiring High Technology Professionals: Trends And Future." **Advances in Global High Technology Management,** Gomez-Mejia, Luis R., and Lawless, Michael E. (eds.), JAI Press. Greenwich, CT. 1992. pp. 91–107.

Cordes, C. L., and Dougherty, T. W. "A Review and an Integration of Research on Job Burnout." **Academy of Management Review,** 1993. 18(4):621–656.

Duchon, D.; Donde, P. A.; and Dunnegan, K. J. "Innovators and Non Innovators: A Comparison of Demographic, Psychological, Behavioral and Role Factors." **Advances in Global High Technology Management,** Gomez-Mejia, Luis R., and Lawless, Michael E. (eds.), JAI Press, Greenwich, CT. 1992. pp. 109–123.

Fairfield, R. P. **Humanizing the Workplace.** Prometheus Books, Buffalo, NY. 1974.

Fancher, R. E. **Psychoanalytic Psychology, The Development of Freud's Thought.** W. W. Norton, New York. 1973.

Fein, M. "The Myth Of Job Enrichment." **Humanizing the Workplace,** Fairfield, Roy P. (ed.). Prometheus Books, Buffalo, NY. 1974. pp. 71–78.

Festinger, L. A. **A Theory of Cognitive Dissonance.** Row Peterson, Evanston, IL. 1957.

French, J. R. P., Jr.; Israel, J.; and As, D. "An Experiment in Participation in a Norwegian Factory." **Human Relations.** 1961. 13:3–19.

Gattiker, Urs E. **Technical Management in Organizations.** Sage, Newbury Park, CA. 1990.

Gellerman, Saul. **Motivation and Productivity.** American Management Association, New York. 1963.

Haire, M.; Ghiselli, E. E.; and Porter, L. W. "Cultural Patterns in the Role of the Manager." **Industrial Relations.** 1963. 2:95–117.

Hall, D. T., and Nougaim, K. "An Examination of Maslow's Need Hierarchy in an Organizational Setting." **Organizational Behavior and Human Performance** (February 1968):12–35.

Herzberg, F.; Mausner, B.; and Snyderman, B. **The Motivation to Work** (2d ed.). John Wiley & Sons, New York. 1959.

Homans, G. C. "Social Behavior as Exchange." **American Journal of Sociology.** 1958. 63:597–606.

James, W. **The Principles of Psychology** (vol. 2). Dover, New York. 1950. (Originally published 1890.)

Katzell, R. "Changing Attitudes Towards Work." **Work In America The Decade Ahead,** Kerr, Clark and Rosow, Jerome M. (eds.). Van Nostrand Reinhold, New York. 1979.

Kaufman, H. G. Relationship of Early Work Challenge to Job Performance, Professional Contributions and Competence of Engineers." **Journal of Applied Psychology.** June 1974. 59:377–379.

Kohn, M. L., and Schooler, C. "Occupational Experience and Psychological Functioning: An Assessment of Reciprocal Effects." **American Sociological Review.** February 1973. 38:97–118.

Krupp, S. **Pattern in Organization Analysis.** Holt, Rinehart and Winston, New York. 1961.

Lawrence, P. R., and Lorsch, J. W. **Organization and Environment.** Harvard Univ. Press, Cambridge, MA. 1967.

Likert, Rensis. "A Motivation Approach to a Modified Theory of Organization and Management. **Modern Organization Theory.** Haire, Mason (ed.). John Wiley & Sons, New York., 1959. pp. 184–217.

Litwin, G. H., and Stringer, R. A. **Motivation and Organizational Climate.** Harvard Univ. Press, Cambridge, MA. 1968.

Maslow, A. **Eupsichian Management: A Journal.** Irwin Dorsey, Homewood, IL. 1965.

———. **Motivation and Personality** (2d ed.). Harper & Row, New York. 1971.

McClelland, D. C. **The Achieving Society.** Van Nostrand, Princeton, N.J. 1961.

———. "Business Drive and National Achievement." **Harvard Business Review.** July–August 1962. 40:99–112.

Mischel, Walter. **Personality and Assessment,** John Wiley & Sons, New York. 1968.

Morrow, L. "What is the Point of Working?" **Time.** (May 11, 1981). 93–94.

Morse, N. C., and Weiss, R. S. "The Function and Meaning of Work and the Job." **American Sociological Review.** April 1955. 20:191–198.

Mortimer, Joylin T. "Changing Attitudes Towards Work." Work in America Institute, Scarsdale, N.Y. 1978.

Murray, A. H. **Explorations in Personality.** Oxford Univ. Press, New York. 1938.

Roethlisberger, Fritz J., and Dickson, William J. **Management and the Worker.** Harvard Univ. Press, Cambridge, MA. 1939. Excerpts reprinted by permission.

Schachter, S. **The Psychology of Affiliation.** Stanford Univ. Press, Stanford, CA. 1959.

Schrank, R. **Ten Thousand Working Days.** MIT Press, Cambridge, MA. 1978.

Sechrest, L. A. "Testing, Measuring and Assessing People." **Handbook of Personality Theory and Research,** Borgatta, Edgar F. and Lambert, William W. (eds.). Rand McNally, Chicago. 1968. pp. 529–628.

Shannon, R. E. **Engineering Management.** John Wiley & Sons, New York. 1980.

Siegel, Alan L., and Ruh, Robert A. "Job Involvement, Participation in Decisionmaking, Personal Background and Job Behavior." **Organizational Behavior and Human Performance.** April 1973. 9:318–327.

Taylor, F. W. **Principles of Scientific Management.** Harper & Brothers, New York. 1911.

Terkel, S. "Working People Talk About What They Do All Day and How They Feel About What They Do." **Working.** New York. (Reprinted from an edition published by Pantheon Books, a division of Random House, 1972).

Veblen, T. **Theory of the Leisure Class.** Viking, New York. 1935.

Vroom, Victor. **Work and Motivation.** John Wiley & Sons, New York. 1964.

Wass, D. L. "Teams of Texans Learn to Save Millions." Reprinted with permission from the November 1967 issue of **Training, the Magazine of Human Resource Development,** Copyright 1967, Lakewood Publications, Minneapolis, MN. All rights reserved.

Weber, M. **The Protestant Ethic and the Spirit of Capitalism,** Parsons, T. (trans.). Scribner, New York. 1930.

Weiss, R. S. "A Structure—Function Approach To Organization." **Journal of Social Issues.** 1956. 12:61–67.

White, R. F. "Motivation Reconsidered: The Concept of Confidence." **Psychological Review.** 1959. 66:297–333.

Whyte, W. F. **Man and Organization.** Irwin, Homewood, IL. 1959.

CHAPTER 7

Putting Theory to Work:
Motivation Prescribed

Case Study:

The Case of the "Silent" Engineer
The "Management Model" that Failed

Cast:

Simon Alinoff: Senior Administrator-Contracts Department
Alice Twofeathers: Procurement Engineer
Mark Bush: Design engineer
Mary Withers: Project Manager

The Johnson Company was a leader in producing low cost, high quality components for the computer industry. The Design Engineer, Mark Bush, was assigned to the company's latest state-of-the-art project intended to develop a new computer chip. Mary Withers, the Project manager, was pleased with the progress that Mark had made in coming up with an initial design that would decrease cost by 20% and still increase the chip's computing power by 50%. She asked Mark to have a sample set of chips made by a subcontractor in order to actually test them out in the new computer frame that was in-work. Mark processed the need requisition to the purchasing department by delivering it to Simon Alinoff. He told Simon that delivery was very important and that he needed the first set of chips within six weeks. Simon said that he would give the job to one of his best Procurement Engineers, Alice Twofeathers.

Three weeks later, Mark called Alice on the phone and asked how things were going.

Mark: Hi Alice, what's the expected delivery for those new chips?
Alice: I'll have to check that for you and call you back.
Mark: O.K.-but as soon as you can.

After a week, Mark had not received a call from Alice. He called her again.

Mark; Hi Alice, it's Mark. Did you get a delivery yet on those special chips?
Alice: Look Mark, I told you that I would call when I was able to. I haven't gotten to it yet.

Mark: (getting excited and letting his voice rise a bit.) Alice, that's just not acceptable. I can't sit around waiting for you to finish whatever you're working on. I need answers and I need them now. What's going on there that's more important than getting those chips? I'll tell you: nothing. If I don't get a call back in half an hour, I'm going to complain to the Project Manager, Mary.

Alice: (shouting into the phone) I don't give a damn who you complain to. These things take time and you'll just have to wait your turn. (With that, she slammed the phone down.)

Mark immediately called Mary Withers and told her about the conversation with Alice.

Mark: Mary, if I don't get those chips on schedule, the whole project will be late. I never liked the general idea of having purchasing getting our experimental materials. I think that we should be able to do that ourselves. And another thing, that "clerk" in purchasing, Alice Twofeathers, claims to be an engineer. Well, she might have gone through engineering college, but she sure doesn't know anything about getting anything done, logically. When I tried to tell her how important those chips were, she just lost her temper and screamed at me over the telephone.

Mary: I understand the problem and I'll take it up with Simon Alinoff. Goodbye.

Mary then called Simon and asked if she could meet with him in an hour. He agreed and wanted to know what the meeting would be about. Mary explained the situation. Simon said nothing. An hour later when Mary entered Simon's office, Alice was there. Mary was surprised.

Mary: Hi Simon. I'd like to talk to you about getting some faster service for my project.

Simon: Hi Mary. I asked Alice to join us since she is familiar with your needs.

Mary: Simon, I would prefer to talk to you alone since this is just a symptom of a larger problem and I'm sure that Alice has other things to do and—.

Alice: [interrupting Mary in a loud voice]: I know what this is all about, your people demand extra service and you always want things done yesterday. In addition, I've heard rumors that Mark is unhappy that I'm handling the project because people tell me that he said that some of us here in purchasing are just incompetent. Because I'm a woman and an American Indian, he dislikes me.

Mary: Alice, this is indicative of other problems we have had in purchasing and I'm sure that your feelings are not the central issue here.

Simon; [sensing a major argument could begin, then interrupted] Both of you are getting excited about this and I think that we should end this meeting and talk about it tomorrow when we can handle it differently.

Mary: [raising her voice] I don't agree. This is an important problem and if we don't get it settled right away, I'll have to take it up with top management. If you like, Simon, you and I can meet again in an hour.

Simon: O.K.-I'll come over to your office.

Mary then turned and walked back to her office. Alice said nothing but remained in her chair.

Simon: Well Alice, what was that all about?

Alice said nothing. Then, without answering, she went back to her office. Simon was perplexed. What was he supposed to do now?

QUESTIONS:

1. What is the major problem here? Are there any other problems that are interrelated with it? If so, what are they?
2. Who is responsible for solving the major problem? Why?
3. Can you suggest anything for Mark, Mary, Simon and Alice? Why?
4. Do you think your suggestion would work? Why?
5. How would you handle the problem of the "silent disapproval" of a subordinate?
6. Can you change people at work? If so, how? If not, why?
7. Has anything like this ever happened to you? What did you do?

1.0 OVERVIEW AND INTRODUCTION

In Chapter 6 we described some concepts about how the model's component *people* were recruited, acclimatized to the organization and possibly could become motivated to contribute. There were numerous ideas and theories intended to describe the thoughts and inner motivations of *people* to work. Many of those theories were not testable in our situation because the test data was the individual's mental state. We can't access that data. But there is an important lesson to be learned and that is: there is *no* best way to motivate anyone. Each theory has its advantages and disadvantages but none have all the answers. They provide us with possibly useful bits and pieces that we might use in our own theories. Perhaps one person is motivated by "needs" of some sort or by a feeling of "equity" or something else. With this knowledge, we are better able to diagnose, if not prescribe. However, we must prescribe if we are to build usable motivational hypotheses.

Accordingly, we can now discuss a generalized prescription. That prescription includes developing practical suggestions for recognizing *people's* work-related contributions and rewarding appropriately when those contributions are delivered. Although these suggestions can never be all inclusive because each of us is different, they seem to be most applicable when all concerned follow a procedure requiring:

1. Setting mutually acceptable work-related goals that are tangible and, consequently, measurable,

2. Objectively measuring achievement against those goals,

3. Rewarding appropriately within the organizational framework.

2.0 SETTING MUTUALLY ACCEPTABLE GOALS

Goal setting for individuals can be difficult in technical organizations because of the rate of change that all of them experience. Individual goals are, therefore, not immutable over time but must be modified as the company responds. This response or change process is difficult for traditionally oriented organizations. When a company is a traditionally organized bureaucracy, it has all the attributes of a classical structure. It has control vested in a formally established hierarchy, very clear job descriptions and performance measurements that are related almost wholly to that position's description. In this situation, the chief engineer probably has well-defined and relatively fixed, specific duties and responsibilities, as do the other corporate managers. There is little attempt to vary the requirements either to fit the individual occupying the job (because there is an assumption that you fit the person to the job) or to respond to changes in the company's economic, political, or social environments (we know what our markets are like). The job is defined and performance is primarily tied to a yearly review that is independent of a natural work cycle of the work group that is being supervised. Reward for achievement usually is increased compensation. That compensation is supposedly based on comparative market worth and the hierarchical position. In my opinion, this is almost a perfect description of failure for the technically advanced company.

When there is an attempt to fit high technology companies into this mold, there is an inevitable eventual stagnation and failure. The world changes. Innovations in technology, modifications in customer requirements and obsolescence of products/processes require more flexible and responsive internal environments. This flexibility is based in technical success and that cannot rely on rigid job classifications. The way the job is done is strongly modified both by the interaction of the behavior of the particular occupant and the response of the occupant to perceived changes in the external environment.

2.1 The Behavioral Approaches

People's motivations are always rooted in self-interest. Unfortunately, it's almost impossible to predict with certainty what an individual's self-interest means. "One man's meat is another man's poison." Consequently, we deal with motivating behavioral changes because we can see and interpret those. Desired behaviors can be obtained when the individual's goals match those of the organization. It is goal perception that must coincide and that can change when the situation changes.

Those of us who have worked in technical companies could certainly support the idea that whoever was our boss at the time affected how we perceived our environment and how the work was done. Therefore, the perceptions of the direct internal organizational environment by the job occupant does affect the nature of the motivation to contribute.

The changes that result from modifications in a company's markets directly affect how people in the company think (Lawrence and Lorsch 1967). The central idea is "differentiation," that is, changes in differences in attitudes among internal groups. This research shows that even though people in groups within the company may think alike, there is a greater amount of different thinking and behaviors *between* groups when there are *fast* external changes than there is with *slow* external changes.

For example, the following statement might be typical in a fast-changing organizational environment: "We in advanced research and development just came back from the world exposition on "turbulent flow" in noxious fluids and learned about recent advances in pumping equipment. Because of that, we have to immediately scrap all our old pump designs and develop new ones. We don't understand why the sales department is not supporting our requests for additional budgets to get this done. They're still trying to sell last year's models. Those are obviously obsolete."

On the other hand, the sales department might be saying, "Why are those guys in R&D always trying to change things just because they went to some technical exposition? Now they want more money because they are trying to convince top management to support their demands to redesign our existing pumps. Sales are still strong, so why worry?" Fast change causes major differences and therefore possibly less uniformity among internal groups, even though there might be similarity *within* groups.

Conversely, in a slower-changing organizational environment, there is more uniformity of thinking with little differentiation among internal groups. Production, sales, and engineering have similar thinking both within and between groups. But even in slower-changing organizations there is always some small differentiation because even those companies do see some environmental changes in today's modern environments. It's the rate of change that is important because it affects individual goal setting.

In a rapidly changing economic environment, successful companies often have to develop *integrators*. These are individuals who understand the thinking of different groups and can interact between them because they have the capability to communicate between these groups. As examples, consider the job of the marketing-engineer or the manufacturing-engineer.

With rapid change, the development of individual goals, assessing achievement, and rewarding become quite complex. The clearly defined world of a bureaucracy with its relatively unchanging budgets as goals, repetitive measurements of monthly expenditures as assessment of managerial effectiveness and yearly reviews of compensation that cannot exceed predetermined percentages are not appropriate in the innovative, high tech company.

In the high technology organization, goals always involve individual creativity because the end result may not be defined precisely. For example, "Can we really invent that new sealed pump to meet new requirements for control of noxious fluids according to the new proposed international standards and have it ready for the spring exposition next year?"

All new product developments may not succeed; thus, we should provide for nonachievement as well as achievement of goals. Technology improvements cannot be exactly forecasted. In addition, goals should be fitted to the individual, if working alone, or to the team, if cooperative work is necessary. Lastly, if the cycle of work doesn't match the yearly review time or the achievement doesn't match corporate salary percentages, these criteria may not apply. Why set goals that are not useful for the individual?

It is the individual, in technical organizations, who must participate in the goal-setting process and everyone may not have similar perceptions of valuable goals. Perceptions can be different (and may be modified with time) as well as the values themselves. One researcher discusses how this is often overlooked when management tends to confuse someone else's (apparent) happiness at work with productivity.

> For over thirty years, . . . organizational psychologists have had to contend with the fact that happiness and productivity may not necessarily go together. As a result, most organizational psychologists have come to accept the argument that satisfaction and performance may relate to two entirely different decisions—decisions to participate and to produce. (p. 41) . . . We need to be reminded that perceptions of job characteristics do not necessarily reflect reality, yet they can determine how we respond to that reality. (p. 46). . . . job attitudes are fairly constant, and when reality changes for either the better or worse, we can easily distort that reality to fit our underlying disposition. Thus, individuals may think a great deal about the nature of their jobs, but satisfaction can result as much from the unique way a person views the world around him as from any social influence or objective job characteristics. That is, individuals predisposed to be happy may interpret their jobs in a much different way than those with negative dispositions. (p. 44) . . . Group-oriented systems may be difficult for people at the top to control, but they can be powerful and involving. We know from military research that soldiers can fight long and hard, not out of special patriotism, but from devotion and loyalty to their units. (Staw 1986, p. 49)

Some organizations attempt to circumvent this problem of goal differentiation among individuals by attempting to obtain an analysis of people's expectations. In many cases, those expectations are discovered through an organizational survey. This laudable intent can fail because of the following reasons.

1. *Ignoring nonresponses:* A very low response rate, may mean that only those who feel strongly about the specific questions are responding. The "silent" majority doesn't answer.

2. *Failure to asses reliability:* Pretesting the survey on a small, but representative, group could mitigate this problem.

3. *Asking nonspecific questions:* There must be an explicit link between the survey question and an organizationally valid process.

4. *Ignoring the results:* Those who respond to the questionnaire expect that there will be some result. If there is none, relations between management and the technical personnel become diminished. (Adapted from Futrell 1994, pp. 65–69)

However even when these pitfalls are overcome, asking the questions may be only the beginning. As noted before, goals may be changed by the organization's needs (i.e., "We have to stop your design of the new pumps because the company has been sold and we're now going to be a valve producer.") or those of the individual.

Typically, people increase their goals after a success and lower their goals after a failure. (p. 207) . . . people with a history of success at similar tasks, are more likely to attribute success to ability and effort, and failure to lack of effort or bad luck. (p. 213) . . . Just as a baseball player with a modest edge in skill will have superior performance over a season, so the individual who sets difficult goals will have cumulatively high performance over a long time period. . . . By achieving incremental goals and many of them in sequence, the nature of the organization (i.e. its level of effectiveness) can be changed. . . . This also implies that managerial persistence in pursuit of some overall plan is essential. (p. 217) (Evans 1986)

Finally, it should be important to establish individual goals that do not become destructive to the organization. Although the primary goals are those of the individual, supplementary (and probably equally valued) goals involving the individual's contribution to the group's achievements should also be defined. As an example, the designer who ignores inputs from manufacturing, procurement, or cost accounting probably will not be able to contribute much to the overall organizational profitability. Salable products often result from a team effort including design, development, production and sales. Although the contributions of the individual are vital, when those contributions do not result in an organizational gain *because of internal competition,* they are not useful.

2.2 Valuing Goals

When compared to the complexity of establishing the value of goal achievement, establishing measurable goals is fairly straightforward; provided both parties agree that those goals are organizationally available and personally achievable. It is the value of the achievement of goals that is often the sticking point. The most obvious example is the value attached to a pay increase. It is measurable and, if the manager is enabled to deliver it, it can be attached to a particular goal or goals. However, a pay increase is not valued equally by everyone. A promotion is also visible but occasionally is not easily delivered and, even when it is, it may not be considered as an unalloyed positive value by the receiver of the promotion. I have seen cases when promotions have been turned down for various reasons; e.g., "I'd have to leave this work group," or "Does this mean I won't be paid for overtime anymore?", or "I see that this does not include a gold key to the washroom."

Appreciation is a most applicable value if one assumes that everyone wants to be admired. Unfortunately, this is also not uniformly true. Some technical contributors may prefer to be just left alone. In any event, although all of these alternatives are available to the manager, there have to be many varied choices that can be more closely fitted to the individual. and that is part of the manager's job.

2.2.1 Pay

An obvious and traditional goal is increased pay or compensation, but this is not always applied appropriately within the organizational framework.

> In order for a pay for performance system to have a chance of success, (1) the merit increase has to match the performance measurement system, (2) the employees must have a belief that the pay program is equitable, (3) the merit component of pay must be significant. As we had previously suggested, expectancy theory tells us that rewards, especially variable rewards, must be significant in order to have impact; merit pay increases are of little motivational value if the merit component of pay is very small. . . . a poorly designed pay for performance system is worse than none at all. (Masters et al. 1992, p. 74)

Even when all these criteria are satisfied, there is the time component and a potential organization percentage limitation. If a pay change occurs that is dependent upon the achievement of some predetermined goal and not upon the "standard" twelve months, there is a strong signal that the goal has been valued. In some companies there is a cultural requirement that pay changes be kept confidential. This is almost an impossibility and is always, in my opinion, a self-defeating requirement. The connection between an achievement and a pay improvement should be an obvious fact as an encouragement for others.

2.2.2 Promotion: The Dual Ladder

Although there are generalities that may not apply to specific individuals, there are indications that technical professionals may have different goals than do managers. Technical professionals possibly seem to view their potential and opportunities for promotion as being supported by important contributions to their fields, thereby extending their reputations. In contrast, managers are supposed to desire upward movement in the organizational hierarchy. The contention is that success for the technician can be partially dependent on status and the respect of colleagues, whereas that for managers can result from approval of their organizational superiors. In recognition of these (very generalized) differences, some companies have established a "dual ladder," with one vertical section providing advancement in management and the other (supposedly equal) vertical section providing equivalent advancement as a technologist. The advancement into management gains additional status, recognition, salary and organizational power. The advancement up the technical section conveys increased autonomy, influence and occasionally salary. The reality does not live up to the promises.

> Many organizations exacerbate these differences by not living up to their promised commitments of creating equal status, perquisites, resources, and other financial and symbolic rewards to those of equivalent levels in the managerial and professional hierarchies. Frequently too, management does a poor job of publicizing the technical

ladder and little observable change takes place either in work activities or responsibilities after technical promotion. . . . the risk is that the technical side becomes
a "parking lot" for bright technologists whose abilities to generate ideas easily
outstrip the capability of the organization for dealing with them.

. . . Finally, there is the inevitable tendency to "pollute" the technical side of
the dual ladder. In addition to rewarding outstanding technical performers who
choose to remain in the organization as individual contributors, the technical ladder
becomes a repository for less successful, unnecessary, and even incompetent managers. . . . Another common practice is to use the professional ladder primarily for
pacifying individuals who are technically competent and who deserve to be rewarded, but who lack diplomatic skills or management ability. . . . (p. 136) Much
of the controversy surrounding dual ladders revolves around the issue of power:
those on the management side have it; those on the technical side do not. (p. 148)
(Katz et al. 1992)

This research may be idiosyncratic, but it seems to support much of the anecdotal
data that I have gathered over many years as a consultant to a broad range of
technical organizations. Therefore, the initiation of a "dual ladder" does not
seem to be an appropriate management goal if the intent is to promote individual
technical achievement. There are a few other observations that I have made and
these are similar to those of others. For example,

The responsibilities of a manager demand political skills and extensive interpersonal
networks. (p. 167) A default option for many technical specialists is to remain
with their firm, or move to a different firm, and adjust their expectations downward
with respect to compensation, status, and technical opportunity. As one project
engineer said - (p. 168)
"The sad thing is, I'm tired of fighting ridiculous battles. Development is too
hard to use so much vital time and energy fighting unnecessary battles. Quite
frankly, I draw the same paycheck whether I bust my butt trying to do something
challenging, innovative and risky, or sit around shuffling papers. Jack, who pio-
neered the display system technology here, left as a result of these pressures. He
had other innovations he wanted to try, and couldn't see going through that kind
of process again. So he left. There are a lot of battle scars, and too much pain. I
never want to do anything like the innovative project just completed again. It's
not worth it." (p. 175) (Page et al. 1992)

There are those who prefer a secure paycheck, the opportunity to pursue
outside interests with normal working hours and have less challenging innovative
opportunities with their concomitant potential for improved status and compensation. They don't want the challenge. They view the necessary power relationships
across groups ("that's organizational politics") and inevitable paperwork ("that's
for the accountants, not me") as distasteful and wasteful, rather than as prerequi-
site supports for organizational coordination. Their attitudes seem to spring
from the social environment they experienced as they developed their technical

expertise. It's independent and somewhat antisocial. In contrast, some types of independence may be rewarded such as the prime example of the academic who is rewarded only for delivering innovative concepts without the necessity to spend any time developing social or interpersonal skills. This type of behavior might be supported in the technical organization if the contributions are really innovative and helpful or when that organization is in rapid change, but those individuals are very difficult to guide. And guidance to meet organizational objectives is quite important.

One suggestion to obtain some semblance of control could be using the group as part of the goal-setting process. For example,

> Group decision making has been shown to be more effective in changing behavior and attitudes than such techniques as the use of lectures or individual instruction. (p. 118)
> 1. Group discussion is superior to lecturing or individual instruction as a means of conveying information;
> 2. the process of decision to perform a specific action raises the possibility that the decision will be performed;
> 3. a relatively public commitment to a decision will be more effective in inducing action than a relatively private commitment;
> 4. a high degree of consensus on a decision raises the probability that the necessary action will be forthcoming. (p. 120) (Golembiewski 1965)

But even this goal-setting or decision-making process, as positive as it may be, can be vitiated by management that is not aware. After the goal is set, in some cases,

> the organization stepped in with the controls. Managers had to consent to even small equipment purchases, group leaders had to approve technical memos and papers carried names of supervisory personnel not involved in the work being reported. researchers could choose which problems to work on (within those available) but managers determined how they were worked on. It should have been exactly the reverse. (Bailyn 1982, p. 45) (my changes in parentheses)

In this example, the setting of goals was useless at best and destructive at worst.

2.2.3 Rewards and Appreciation

Returning to the supposedly highly valued reward of pay, the notion that pay is a prime motivator in technical organizations has been questioned by many researchers (Masters et al. 1992; Gattiker 1990). Compensation levels are often established through comparison with (ostensibly) equal job categories in other companies. When certain job skills are scarce, the law of supply and demand applies and wage scales can increase. In recent times however, with restructuring and downsizing, it is occasionally likely that companies would release technical personnel rather than increase compensation and become noncompetitive in world markets (or so they say).

In some cases, if neither pay nor possible promotion to management are possible as potentially prime values in technical organizations, other more personal goals seem to be more applicable and have established great prominence recently.

Public recognition, achievement awards, gift certificates, larger offices, private parking spaces, time off, special company clothing and cars have all been suggested as potential noncash rewards for jobs well done. Other possibilities are:

> tickets to sporting or cultural events,
> letters of recognition,
> appreciation lunch,
> use of individual computer at home,
> round of golf,
> club membership, etc. (Nelson 1994, pp. 16–17)

The difficulty for us, as managers concerned with goal setting and delivering value for goal achievement, is how to select the appropriate value that fits the particular achievement that we want to have accomplished. The person might feel (justifiably) that a public pat on the back at the annual company picnic is hardly a sufficient reward for developing the latest "widget" that increased profits by 200%. On the other hand, gifts that are related directly to performance (and not necessarily to a yearly review) such as special company wristwatches, balloon trips, use of company-purchased desk computers at home, expenses paid to attend special technical conferences and extra vacation days would work well. However, both the person and the manager must understand what the range of values can be. If someone expects to receive an original painting by Michelangelo for a design job well done, that would be an unlikely value, to say the least. Perhaps, a year-long membership in the local art museum would be more practical. But there are alternatives that we can more easily offer. For example,

> 1. Make sure each scientist and engineer has a chance two or three times a year to tell a gathering of colleagues about accomplishments, work in progress, and future plans. In meetings to review progress on projects, allow the individual, not supervisors, to explain what has been done.
> 2. Actively encourage the publication of papers and reports and the filing of patents.
> 3. Let the individual sign design documents produced.
> 4. Break a big project into meaningful segments so that progress and accomplishment of each segment are visible to its workers.
> 5. Show intense interest in each project the person is working on.

6. Assign some administrative responsibility as soon as possible so that the individual can deal meaningfully with people and become involved in decision making. The results in a tangible recognition of responsibility.

7. Insofar as possible, allow the scientist or engineer to work on projects of greatest interest. This results in personal commitment and involvement. (Shannon 1980, p. 199)

2.3 The Cafeteria Approach

In the cafeteria approach, the person is involved in the goal definition and also has available options for selecting the values to be received. (This discussion is based on a modified management-by-objective concept.) The goals are a joint determination between the person and the person's direct technical manager. However, because that person's values may change between the time that the goals were set and when they were achieved, the rewards do not have to be rigidly fixed in the beginning. Within the range of rewards that the organization can deliver, it should be possible for the person to be able to select the reward that is most appropriate when payoff time comes.

Assuming that the basic requisites of satisfactory compensation and security are achieved, it might be appropriate to design a plan to provide other rewards. Then, if the person can pick an appropriate reward that she or he considers justified for reaching a predetermined goal, a motivation formula is achieved. Examples of these rewards, as noted above, might be getting additional vacation time, attending technical conferences, getting a bigger office, adding to one's medical insurance policy, or even receiving a reserved parking space. The medium of exchange could be any kind of company scrip and the selection could be from the company "catalog" of rewards. The scrip would be "spent" for the reward. (I say again, however, that this does not eliminate the need for satisfactory compensation and job security.)

3.0 GETTING STARTED

The "cafeteria" approach is not new; it's been available for many years. According to Perham (1978), there are companies that have implemented this cafeteria approach to rewarding people, but so far they have limited it to selection of fringe benefits. People select the benefits they want from a catalog of available alternatives of medical, disability, vacation, savings, and pension plans. They use "company dollars" to buy them and may change the mix they choose every year. Those company dollars are equivalent to a medium of exchange and are distributed in accordance with salary level, tenure, technical achievement and other variables of employment. The employees "buy" different packages of benefits as they wish, with only a minor stipulation by the companies that certain minimums in insurance and personal protection must be maintained. As a person's circumstances change over the years so can the benefits selected.

This concept should be equally applicable to goal setting and achievement, and would be particularly applicable in a technical organization, where effort and productivity are difficult, if not impossible, to measure objectively. In that situation, the end results could be predefined and the general methodology chosen to achieve them. A program of implementation might involve the steps of goal setting, value setting and measurement, and payoff. These steps are discussed below.

3.1 Step 1: Setting the Goals

The problems are no different from those in any goal-setting program. They include,

    1. Arriving at a definition that is mutually agreed upon.
    2. Determining the measurement of achievement and allowing for variances both above and below goal achievement.
    3. Setting the methodology, including resources available and procedures to be followed. (McGregor 1972)

3.2 Step 2: Setting and Measuring the Values

This step can be divided into the following two parts.

1. Determining the values available in the organization: Those values can be almost anything that is within the management prerogatives of the manager concerned. (This is not a straightforward task. I remember one manager who confidentially told me that he had no control over salaries, positions, hiring and firing, desk location, or other rewards. These rewards were all controlled by company policy. Fortunately, that manager was able to influence his people through a pleasant personality (he said) and an obvious superior technical knowledge about the tasks that they were to perform. He acted as both the technical and the social leader.)

2. Defining how the value shall be connected to the goal achievement: As a trivial but illustrative example, "for every hundred valves assembled during the next three months in your assembly department, there shall be a maximum of three functional failures at final test, using test procedure 8364-B. This goal shall be worth fifty points. Each failure over three shall lead to the deduction of twenty points, and each failure less than three, to a gain of twenty points, e.g., no failure = 50 + (3 × 20) = 110 points. These points can then be spent any way that you like when using the company catalog."

3.3 Step 3: Payoff

Define the value of the "points" received toward the potential rewards; e.g., at that person's grade level, fifty points may be worth an extra week's vacation, a raise of a predetermined amount, or some other reward. Of course when all that can be given is the salary increase, the "cafeteria" approach begins to resemble closely a management-by-objective program that is completed honestly and paid as salary increases.

In some cases where change may affect the end goal, it might be reasonable to set interim goals and equivalent values. Then the person is not penalized if a change occurs that is beyond his/her responsibility. "OK, sales wants to drop the whole design-development program for our new line of "widgets." Although you're on schedule and the whole program was valued at 100 points, at this 50% completion point, you'll receive 50 points even though the job was canceled."

Keep in mind that compensation rewards, such as a raise or bonus, do not necessarily change the person's *intrinsic* interests; that is, the self-rewards in getting something difficult done right. However, if the *extrinsic* rewards are inappropriate in the person's situation, there would invariably be a negative result. As noted before, "A pat on the head might be appropriate reward for a dog that has learned a new trick but not for an employee who has brought the company a multi-million contract" (Gattiker 1990, p. 87).

4.0 ADVANTAGES AND DISADVANTAGES

There are both advantages and disadvantages associated with the cafeteria approach. The advantages are:

1. Definition of needs: There is no necessity to define the status of the person's present needs or historical antecedents. Only those defined now by the person in making choices of values are relevant.

2. Measurement: There is either an objective measurement of achievement (e.g., "You really delivered that design on time.") or else an agreement as to the value of the "ordinal" measurement (e.g., "If it's done by Thursday, that's worth a score of 10. For every day it's late, we'll deduct one point.").

3. Payoff: Even when the goals are defined, measured, and are within the prerogatives of the manager to deliver, everyone understands *when* the payoff will occur.

These suggestions utilize general motivational concepts as raw material for goal and value setting. A cafeteria approach helps the person to coordinate personal as well as organizational goals and helps the person to control the job. The disadvantages of this approach are:

1. Openness and trust: This requires an environment quite difficult to achieve in many industrial organizations. Machine-like organizations that are concerned primarily with high production and very tight product quality controls might not respond well to creative kinds of organizationally open activities. Conversely, a typical advanced research and development group could not operate without it.

2. Value statement and goal setting: Although this is intended to be a one-on-one process, the group and cultural environments often provide an all too effective limitation on it. A culture that is tied to a rigid company policy and not supportive about viewing people differentially will not allow this approach.

3. Administration: This system is more complex and costlier than adopting a uniform or an anniversary-based performance review system.

An "expectancy"-contingency cafeteria approach seems to resolve the problems of internal motivations and their change over time. It promises to be a self-fulfilling prophecy if limitations on achievements and rewards are well defined in the context of the situation. The responsibilities for implementation are clear: the person achieves and the organization delivers. Both sets of implementation tasks require definition, and although those definitions are difficult, they are not impossible. For example, can there really be an equivalent definition of the value of an extra week's vacation or, say, a larger office or a private parking space? The first stage in the solution of any problem is definition. Implementation depends on the motivational hypothesis selected, and that is up to you to determine.

4.1 Creativity and "Dis-creativity"

An incentive program that includes the rewards that the individual values is almost a sure way, I believe, to support creativity and motivate a person to meet predetermined, organizationally relevant goals. On the other hand, a great deal of thought is required before beginning such a program and what the program will and will not do must be clarified exactly. Otherwise, I believe the program is a sure way to demotivate a person. The procedure is quite straightforward to understand (What do you, as the manager, want and what can you offer for it?). It is very difficult to implement on a company-wide basis because each group has different social mores and expectations.

5.0 REVIEW AND PRACTICAL TIPS

This chapter was intended to build upon the various theoretical approaches reviewed in the prior chapter with the intent of assisting you to develop your own theory of motivation and provide possible hypotheses to test that personal theory. We recognize that each individual, group, and organization is unique, so the applicable theory probably will be as unique as the situation to which it is applied. The building blocks in Chapter 6 included both universal and situationally interactive descriptive frameworks. In many cases, their predictions were not testable and therefore less than useful. This chapter provided specific prescriptions and suggestions for building your own applications. Pay, promotion, and social rewards were proposed as possible motivators.

The general conclusion is that no one can ever be sure that someone else is completely motivated at work, as it is literally impossible for any manager to define for any other person the variables that will effect positive change. Motivation comes from within the person; it is an urge to accomplish something, to do something that will help to reach some predefined goal. Managers, therefore, can only manipulate values to be received or rewards to effect behavioral change; they cannot change internal motivation directly. We can see the output of motivation as the person's behavior, but not the motivation itself. It becomes even more difficult to approach motivation when we recognize that we are part of the process. After all, it's our interpretation of the behavior observed that may or may not be faulty. In effect, the questions is always: Why did that behavior occur (in my opinion) and how can I either cause it to occur again or conversely, stop it? Finally, the rewards any manager can offer are relatively limited: money, promotion, social values and a few others. It depends on how much the person values those rewards. However, I suggest that you consider the following.

1. Any method for changing the behavior of the person requires consideration of the interaction between the person and the situation. Universal theories are just too general to be of any use.
2. Using expectancy theory, attach a reward the recipient values to mutually agreed-upon, measurable goals. Then behavior can be modified.

It's up to you to determine initially what behavior is expected and the interpretation of the behavior that does occur.

6.0 "ONE-SENTENCE" SUGGESTIONS

1. Develop a tentative motivation program on paper. Then get senior management approval of the values you can offer; e.g., can you deliver a raise in pay, a change of office, a promotion, a trip to a conference, etc?

2. Don't promise more than you *know* you can deliver—without a senior manager's approval.

3. Provide interim values as interim goals are achieved.

4. If there is a change in goals, pay off on the interim basis and restructure the new goals and values. It's a new program now.

5. Make all changes only on a mutually acceptable basis. Both you and the person must be satisfied.

And finally:

The more clearly we see the best method for achieving a goal, the easier it makes doing so. (Anon.) and when developing your motivation program, remember there are three typical reactions. 1. That won't work-it's impossible. 2. We tried something like that last year and it failed. 3. It's OK because I said it was a good idea all along.

SUGGESTED ANSWERS TO CASE QUESTIONS

1. There seems to be a personal problem that Alice has in dealing with some of the people in the company. Either her job and responsibilities have not been defined specifically for her or if they have, she apparently doesn't feel motivated to meet the goals. Without dealing with any of Alice's personal problems, she should have her responsibilities to her functional boss and to the project people outlined clearly. Because this could also be a problem in delegation, there is high uncertainty here for Simon. Therefore, he should schedule frequent meetings with Alice to obtain ongoing feedback about her performance. This should not negate a parallel investigation of her comments about being treated unfairly by others.

2. Who is responsible for solving the major problem? Why?

   Simon is. Because his uncertainty level has risen, he should take immediate steps to investigate, develop a problem solving hypothesis (as suggested before: a faster feedback from Alice and an investigation of her conflicts with others) and solve the problem, thereby decreasing uncertainty.

3. Can you suggest anything for Mark, Mary, Simon and Alice? Why?

   After investigating Alice's claims and instituting a feedback schedule with her, the next step would be to call a meeting and require all participants to develop an improved operating process that they can accept. It would be their responsibility to do that, not Simon's.

4. Do you think your suggestion would work? Why?

There is no way to be absolutely sure because, in management, we deal with people and that means there are always more variables than we can account for. Because of this there should always be a "backup" plan. In management, there is rarely "one" answer. Therefore, we always need some idea of what to do next if the initial solution or hypothesis cannot resolve the problem.

5. How would you handle the problem of the "silent disapproval" of a subordinate?

State the obvious: mature individuals are those who able to communicate. Individuals who can't or won't will probably not succeed in any organization.

6. Can you change people at work? If so, how? If not, why?

You cannot change people easily because you, as the manager, are only perceiving a small part of the individual's behavior. But you can change how they behave in the workplace by changing their jobs. That will change their behavior and if that changed behavior is acceptable, eventually they will change how they think, to some extent.

7. Has anything like this ever happened to you? What did you do?

Do you agree with my answers. What is different between yours and mine if you disagree with them?

REFERENCES

Bailyn, L. "Resolving Contradictions in Technical Careers or What if I Don't Like Being an Engineer?" **Technology Review,** November-December 1982. 85(8):42–47.

Evans, M. G. "Organizational Behavior: The Central Role of Motivation." **Journal of Management, 1986 Yearly Review of Management.** 1986. 12(2):203–222.

Futrell, D. "Ten Reasons Why Surveys Fail." **Quality Progress** (April 1994):65–69.

Gattiker, Urs E. **Technical Management in Organizations,** Sage, Newbury Park, Cal. 1990

Golembiewski, R. T. "Small Groups and Large Organizations." **Handbook of Organizations,** March, James G. (ed.). Rand McNally, Chicago, IL. 1965. pp. 87–141.

Katz, R.; Tushman, M. L.; and Allen, T. J. "Managing the Dual Ladder: A Longitudinal Study." **Advances in Global High-Technology Management,** Gomez-Meija, Luis R., and Lawless, Michael W. (eds.). JAI Press, Greenwich, CT. 1992. pp. 133–150.

Lawrence, P. R. and Lorsch, J. W. **Organization and Environment,** Harvard Univ. Press, Cambridge, Ma. 1967.

Masters, M. F.; Tokesky, G. C.; Brown, W. S.; Atkin, R.; and Schoenfeld, G. "Competitive Compensation Strategies in High Tech Firms." **Advances in Global High-Technology Management,** Gomez-Meija, Luis R., and Lawless, Michael W. (eds.). JAI Press, Greenwich, CT. 1992. pp. 59–87.

McGregor, D. "An Uneasy Look at Performance Appraisal." in **Harvard Business Review.** September-October 1972. 50(5):133–139.

Nelson, B. **1001 Ways to Reward Employees.** Workman Publ., New York, 1994.

Page, R. R.; Stephens, Gregory K.; and Tripoli, Gregory K. "Traditional and Entrepreneurial Career Paths: Variations and Commonalities." **Advances in Global High-Technology Management,** Gomez-Meija, L. R., Lawless, M. W. (eds.). JAI Press, Greenwich, CT. 1992. pp. 155–175.

Perham, J. "New Life for Flexible Compensation." **Dun's Review** (September 1978).

Shannon, R. E. **Engineering Management.** John Wiley & Sons, New York. 1980.

Staw, B. M. "Organizational Psychology and the Pursuit of the Happy/Productive Worker." **California Management Review.** Summer 1986. 28(4):40–53.

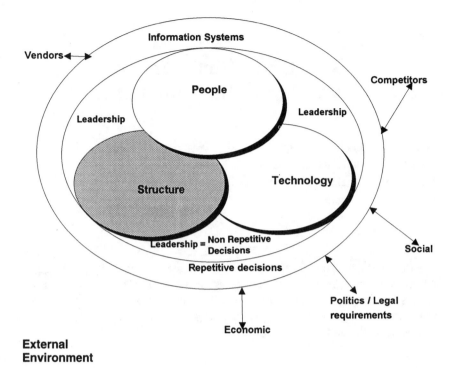

Information Systems

Vendors

People

Competitors

Leadership

Leadership

Structure

Technology

Leadership = Non Repetitive Decisions

Social

Repetitive decisions

Economic

Politics / Legal requirements

External
Environment

CHAPTER 8

Structures: Description and Prescription Semi-Permanent Teams: Functions and Survival Today

Case Study:

The Rigid Structure of Corvis Manufacturing

Cast
George Mulvaney: Sales manager
Mike Casey: Manufacturing manager
Albert Halloran: Chief engineer
Don Corvis: President

The Corvis Manufacturing Corporation had been started about fifty years ago by Will Corvis, Don's father, as a producer of high-quality children's toys and doll houses. Will had just retired, and although Don had been general manager for ten years, this was his first staff meeting as president after his father's retirement. All four officers attended.

Don: According to our agenda, we have several problem areas to review. Decrease in sales of "Missy" dolls is continuing. Manufacturing costs for our new doll house assembly line have not decreased to meet the forecast when we approved the budget for it last year. We might be sued because one batch of doll dresses we shipped last year was not treated for inflammability, and we have several reports of children being burned last Christmas when the clothes caught on fire. Who has some answers to these problems? Anybody ready to start?

George: As I see it, Don, the problems are all caused by ourselves. Let me give you some background that you might not have had as general manager, as you weren't handling sales then. Although we have the best sales force in the field, cheaper prices and some product innovations from competition overseas are killing us. More life-like, flesh-feel plastics and cheap imports are affecting our volume. We need new, less expensive, and more attractive products, and we've got to take more advantage of our long relationships with our customers to give them everything they want from one source of high-quality, low-price products. If we can provide all their needs, they won't have to buy imports. Imports are an administrative bother, even if they are occasionally less expensive. We've got to expand into allied areas. Dolls and doll houses are not enough. We've got to sell the whole market of children's recreation: dolls, toys, books, and games. Give me some new products.

Don: OK. Sounds good so far. Anybody else?

Mike: Look, fellows, we've been trying to keep costs as low as we can. The new doll house assembly line was expected to lower costs, but if you keep changing the designs, there's no savings left because of the increased costs we incur when we break down and set up a new product. We can't break setups and produce cheaply too.

Don: OK, what do we do now? Sales wants new and different products at lowered costs and manufacturing can't produce at the costs we need unless we keep our products fixed long enough to get some economies from long runs.

Albert: Our engineering department is set up to design toys the same way that we always have. Before we used any new materials, we always tested them thoroughly in our own materials lab. Every plastic, fabric, and metal was checked against our standards to ensure that no problems would occur in the field. The one time that we delegated responsibility to manufacturing to get materials to meet our fireproofing standards, a batch of nonstandard doll clothing slipped through. This company has always been organized on a product line basis and engineering has assigned technical people to each line. Dolls and doll houses are almost separate departments. Why don't we become a bit more flexible and change the way we're organized?

QUESTIONS

1. What would the organization look like if George Mulvaney were running it?
2. What would it be like if Mike Casey were running it?
3. Are these managers responding to the questions that were asked? Why? How would you explain this? Correct it?
4. What can be done to resolve the questions on the agenda?

1.0 OVERVIEW AND INTRODUCTION

We begin by describing applicable theories and then suggest applications of those theories (as influenced by personal experience) for you to consider as part of your own unique management theory. After discussing general ideas about how to develop and use management theories in your own organization, we now go on to deal with the second most important component, *structure*, which is the repetitive and predictable behavioral patterns that people in the company exhibit in their daily interactions with each other. Structure includes both the formal and the informal interactions. In other words, it's both the organization chart and the real way things actually get done. This chapter concentrates on the structural interactions of the permanent groups in the organization, the *functions*, and provides suggestions for optimizing those interactions. Functions are those "permanent" groups that go on "forever," although the people in them may not. For example, those groups might include engineering, research and development, manufacturing, and quality departments. Chapter 9 deals with another organizational structure called *projects*. These are the increasingly important, but temporary, technical organizations that are intended to be short-term groups, assigned to achieve specific, limited but very important goals.

2.0 THE MYTH OF OVERALL CORPORATE PLANNING—CREATING THE FUTURE?

The design of the long-lived *structure* is usually discussed in many general management texts first under the development of an overall corporate strategy. Then these texts deal with planning for groups (i.e., structure follows strategy). That idea is logical and consistent for senior management, but most technical managers are not concerned directly with overall corporate strategy. They're concerned with how their departments will fit into whatever strategy is developed in those upper echelons. Therefore, we will move directly to planning for groups within the corporation because
   We should note that:

1. Recent research indicates that the current method of overall corporate strategic planning is not very useful (Mintzberg 1993).

2. In my experience, as noted above, the managers in most technical organizations are not consulted about overall strategy but are simply requested to fit into that strategy.

3. As this book is intended to be a pragmatic guide that assists your own management theories, the planning of technical structures will be limited to those alternatives that seem to apply best based on experience and can be easily used.

Most overall corporate strategic planning seems to be based on a belief that the future can be controlled. This is obviously a fallacy. The future can only be predicted and action can be taken *today* with a plan for taking action tomorrow, but all the plans for tomorrow must be subject to a periodic review of how the future is really developing.

All plans are based on a large number of assumptions. . . . It differs from a forecast. For example, we carry a spare tire in our cars because we assume we may have a flat tire but we do not predict that we will have one. (Ackoff et al. 1984, p. 7)

In too many cases, assumptions are included as if they were reality and resources for that year are made available to meet those assumptions. That occurs when a yearly corporate plan is intended to be untouched, and when the future turns out not to match those forecasts, it is the managers who are expected to explain the variances. The forecast, of course, was initially flawed.

When forecasts are based on expectations that something will happen at some future time and resources are provided to deal with that, variances will always occur and then there is often an attempt made to fix the blame rather than fix the forecast. There really are two alternatives when these variances happen.

1. Change management actions to fit the predetermined plans if the variances from the plan are minor and it is appropriate to do so.

2. Change the plans.

If these two alternatives are equally usable, planning becomes a reasonable process. As a consultant to major corporations, I have found that forecasts are sometimes regarded as fixed, immutable plans that each manager must follow. The process of planning then becomes an end in itself with little relevance for reality as the future becomes the present.

A good deal of the corporate planning I have observed is like a ritual rain dance; it has no effect on the weather that follows, but those who engage in it think it does. Moreover, it seems to me that much of the advice and instruction related to corporate planning is directed at improving the dancing, not the weather. . . . The better we can adapt to what we do not control, the less we need to control. (Ackoff 1981, p. ix)

For many years, organizational planning was only a top-down operation. In more recent times, it has become a participative and more effective task. The original concept behind planning was that it really was possible to optimally forecast and control the future decisions for the whole company.

Perhaps the clearest theme in the planning literature is its obsession with control of decisions and strategies, of the present and the future, of thoughts and actions, of workers and managers, of markets and customers . . .

. . . An obsession with control generally seems to reflect a fear of uncertainty. Of course, planners are not basically different from anyone else in this regard. We all fear uncertainty to some degree, and one way to deal with a felt lack of control, to ensure no surprises, is to flip it over to seek control over anything that might surprise us. At the limit, of course, that means everything—behaviors as well as events and some planners at least give the impression of wanting to approach that limit. In a sense, reducing uncertainty is (or at least has become) *their* profession. . . . (Mintzberg 1993)

I believe that planning can be useful provided that the planner understands the process and its limitations. Planning consists of three interrelated parts.

The first is the *forecast*; that is, what the planner expects to happen or to be accomplished including resources, time scales and outputs.

The second part is a *measurement system* that provides objective feedback to the responsible manager receiving reports of reality against the forecast.

The third part is my definition of *strategy*, which is the expected activities that the responsible manager can follow in order to deal with any differences between the forecast and reality.

When "strategy" indicates that the forecast cannot be achieved, it is time to reforecast. The original forecast now has only historic value, if the planners assume that the future is related in some definable way to the past. "Strategy" is the repertoire of actions available that managers can take to meet predetermined plans and most important, the amount of variance (between the forecast and the actuality) that requires a completely new forecast. As an example, when the variance exceeds 5%, the manager might follow corporate standard procedure "#62" (or whatever) to correct the problem. If the variance exceeds 10%, the manager will draw up a new forecast because the original one no longer applies. And so on.

A plan should be an analysis of where the organization presently is against where the planner thinks it should be now. (*Now* is the operative word!) Then the planner can attempt to define how the organization should act to get closer to where he/she thinks it should be. All forecasts, of course, should be "rolled over" at predetermined time periods (say, every quarter) or as part of "strategy" when the variance exceeds the strategic limit. Too often yearly plans become destructive because situations change over time and an unchanged yearly forecast doesn't allow for corrective planning. Another problem (that is fortunately quite rare in my experience but one that you should be aware of) is the practice of setting overall corporate or group goals that the planner knows cannot be reached reasonably. The idea is "Well, let's stretch them a bit!" This is one of the most egregious faults that any management can be guilty of. No manager can make the psychological commitment to attempting a goal that he/she thinks is not reasonable (Steele 1989, p. 85).

However, with or without a grand corporate strategy, the technical manager is responsible for optimizing the way repetitive human interactions in the technical group take place. Some companies attempt this by developing standard operating and design procedures. These are invariably based in the corporate environment so we will explore that next beginning with some history.

3.0 A BIT MORE HISTORY

It's obvious that no organization can exist without a people component (i.e., individuals), but this emphasis is relatively modern. As noted before in Chapter 4, only in recent times has there been any appreciable management attention paid to people as individuals and not merely as economic resources organized in a machine-like structure. It is difficult for a modern manager to believe that, in the past, most organizational managers considered people to be no different from other raw materials.

Previously, management effort concentrated on the optimum acquisition of inanimate resources and facilities and minimal effort was exerted to develop the human organizational structures. Machines, and not the machine operators, optimally transformed resources into revenue. The organizational structure was unimportant because people were considered to be interchangeable. If an employee didn't produce, another one was ready to take that one's place. Manufacturing plants, cash, tools, materials, and processes were important and highly valued. They deserved close attention from management. The relatively inexperienced, interchangeable and inexpensive labor had to fit an extremely limited number of organizational designs. This made any extensive attempts to develop specialized or unique organizational structures unnecessary and even wasteful.

During those times, there were rational reasons for this management thinking. In addition to the stable social culture, there was a much slower rate of change in public environments and customer needs. Market demands for products were slowly increasing. A product that served a market need had a reasonably long life because obsolescence was not a major factor. The organizational structure could meet current demands from growing markets due to slow changes in the public environments and relatively interchangeable people. Any resources spent on improving human interactions were wasteful because those interactions contributed so little to overall productivity.

The organization design of that time mirrored the existing society. That design was primarily a single proprietorship organized like a patriarchy. Although there were minor differences, even the larger corporations were organized internally as if they were single proprietorships. Henry Ford never accepted the idea that he was not the sole proprietor even when the Ford Company became a publicly owned corporation. Also, within the company, the manager or the foreman of each group was really the boss; he hired, fired, directed the work, and fixed the wages.

Even different industries used the same generalized, patriarchal structural model. Therefore, organization design alternatives were obviously not very important. The important decisions were about machines. If production equipment efficiency were improved through redesign, that would be important! The people could easily be replaced with others, and they often were if production quotas were not achieved. At that time, it was a rational or logical point of view to have the organization's structure directed by top management. Concern for individual human motivation was not required. Productivity was built into the machinery. Human beings were intended to supply and service the production equipment as that equipment transformed raw materials and power into finished products and, eventually, organizational revenue.

The logic and efficiency of this machine-archetype was easily understood. Management first designed the machines and facilities, then little attention was paid to designing the organizational structures. Today environments and markets change rapidly and technical industry has moved from the production of stable, unchanging goods to the production of innovative products and specialized services. In technical industry today, people have displaced the machine as the organization's capital resource. The people are no longer interchangeable, because the process of developing and producing the innovative products and services that support the organization's growth is a creative one: one that is unique, not interchangeable. (Human interchangeability, of course, never really existed.) It is creativity resulting in innovation that drives profits today and creativity is often in short supply.

This is supported by the notion that there is *not* an even distribution of creativity in the technical population, when measured by the output of research personnel (Shockley 1957). Each one of us, as a knowledge worker, is unique and because people (with their creativity) have displaced the machine as a major source of organizational growth (and consequent revenue), the attention paid to improving "people's" effectiveness has increased in the past few years. This means increasing the effectiveness of the organizational structure in order to support "people's" creativity. Creativity is an attribute of the individual and therefore structure should be designed primarily to minimize individual behavioral constraints. Uniformity is the enemy of creativity. Although there are always behavioral constraints imposed by external legal or social requirements, the organization should limit any additional ones to product and process design standards for minimizing costs. The knowledge workers should be allowed to work out their own informal organizational structures whenever possible.

The historic bases for some present technical organizational structures have been discussed in previous chapters. This chapter moves on and extends into descriptions of various modern structural designs and how these can be applied to technical organizations. As before, description will be followed by prescription, which you should modify to fit your own situation. We first cover structures that are supposedly universally applicable to all types of organizations [i.e., the recommendations for the "one best way" that Taylor (1911) developed, which in some cases still are used today]. Then we cover those structures that are situationally based and designed according to existing contingencies in today's technical companies. Contingency theory will also deal with how structures change as the organization itself develops and changes.

An obvious example of development could be that of the small, start-up technical company organized around the creative genius with a structure that becomes more bureaucratized as the company grows and develops different products to meet the demands of different markets.

3.1 Subjectivity in Research on Formal Design Elements

Most successful modern organizational structures are based on situational contingency models, even when the designers are not aware of it. These relatively new structures recognize that there are many ways instead of just one best way to organize, and that the optimal design depends on the particular contingencies or situation. These structures include the variable of "time" and are based on an if-then kind of design (Osborn et al. 1980). In other words, if "this" happens next year (e.g., a change in customer needs), "then" we will re-organize on a project rather than a function basis. Contingency models could almost be analogous to the models of modern physics, as they now include uncertainty rather than Newtonian certainty. The contingency models reflect the changeability and the consequently primary importance of people to the organization. A familiar word of caution though: some of the concepts that we will cover include nonobjective, and perhaps biased, assumptions that are inherently normative and therefore depend on the particular researcher's point of view. Normative theories suggest that there are approachable universal ideals like participation, openness, and conflict confrontation, whereas contingency theories suggest that people's behaviors depend upon both the environment and the participants' characteristics (Beer 1980). Unfortunately, management research has not reached the heights of scientific objectivity as in physics or chemistry; thus, there are always some personal values in the data being presented.

Some structures, such as those in process industries like refineries and chemical processors, cannot stand very much openness, participation, or confrontation of conflict. By necessity, they must have well-documented procedures that direct and control human behavior. It's not a good idea to have process control operators in a pharmaceuticals plant think about "creatively" modifying the chemical processes that they are monitoring. Therefore, the normative viewpoints suggested by supporters of research theories that support complete openness and creativity do not always apply. However, these viewpoints can be a starting point, when there are no other data for a manager to consider. ("I think that's the way it should be because there are no relevant findings that I can use.") But that situation is very rare.

Since we do actually have some research findings to start with, we begin our design by describing those structural building blocks suggested by various researchers that you might decide to use to build your unique organization (but remember the source of the blocks). The formal model that you design may typically include blocks such as technology, organizational history, markets served, background of organizational participants, legal restraints, and/or product life cycles.

3.2 Overall Direction Defines Structure: in Addition to Correcting Forecasts (But It's Tactical, Not Grand Strategy)

If there were only one factor that affected the tactical organizational structure, it would be that structural design is determined by the market served. It is the rare chief executive who can understand the needs of the markets and then develop a perfect company strategy and supporting structure to meet those needs. However, if this rare situation occurs, the structure selected to support that strategy is a major tool to convey that strategy to organizational participants. The structures are the documented decisions intended to guide how people should interact in the future as part of the firm. Unfortunately, such chief executives are scarce, but they do come along once in awhile. In my experience, most of the time, the chief executive relies upon special planning groups with their supposed expertise in forecasting.

A positive example was the deliberate planning of Alfred P. Sloan, the President of General Motors in the 1920s, that transformed General Motors from "an agglomeration of many business units, largely automotive, into a single, coordinated enterprise" (Chandler 1962, p. 130). Sloan set up uniform accounting procedures for all the divisions but supported decentralization to serve separate segments of the automotive market. The central offices were intended to define lines of organization and maintain communications between divisions, but each division had its own authority to operate autonomously to service its markets and, in some cases, even compete for markets with other divisions.

This structural design quickly impacted the market positively and the industry leader of that time, the Ford Motor Company, negatively. General Motors quickly passed Ford and became the industry leader within a relatively short time after the new structure was implemented. The corporate officers at General Motors made policy and coordinated the operations of the various divisions within the corporation. The separate divisions were encouraged to develop their own *unique organizational structures* to meet their perceptions of the needs of the marketplace. The structure of the Cadillac division varied from that of the Chevrolet division, even though both were within the General Motors framework.

However, despite how important the "market planning" building block is, it is still only one foundation building block. There are others that not only determine the rest of the foundation but also the structure built upon it with stated organizational goals.

4.0 THE VARIABLES AFFECTING STRUCTURE

Because all management theories have parts that interact with many other organizational variables, it is impossible to evaluate variables independently of one another. We discussed this in prior chapters on theory building. Therefore, some of the variables upon which structure is based are covered in different sections of this book. The work of Lawrence and Lorsch (1967) on the interactions of rate of marketing environments change and people's thinking processes is an example of this. That work is discussed in several chapters because the implications are different for different parts of the management model. That multiple discussion will also apply to the work of the next theorist covered here: Joan Woodward. Her ideas are briefly covered here. For other detail, see Chapter 10.

Woodward showed that one possible cause of structural differences was technology (Woodward 1965). This was an intervening variable (a variable between organizational structure and economic success) that was poorly understood and seemed to be partially responsible for organizational success, whereas others were not. This research partially explained why different types of structures could support economic success. There was a place for the rigid, classical, top-down structure as well as the group-oriented structure. In Woodward's research, structure was defined by technology. She defined technology in a limited way as the manufacturing processes used by a firm. According to her findings, the organizational structure of classical and human relations theory, with fixed, machine-like, formal relationships, is only one alternative in the range of structures. If the firm is in large-scale production, defined as the production of fairly large-sized lots of one kind of product, then classical theory, with the familiar pyramidal structure, is the optimal choice. (Sometimes it does work.) However, when small lots or individual units are produced (such as in tool and die shops) or production is continuous (as in process production in refineries or pharmaceutical or chemical plants), classical structures are not appropriate. (Sometimes it does not work.) In small lots, the skill of the people determines productivity, whereas in process production, the design of the plant does. (More pieces of the puzzle are falling into place!)

We now have more structural design alternatives. For example, depending upon the organization's variables, we can apply the following guidelines.

1. If we need to minimize creative behavior to maintain order and direction, we adopt a structure that fits rigid classical theory.

2. If we need to coordinate groups that produce products that may change with time, we adopt less rigid human relations theory to manipulate those groups' goals.

3. If we need to respond to manufacturing technologies, we can adopt three different structures dependent upon unit (one of a kind), mass (large lots), or continuous (nonstop) production. (See Chapter 10 for descriptions of "Unit," "Mass" and "Continuous" production.)

But this approach provided only a partial explanation of contingencies for organizational structural design. Other researchers had developed situational factors that suggested new and different structures. These approaches are studied next, as they are more adaptable for us in our development of personal theory. They provide a further development of this personal management theory using a modern, more adaptable contingency framework, rather than a universalistic one. We begin to understand that: If A occurs, do B, rather than the universal direction of: This is the best and only way to do it.

4.1 (Situational) Contingency Theory: The Uncertainty of It All

Woodward's work describes the situation in terms of the technology of production. Lawrence and Lorsch's work (1967) describes the situation in terms of the attitudes of organizational participants as affected by rate of product development or rate of change and economic uncertainty. These are more limited, less universal and therefore more applicable design criteria for us than the grand machine prescriptions of classical theory or the conflict resolution methods of human relations theory. Woodward's work is explored in detail in Chapter 10, so we will next move on to a further exploration of Lawrence and Lorsch. (By the way, as noted before, discussing a research finding in several areas is indicative of the cross-fertilization of modern multidisciplinary techniques.)

Lawrence and Lorsch attempted to find out why the division of tasks and responsibilities among organizational departments did not always increase productivity, as predicted by classical theory and the ideas of bureaucracy. In fact, there was even a conflict with human relations theory here, because sometimes misunderstandings and conflict increased even when the informal structure apparently supported the formal one. There were even departments in the organization that obviously agreed with the goals of the formal structure but exhibited behaviors that conflicted with other groups within the same formal overall organization.

The researchers felt that insufficient attention had been paid to the relationship between the structure (i.e., of groups) and the external environment (i.e., the situation). They recognized that organizational participants, the managers in different jobs, had not only different informal organizations but also different personal orientations toward particular goals. The research, therefore, parted from classical theory, because that theory had no provisions for differences in managers. It was also different from human relations theory in that it accepted differences between groups of managers that could not be (and should not be) adjusted through counseling or organizational modifications. The researchers found that cognitive differences both between managers in different groups and in the design of the organizational structure were important if an optimal match among formal, informal, and environmental factors was to be achieved. The informal organization, of course, is not uniform across the total organization but changes according to the needs of smaller groups within it.

The relatively uncontrolled variable, the environment, was defined from the perspectives of the organizational participants as they looked outward. The researchers defined three main subenvironments (see Figure 8-1): the market, the technical economic environment, and the scientific environment. The market was concerned with customers, sales, and service; the technical economic environment with production; and the scientific environment with research and development or scientific functions.

Managers in three different industries, plastics, food, and container manufacturing, were questioned about how they viewed these three subenvironments and what was important to them in each. As predicted, the orientations for each industry were different and the orientations of departmental managers within companies in the same industry differed from each other. "Sales personnel . . . indicated a primary concern with customer problems, competitive activities, and other events in the marketplace. Manufacturing personnel were all primarily interested in cost reduction, process efficiency, and similar matters" (Lawrence and Lorsch 1967, p. 37).

There were definite differences in cognitive processes. The cognitive differences were defined as differentiation: ". . . the differences in cognitive and emotional orientation among managers in different functional departments" (Lawrence and Lorsch 1967, p. 11).

This differentiation was measured by various research questionnaires. The variables for individuals were:

1. The time orientation of the managers. Do production executives have a shorter-range viewpoint than design engineers, who are concerned with longer time spans?

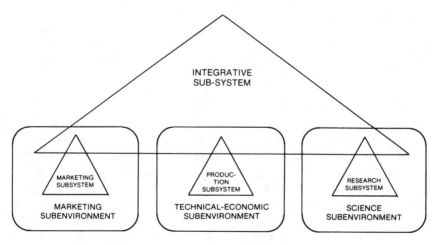

FIGURE 8-1: Differentiation and Integration

2. Interpersonal orientation, or how people relate to each other. Are there different task and relationship patterns?

3. Orientation toward goals that are a function of the particular department. To what extent are sales managers concerned with different objectives (sales volume) compared with production executives (units produced)?

Another variable for groups is of direct concern to this chapter, *formality of structure*; i.e., are there different formal supervisory levels in different departments of the organization? This last variable was measured by the researchers themselves and was not part of the questions they asked of the organizational participants. Resolving interdepartmental conflicts and coordinating the various organizational structures was accomplished by a process called integration: "the quality of the state of collaboration that exists among departments that are required to achieve unity of effort by the demands of the environment" (Lawrence and Lorsch 1967, p. 11).

Classical theory provided for integration through enforcement of the rules and procedures that governed the behavior of organization members. With a stable and predictable situation, these mechanisms are often effective. With rapid change in the environment and/or in the people in various organizational positions, the effectiveness of these rules and procedures declines.

However, this research by Lawrence and Lorsch showed that managers in different departments really did have *different* thinking processes. These processes were not modifiable by the overall formal organization to any great extent; the managers were not trying to be difficult, they really were different! (What does this say about trying to develop an overall strategic structure that applies to every department in a technical organization?)

With these different viewpoints (differentiation), there is a greater need for interdepartmental conflict resolution (integration). Stability in the environment, with little change in the structure, supports the idea that rigidly defined rules and procedures for all can help in getting work done in these kinds of environments. Conversely, environmental instability and fast change requires the use of more flexible, changeable structures and continual coordination of very different kinds of people.

One conclusion of this work is that there is no *one best way* to organize, at least in technical organizations. Therefore, *within the same company*, there may be one functional structure that is appropriately closed and bureaucratic (e.g., quality control, standards design, computer operations) and one that is open and relatively unregulated (e.g., quality assurance, advanced design, computer systems development). When this happens, organizational conflict is neither good nor bad but an expected part of the structural design. Uncertainty about the future and in decision making is a major consideration that determines how bureaucratic (low uncertainty) or freewheeling (high uncertainty) the structural design should be.

The more differentiated an organization, the more difficult it is to resolve conflicting points of view and achieve effective collaboration (p. 108). . . . organizations will tend to elaborate and subdivide units that cope with the more problematic or uncertain sectors of their environments (p. 100). . . . in addition . . . the organization must fit not only the demands of the environment, but also meet the needs of its members (p. 55). (Lawrence and Lorsch 1967)

Although I have taken certain liberties with the sequence of the authors' ideas, I have tried not to tamper with their intent. Surely these are changes from the relatively stable and fixed formal organizational structures of classical theory. An unusual thought to any traditionalist: the people in the organization are the sources of internal differences, and the greater the internal differences among groups, the easier it is for the organization to adapt to rapid change and, if necessary in that situation, to grow and prosper.

This also leads to a familiar and obvious conclusion. Because each organization's rate of change might be different, each organization will certainly be different from its neighbor, not only because of this differing rate of change but also because of the different perceptions of the participants in the groups within the organization.

Therefore, designing an organizational structure for your group that is better than the one you have becomes a very interesting problem. It requires selecting the right interpretation of the rate of change of the environment and a formal structure that matches the differentiation or differences among groups in the informal structure. For example, fast change means wide distribution of power and of the decision-making process. Slow change means consolidation of power and of decision making. The social and psychological variables seem to be almost as important as the main purposes of the technical department.

This may be quite a bit more complex that the prescriptions of prior theories, but that's reasonable, too. More complex situations sometimes require more complex explanations. A simpler answer to a complex problem may be desired, but will it work as well? As we go further with our prescriptions, we must consider the amounts of differentiation among departments and the consequent amount of integration that will be needed to coordinate these different thinking groups. The design variables will depend on the amount of perceived change in the organization's environment by the participants who will resemble others in their particular group but will be different from other groups. This legitimizes the common sense idea that everyone isn't supposed to be like everyone else or be treated like everyone else. It might even be possible to have a company policy that says, "There will be no overall company policies except those required by contract or law. Everyone is be treated differently, as individuals, or at worse, as members of their own particular group within the organization." In the final analysis, creative people who are very different from each other do not seem to respond well to uniform policies, even if those policies are egalitarian in intent.

These design concepts have been supported to some extent by other research findings. For example, there are those involving the attitudes of managers (Morse 1970), which show that some managers "like" to be in well-defined structures and some don't. Those that like it seem to fit better into production-oriented organizations; those that don't, into less restricted groups. There are other prescriptions that research can offer to us.

1. The relationship between functional departments is the most important variable and the various departments' interdependence determines effectiveness (Galbraith 1970).

2. The "linking pin" concept (Likert 1961), in which each organizational function has members in it who are simultaneously members of both upper and lower groups in the structure, is most important for greater cooperation and clearer communications. This might be analogous to the "integration" of Lawrence and Lorsch.

3. The maturity of participants (Argyris 1967) is the major determinant of the effectiveness of the organizational structure, and allowing workers to develop their own methods and procedures improves this structure. This is a bit away from the differentiation-integration scheme but it does support the idea that cognitive differences among participants (such as maturity) is important.

4. There is also the concept that the distribution of power within the organizational structure is the prime consideration (Kotter 1978). Power in a slow changing organization is not distributed widely and is usually held at the top.

4.2 Developmental Theory—Time

In addition to the human variables of the organizational participants that affect the structure design, there is the one very important variable over which we have no control: *time*. Some applicable research on the relationship between time and organizational success was completed by Filley and House (1969). Organizations are affected by time just as participants are. Organizations grow, shrink, change their borders, become centralized or decentralized, and definitely are changed over the passage of time. One of the reasons for this change is the incorporation of modifications to the structure that resulted from repetitive management decisions. A nonrepetitive decision that originally resulted in a unique solution has been repeated on several occasions, and the structure is now changed to institutionalize that decision. For example, why decide on a method to place a purchase order for the tenth time if you can solve the problem once by writing a purchasing procedure that the new purchasing department personnel will be able to follow without your intervention? The method of purchasing is no longer a nonrepetitive decision. It has been institutionalized and the organization's structure changed. However, when nonrepetitive contingencies are incorporated, one possible result is excess capacity over that needed for the major existing organizational goals (Thompson 1967, p. 47). Changing the structure may not give you a perfect fit between present expanded needs and the new organizational design.

For example, now that you have written the purchasing procedure, what was once a part-time assignment for you as a technical manager has grown into a full-time assignment for a purchasing manager, a clerk, and an expediter. Very few things are free, especially organization designs. Even though all you needed was two and a half people, people come in whole units, and occasionally overcapacity results. In this example, it's best to change the organization slowly and keep it a bit lean.

Time does other things to organizations besides institutionalizing nonrepetitive decisions. One approach (Filley and House 1969) that deals with the whole organization proposed a model of growth and change that could fit many of the overall needs of technical organizations. See Figure 8-2. However, this model applies only to firms that begin their growth with a single product base. That is a typical beginning phase for many technical companies; thus this model could be very appropriate. It provides a general yardstick to measure where your organization is against where it should be. The vertical measurement, Index of Growth, of course, is an ordinal measurement. Time, as the horizontal measure is a ratio measurement (at least in the dimensions that we work with). The single-product model describes three general stages of growth: the traditional (or craft) firm of Stage 1, the dynamic firm of Stage 2, and the rational administration of Stage 3.

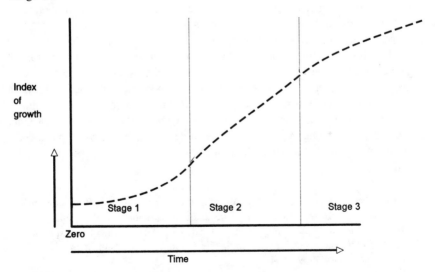

FIGURE 8-2: Three-Stage Model of Growth and Change (Courtesy of Filley and House, *Managerial Process and Organizational Behavior,* Scott, Foresman, Glenview, IL, 1969, p. 442.)

The majority of small technical businesses seem to fit into Stage 1. The structure is simple and direct; it consists of the owner manager and the individuals who report to her or him. There is little or no documentation of the formal organization because the personal views of the owner are well known to everyone in the firm. The future is judged to be like the past, so there is little need for planning, and because the organization is relatively stable and unchanging, there is no justification for using planning to modify the organizational structure. Conflict is minimal and communications are good. Except for occasional innovations, any functional activity that will not contribute to the bottom line is not developed. With no planning or marketing staff, the organization is very flat.

Stage 2 occurs when innovation brings growth and change and the owner has to cope with the much greater uncertainty these things bring. The organizational stability of Stage 1 gives way to greater managerial differentiation, with an increase in internal conflict as departments (and their departmental managers) rise and fall in importance. Growth goals are forward looking as the future will not be a repetition of the past. The technical staff either comes under the direct supervision of the organizational chief or else is given a free hand as long as the conflict among departments is acceptable and new products and services that are profitable are being created. Planning is used to integrate and coordinate the more diverse departmental activities. The organizational structure is not as flat as it was in Stage 1, because staff functions such as finances, marketing, and planning have been added.

At Stage 3, the firm is much larger and more complex. The owner and the entrepreneur have been replaced by trained executives who "plan, organize, direct, and control" (whatever that means to them). The organization often becomes the familiar triangular or pyramidal shape. Action is taken only after careful analysis and prediction of probable consequences. Rationality and formal organizational structures are guides for solving typical structural problems such as the "correct" span of control and amount of responsibility delineated for each position. The structure is minimally adaptive and not very innovative. Technical "experience" is valuable and "innovation" is very carefully considered (if not distrusted) before being implemented. The "not invented here" syndrome appears very forcefully to obstruct the adoption of any ideas not developed internally. At this stage:

> The firm becomes a collective institution rather than a one-man show. Its institutional objectives become separated from the needs of its participants and organizational efforts are directed to the changing needs of a defined market, rather than exploiting the short-term advantages of the initial innovation. Its mechanism is adaptive. (Filley and House 1969, p. 449)

Some organizations have reached Stage 3 in total size without attaining the more or less collectively rigid mental set that is described. They have done this by dividing their companies into many smaller ones and allowing extensive decentralization, an excellent form of differentiation. The corporate headquarters serves only as a resource supplier when the smaller divisions cannot provide sufficient funds, personnel and/or facilities to sustain their own growth.

4.3 Contingency and Uncertainty

Many of the technical organization structures that we discussed could fit into this model. The contingency, in this case, would be "time." In other words, taking the formal structure as one element, Stage 1 could be classical theory, with the owner manager acting as the organizational patriarch. Stage 2 shows the high differentiation and integration patterns of the contingency theory of Lawrence and Lorsch, and Stage 3, possibly the "linking pin" structure of Likert (1961) with each group being semi-independent with "links" between each organizational level. This structure might even be modified internally using some of the manipulative structures of human relations theory, with its emphasis on internal harmony.

The point is not to determine the exact theory that applies to your situation, because no theory will fit exactly, but to be sensitive to the variables and how they relate to you in order to build the best structure that fits your situation as you then perceive it. It's not a onetime job, because the situation is always changing. As an aside, organizing your group's structure doesn't necessarily require a top-down approach. After coordinating with your subordinate managers, as the technical manager, your structure can follow the design that you select.

5.0 ORGANIZATIONAL PRESCRIPTIONS

We now move into other, more direct prescriptions that apply to the unique needs of the technical organization. These prescriptions assume that the mission of the technical organization in your company is fairly well known to you. The next section prescribes several structures that are more specific.

Because the environmental situation is more of an independent variable than the formal organizational structure, the logical design steps would be to define:

1. The *situation* as it presently exists,

2. The *appropriate organizational design* that you believe will best fit it,

3. The *existing design*,

4. Some methods for *moving* from the existing design to the desired one.

All these steps require definition and ordinal measurement. You, as the designer, establish the values and the scores.

The design that works best should have less difference between "what is" and "what should be" than other designs. It also requires a close match between the designed formal structure and the informal climate, because our organizational participants are "knowledge workers" who control the major parts of the work quality and quantity. For example, if top management operates in a bureaucratic, top-down manner and you feel that a cooperative, supportive approach should be used in engineering, I can predict major conflicts (and in the short term, you'll probably lose).

Even so, all this does not promise perfection; we are dealing with perceptions in all cases and in prior chapters we've reviewed some of the problems with those. So although it may be possible that the organization built will get the job done or satisfy the needs of the people, achieving either goal may mean compromising the other (Morse 1970, p. 85).

The message seems clear enough. We will try to develop a formal organizational structure that matches the needs of the overall management environment first and the informal structural situation second. The environment is always the prime variable. The closer the match, the more effective the structure will be in assisting managers to get their jobs done, because the environment will provide economic rewards, and the informal structure, personal ones. But there will always be compromises that the formal structure must make, depending upon the importance of the economic and personal goals. The closer the goal congruence between formal and informal structures, the more effective is the intended overall organization structure that includes them both.

5.1 Making Design Goals Operational

However, we can face difficulties even beginning the design. Occasionally, the attempt to define the limited goals of the technical department may result in those that are different from overall company goals. The definitions are always operational; they refer to what is being measured. For example, "intelligence" is measured by the score on the intelligence test; a "successful" organizational structure provides more economic and social rewards to organizational participants than an unsuccessful one does. (Operational definitions are always a bit circular.)

I believe that most technical managers know the desired goals or definitions of their groups; e.g., "developing the best products in the gas transmission industry as measured by industry acceptances of these designs in the new pipelines." That's more limited and somewhat different from the company goal, "to make the best return for our stockholders and employees." But even that limited technical goal has not been made useful enough to be used as a basis for organizational design.

The answers for a basis for designing the organizational structure often result from proposing a lot of questions that begin with "how." For example,

How will it be done?

How will we know that it has been done when it is supposed to be finished?

How will progress be measured? And so forth.

These are typical questions to be answered if the technical goal is ever going to be satisfied. The concept is no different in answering these questions for the structure than in answering the same questions for an individual in a personal motivational system (see Chapter 7). When goals are made clear as, for example, developing a line of valves for the gas transmission industry to meet the new functional specification of the latest convention standard that weigh half as much as our present designs, and present no cost increases, and doing it by the end of this year, we can begin thinking about the types of repetitive behaviors or structures needed to reach them. This could include the differentiation/integration thinking or the technology of manufacture or the time variables or whatever the design hypotheses use. After goals definition, we can deal with defining the various structural design hypotheses that are easily available. Our further discussion here is limited to the technical operations, not those of the company as a whole.

5.2 Functions: Definitions and Applications

The term *technical organization* is intended here to include both the scientific-engineering and the production-manufacturing departments of the company. Most technical organizations, such as in engineering and manufacturing, are organized either on a functional or a project basis. Functional structures are *continuing* structures with group goals that are defined within more or less constant, overall company goals. These functional structures are desirable in stable situations, are survival oriented, and are intended to be "immortal"; that is, they continue to exist relatively independently of the tasks assigned to them.

These tasks are often not very long lived. They arrive and are generally solved quickly and rarely exceed the time limits of the company's financial reporting periods, in most cases, a year. The engineering organization itself however, for example, goes on as long as the company exists, and though departments within that organization, such as the electronics or hydraulics departments may change with time, those changes are usually quite gradual. But even though the goals of a particular engineering department or even the whole engineering organization may not be the same as overall company goals, they are usually constant over time. Someone may decide, "Well, we ought to strengthen our field support of the new stainless steel widgets because they seem to be having some installation problems that will require continuing education of our customers." That particular structural change will probably result in the implementation of a support department or group in engineering that will continue as a more or less permanent one as long as those stainless steel widgets are in use. Functional structures emphasize cooperation, harmony, and logic. They are less expensive to operate than project structures because they are stable and the people can be trained consistently and procedures can be applied (in a decision matrix) without a great amount of structural change needed over the passage of time.

In comparison, project structures are usually single-purpose designs that have a definite time limit. They are desirable in rapidly changing situations, have limited, specific but very important goals, and, when they achieve those goals, the particular project structure is dissolved. The people in projects usually perceive a more turbulent social environment than those assigned to functions. Conflict is endemic around projects because people are required to interact with different "others" and have to learn new procedures that may be different for each new project. Organizational procedures in projects often depend upon the project manager and the team leaders; thus, they are rarely applied consistently for more than one project. In projects, modification of past decisions and behavior or the creation of new decision matrixes is more or less required each time. Conflict is also caused by occasional confusion when the projects cut across well-established functional boundaries. The term "project" used here includes task forces, matrix groups, and other limited life structures.

These two structures, functions and projects, are not an either-or design in some technical organizations. Functional organizations divide the work among them. As each group's work is completed, if there is more to be done by another group, the work and the responsibility for the next sub-task is passed to them. As examples, there are the engineering, quality assurance, purchasing and manufacturing groups. These groups process and are responsible for separate parts of the work. Projects, on the other hand, do not divide the management of the work. They manage the whole task, in some cases by using personnel drawn from functional groups. Although functional organizations may have some small internal project-like structures where the work is passed from an individual to another one, the responsibility for the work doesn't cross the group's organizational boundaries (e.g., work done by mechanical, electrical, and hydraulic designers within the engineering group). Similarly, a project organization may have structures that may seem to be similar to functions when it organizes the work to be handled sequentially by internal groups within it such as design, development and testing. But the central criteria that differentiates functions from projects is a limited life span for projects as compared to the maintaining of the work within the group boundaries for functions.

Functions are designed to exist indefinitely and their work is contained within their organizational boundaries. There are some project structures that may seem to be like those of functions when those projects have an extended life over several years. Therefore, this classification of tasks as either functionally based or project based is contingent upon the particular company culture to some extent. There cannot be a single model for all times and all companies because structure evolves in response to market and environmental changes.

5.3 The Functional Model: Functions Are "Forever"

Now that we have classified the two major structures of functions and projects, we begin with the prescriptive designs available to functions. Projects, being more complex, are discussed in Chapter 9. In functions, the least complex formal organization design is the triangular or pyramidal model. It's a familiar model based in classical management theory and it can be either discipline or product oriented. This triangular design is effective because it fits the requirements of a relatively unchanging situation. Uncertainty in both the external environment and internally in the organization is relatively quite low. Classical theory, the Stage 1 development company, and the Lawrence and Lorsch production-oriented container manufacturing company with low differentiation and high integration are all examples of this very cost-effective design. It fits bureaucratic descriptions very closely.

It is perhaps the oldest design available, possibly having been described in Exodus 18:25 when Moses, on the advice of his father-in-law (could he have been one of the first management consultants?), selected one able leader to rule tens of people, then from the leaders of the tens of people select other leaders to rule the hundreds, and finally from the group of leaders of hundreds, other leaders to rule the thousands. With such leaders being responsible for the limited span of control of ten people, it was possible to communicate to 10,000 people in just four steps. (As an interesting intellectual exercise, consider how many levels of organizational structure your company now has and how many it would need with this span of control of ten people. Do you really think that all those managers are needed?)

A pyramid permits rapid communication and it becomes more effective as the shape becomes flatter (longer span of control), because there are fewer levels through which to communicate. This structure has tremendous potential to use power as a command instrument as it is relatively easy to obtain feedback, determine which level of the organization has not responded, and take corrective action. In many cases, there are two major divisions of this design: organizing by discipline (see Figure 8-3) or by products and services (Figure 8-4).

5.3.1 Organizing by Discipline

5.3.1.1 Advantages. Grouping of personnel and facilities around a particular discipline provides the organization with expertise that can be applied horizontally

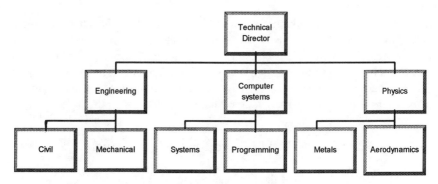

FIGURE 8-3: Functional Organization by Discipline

across the total structure. For example, anything that engineering knows about "corona loss in wires" can probably be found in the electrical power transmission group of the electrical division of the engineering department. When all personnel with particular skills are grouped, it is possible to hire experts, train technical specialists to achieve the expertise needed, or even to minimize the number of people needed to supply that expertise, as it will be available to the entire organization. Skills can be shared or pooled as required.

There is a clear promotion path for individuals upward that is dependent upon the personal acquisition of professional competency within a vertically defined specialty. Problem solving and decision making are quite straightforward if the problem is classified to be within a particular technical specialty. It is likely that the latest state of the art will be applied to problems, because all the specialists within a discipline are located together. Errors and/or breakthroughs are easily communicated. Positive and negative experiences are directly available to all interested parties.

It is also possible to maximize the use of expensive facilities and equipment and to justify their procurement. As an example, it is easier to justify the installation of a large computer-assisted design facility if many people will be using it continually and the experts are right there either to operate it or to assist the user in operating it.

This structure has another very important advantage: the informal organization tends to be very similar to that of the formal structure. Being trained in a particular discipline for years generally develops behaviors in individuals that are recognizable and accepted by their peers. Joining a structure in which those behaviors are encouraged and specialized technical expertise is respected further supports social behaviors that match the organization's culture. In this case, the culture tends to match the formal structure and formal authority (the power assigned to a particular position) tends to match influence (the power that accrues to a particular person). Social groupings (one aspect of culture) parallel the formal design, as specialists tend to develop recognizable behavioral patterns that are accepted by their group. This explains in part why specialists socialize with each other in after-work activities. The lettering on the tee shirts of the interdepartmental bowling or golf teams often describe a functional discipline such as the "Power Group" (i.e., the electrical transmission department).

5.3.1.2 Disadvantages. When a functional discipline is emphasized, the integration or exchange of ideas across the organization can suffer. Cooperation is difficult to achieve, because vertical relationships are emphasized over horizontal ones. Achieving integration (Lawrence and Lorsch 1967) becomes more time consuming and expensive. The development of vertical promotion paths rarely provides the management ability that managers who coordinate the activities of others must have. Additionally, the common professional and social bonds among technical specialists can develop extraordinary resistance to any change in the status quo. In some cases, self-perpetuation and improvement of narrow expertise support the mental walls that are typical of the "not invented here" (NIH) syndrome, which acts as a strong deterrent to any creativity that is not generated internally. The state of the art can be applied but it had better come from within the group and not by any "outsider."

5.3.2 Organizing by Product

5.3.2.1 Advantages. The organization structural model that is product oriented groups all personnel and facilities according to a particular product, service, or market. This coordinates organizational expertise in much the same way as the discipline-based model did, but whereas the discipline model applies specialized knowledge across all products, markets, and processes, the product-based model concentrates in depth on a limited number of products, markets, and processes. For example, the product orientation of the power transmission tower group could include the skills needed in structures, metallurgy, electrical and civil engineering, and fabrication.

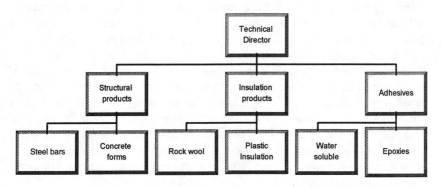

FIGURE 8-4: Functional Organization by Product

Therefore, solutions to problems that are concerned with different aspects of a *particular product line* are concentrated. Expensive equipment and facilities are harder to justify because the equipment and facilities now apply to more generalized areas, such as industries, and their use might therefore not be as important as in a technical specialty. A high-frequency voltage tester that was, for example, needed by a discipline concerned only with high-frequency electrical transmission might only be marginally justifiable for a whole product line that services the electrical transmission industry.

In a product-oriented structure, the state of the art is less likely to be as current as in a discipline-oriented structure, because all the specialists in that art are not always grouped together. Both positive and negative experiences are still directly available to interested parties in the rest of the company, but they are transmitted only as they apply to an industry or product line. There probably is a higher level of technical uncertainty in product-oriented structures than in discipline-oriented structures, because changes in products and services often occur much faster than changes in basic disciplines.

Although the informal organization still parallels the formal organization, it is not quite as congruent in its goals as in the discipline-oriented structure. One broadens one's goals in moving from the specialized discipline orientation to the broader product orientation. Although the product orientation may include fewer disciplines than the discipline orientation, there seems to be a wider range of work and problems in product organizations. Another difference is that discipline-oriented groups seem to have higher integration within their own department but more differentiated from the rest of the company. In other words, they're more like each other within groups but more different from other groups. They therefore require more integration among various technical groups. Product-oriented groups, however, are less differentiated and can work better with other organizational groups such as accounting or marketing. This design requires less integration across groups.

Promotion paths in product-oriented groups are broader than those in discipline-oriented groups. They not only go up the organization, but might also go into different groups of the same organization. Additionally, the cost of the technical operations should decrease because very highly specialized (very costly because probably less than fully utilized) technical talents are rarely needed on a continual basis. Although it is possible that product-oriented structures can limit participants' viewpoints as discipline-oriented structures can, there is increased product or industry knowledge as a powerful justification for the individual's promotion into higher levels of management. In product-oriented organizations, the scope of the problems increases. It is one of those "necessary, but insufficient by itself" criteria that often appear in textbooks. A knowledge of the market environment is an absolute necessity if the individual is to succeed as a manager, and product-oriented structures encourage that.

Cooperation and coordination across product or industry lines within a company are probably easier to achieve, as differentiation is lower. The "not invented here" syndrome could be almost as applicable here as in a discipline-oriented organization, but a new idea is more difficult to kill because products cover broader areas than disciplines and are therefore not as susceptible to an expert who says that "it won't work."

5.3.2.2 Disadvantages. The major disadvantage of using this design is a loss of specialized and in-depth technical-professional skills. We all lose the ability to apply learning quickly unless we use it all the time. For example, when was the last time you used the partial differentials that you had to learn in the classes in advanced calculus? How much trouble would it be to dig out your old texts and relearn them? (Do you even know where they are?)

5.4 Summary: Designing the Function

The alternatives for functional structures that were mentioned above are useful when you have generally defined the situation as you would like it to be and have compared it to the present situation. They are helpful in that first stage of selecting the structural pieces with which to change the "is" to the "should be." When and how do we get that change started? We start with the "when."

> . . . the organization has two choices. It can adapt continuously to the environment at the expense of internal consistency—that is, steadily redesign its structure to maintain external fit. Or it can maintain internal consistency at the expense of a gradually worsening fit with its environment, at least until the fit becomes so bad that it must undergo sudden structural redesign to achieve a new internally consistent configuration. In other words, the choice is between evolution and revolution, between perpetual mild adaptation, which favors external fit over time, and infrequent major realignment, which favors internal consistency over time. (Mintzberg 1981, p. 115)

I agree and I believe that "perpetual mild adaptation" is the better one. Obviously, when there is a major mismatch between the situation and the goals of the organizational structure, there will be a negative economic impact on the organization. That mismatch can occur slower and more insidiously in the technical group than within the overall organization, technical managers should follow an organizational redesign process that is somewhat different from any other ongoing corrective process in other departments. But when there are continual changes in scientific advancements, product innovations and technological improvements, the technical group is often faced with new requirements that do not come from the organization's top levels. New knowledge must be acquired and, unless there is a plan to change with that acquisition, technical groups can face scientific obsolescence. The "when" to change structure therefore is obvious: at regularly scheduled internally planned intervals, rather than waiting for top-down directions. In other words, when the question becomes: Why change something that works?, the answer is: If you don't, you will become obsolete.

The "how" to do that re-evaluation is not that obvious because there is probably a record of technical past successes that was the basis for the existing structural design. "We know that what we have works. Will the new structure work as well?" The answer is, "We're not sure but can we afford to remain as we are seeing that we are not alone in our markets?"

Uncertainty can always be defined as a *nominal* or an *ordinal* number, i.e., it is different from something else. In the game of chess, a red queen is different from a black queen, no better, no worse, just different. A specific decision under uncertainty can be defined as an *ordinal* number, because it is not only different from other decisions, it is also defined as being larger or smaller than other decisions. Using this example in structural design, we might say that the proposed structure differs from our present structure and has a *subjectively determined* higher potential for success. In other words, it is our opinion that is now measured.

5.5 Measurements—A Brief Review

As an interesting point, most of our physical measurements are interval based, such as the example noted before of degrees Fahrenheit used to measure temperature. Because there is no absolute zero in interval scales, the numbers in them are not multiplicative; e.g., 60 degrees Fahrenheit is not twice as hot as 30 degrees Fahrenheit. It is just 30 degrees hotter. It we could find an absolute zero for our measurements, we would be able to use ratio measurements, in which there is an absolute zero and the numbers are multiplicative. The Kelvin temperature scale is an example of this. We have no such numbering system available in management because we have no absolute basis for our decision making and organizational design.

We will, however, be able to use ordinal numbers qualitatively in developing a starting point for our organizational design; i.e., "This proposed design is going to be better than the one we had and I estimate that its successes will increase our technical output by about 25%."

As an extension of my suggestion to assign factors to a potential scenario subjectively, you might consider each of the elements of the structure, such as job specialization and behavior formalization, or whatever you wish to use as a measurement, and subjectively "score" them. Then you, as the designer, can come up with a total "score" for the structure that exists now. The assignment of the subjective factors can be done by using a personally designed evaluation scale for each element. See, for example, the following Table 8-1 which uses five different measurement points.

Table 8-1. Evaluation Scale

Score	Description
1	Very low
2	Low
3	Average
4	Above average
5	Very high

Table 8-2. Results of Element Scores

Elements	What Is	What Should Be	Differences
1. Fast response to market	2	4	2
2. High voltage line expertise	5	5	0
3. New product output	1	3	2
4. Conform to company policies	3	2	−1
5. Other	?	?	?
Total	11	14	3

When you have assigned all the scores for **what is**, you can do another score for **what should be**. The difference between the scores is an indication of the amount of structural redesign needed to fit the structure to the situation, as defined. The differences between the scores on each element are indications of specific corrective actions that should be taken to modify the structural design. It should always be understood that these numbers are ordinal, that is, subjectively assigned.

Just as an example, let's assume that we have subjectively decided to use these typical elements to measure our organizational structure. Using that subjective scale of 1 to 5, we arrive at the results shown in Table 8-2.

The results might indicate that there should be a concentration on two elements—1 and 3—with a consequent structural redesign toward a *project* orientation (notice the positive score on faster response to market and the negative score on conforming to department policies), using people who are highly qualified in a particular discipline such as high-voltage lines (new product output) and are willing to act independently of department policies. The methodology suggested above is not the only way, there are probably many other ways. In addition, this way been tested extensively in controlled research environments. That's probably impossible. Some of the problems to consider in using this particular method are that you might not be able to attach numerical values to elements or they may not be additive or the values may not be equivalent. (Is a "2" for element 1 above equal to a "2" for element 4, or is a "4" at least two units better than a "2"?)

However, whether the methodology has been tested or not is really not as important as developing some kind of useful tools that result in a method for you to use in organizational design. What is important is that somehow you absorb uncertainty and move measurements, *documenting* your thought processes, as you propose your hypotheses for a new design. The documenting *process* helps you to learn. This approach potentially includes all aspects of the organizational structure.

Other methods of organization design are more qualitative than those noted before. The answers to the questions below could be the design criteria or a changed structure, for example:

> Means-end schema: What business are we really in?
> Authority structures: Who bosses whom on what matters?
> Job-slot pyramid: What is the extent of individual responsibility?
> Communication network: Who talks to whom, when, and about what?
> Group linkages: What teams coordinate, and how?
> Decision rules structures: When internal conflict threatens, what gives?
> Program inventory: What tasks are we proficient in?
> Spatial organization: Where do we operate from?
> Style: What special character shall we assume?
> Socio-political environment: What kinds of public service are worthwhile to the organization? (Meir 1967, p. 477)

It is then possible to use a similar personal score here for "what should be" and "what is" and again work from that position.

Another method of coming up with an organizational design involves charting the various responsibilities of managers. This charting process enables managers from the same or different organizational levels to participate in identifying their roles as well as the roles of others in making particular decisions. The process assists the organizational designer by clarifying roles and relationships, both formal and informal, as perceived by the managers themselves. The way it's done is to have the respective managers complete a form (see Table 8-3) that lists all their decisions in a column on the left side of the page. Against each decision, in a horizontal row, they score whether their participation is absolutely required (score = 3), required for approval (score = 2), required for information (score = 1), or not needed (score = 0) (Galbraith 1970). Management loads are distributed according to the numbers of 1s, 2s, and 3s. According to our chart, Sam seems to be the major decision approver with the most 1s, Mike may be able to take on more work, and Charles could be the important decision maker. This is very subjective, as are all the other measurement schemes.

We have reviewed several paths to use in structural design and the one that

Table 8-3. Subjective Measurement of Responsibilities

	Charles	Mike	Sam
Budget/cost decisions	1	0	1
Recruiting decisions	2	1	2
Technical decisions	3	0	1
Totals	6	1	4

you decide to choose may not be that clear to others in your organization. However, some type of selection must be made, because if you don't (or can't because of other duties) accept the idea of periodically scheduled redesign, the "when" to design will quickly select a time of its own. That will be when the overall organization suffers economic losses as the fit between it and the situation deteriorates with time changes in technology or the effects of competition, and your mismatch contributes to those losses. Sooner or later, something will give. If the redesign direction comes from your own evaluations, the task will probably be much smaller than if it comes from the external economic or social environment. In any event, that environmental change cannot be ignored for long; eventually, something must be done. If you wish to start slowly at your own group level, here are some potential questions to be used in determining the need for organizational redesign before it begins.

1. What is the most important contribution of your group?
2. What is the most important problem?
3. If your group were eliminated, what would the result be?
4. How can your group easily be improved?
5. What recommendations for improvement would you be comfortable in suggesting today? (Galbraith 1970)

Another way to start a redesign could be to start with your major variable: *people*. Table 8-4 shows some of the assumptions about people in several very generalized structural designs. If you see a fit, you may have a starting point. People are at the basis of any organizational design, and these assumptions should be understood before attempting to use the designs shown in Figures 8-3 and 8-4, or any other designs. Scoring of the elements selected can be done for your organization as suggested before, by assigning subjective values to all the statements for "what we have," then doing it again for "what we want." The arithmetic differences should indicate where to redesign.

The idea that there is no "one best way" to manage has now been expanded to include no "one best way" to improve the organizational structure. Moreover, designing an organizational structure that is applicable to the technical department

Table 8-4. Assumptions About People in Various Designs

Design	Element	People
Cottage industry	Paternal, familial	Children, need guidance
Classical theory	Hierarchical power, limited span of control, owner is chief	Have only economic needs
Human relations theory	Social concepts, conflict resolved through managerial solicitude	Satisfy social needs on the job, cooperate with group norms
Technology	Knowledge workers control the design	Select their own jobs, needs for guidance or freedom
Bureaucracy	Discipline, professionalism, documentation and procedures	Highly trained, experts in the job
Project/matrix	Authority follows responsibility, constant restructuring, endemic conflict	Need for achievement, self-actualization

presents a greater problem for the structural designer than perhaps any other department of the company, because the repetitive decision making (that the formal technical organization is supposed to handle) is often split into two orientations: functional and project-matrix. We do, however, have some guidelines.

Determining whether to select the functional or project-matrix structure (see Chapter 9 for details) depends first upon the importance of the tasks to be accomplished, than upon the time constraints and finally upon the cost factors. A functional structure is better when a continuing or never-ending series of smaller tasks. It's best described by the highly integrated, thoroughly controlled, mechanistic model of an organization involved in mass production. This is the least costly structure to operate and to control, as it is a top-down directive structure that does not (and cannot) consider major differences of the participants. It assumes that their personal welfare coincides with that of the organization. Interestingly, effective participants in those organizations usually agree with that (e.g., the company man). This structure tends to prevent dysfunctional change and emphasizes harmony and cooperation.

On the other hand, project-matrix structures are specific, nonrepetitive, and short term. They have major and fairly well-defined tasks to complete or accomplishments to attain. They are more costly to operate than functional structures, as they tend to be bottom-up structures, incorporating differences among participants. They gain the advantages of creativity but also experience the disadvantages of potentially increased organizational confusion. The structure itself may vary from loosely organized task forces through various modifications, finally becoming a separately housed project concerned with extremely large, costly, and relatively lengthy tasks, but

. . . (there) is always a time limit on the life of the project structure. One definition is ". . . an organization designed to accomplish a specific achievement, created from within a functioning parent company and dissolved upon completion of that achievement" (Silverman 1967, p 1).

6.0 THE FUNCTIONAL MANAGER'S PRIMARY JOB—TRAINING

The structure will survive and adapt only as the people within it are trained. Although the idea of survival may seem obvious because structure by definition is the repetitive ways in which people interact at work, unless there is at least an ongoing training program to teach people about the formal organizational processes and procedures, the informal structure will develop its own procedures. This may be antithetical to what you want. Adaptation is necessary because of the never-ending changes in technology that you need to consider in order to maintain and improve company products and/or process designs. When training is delivered in an organizational context, it can be defined as any procedure intended to foster learning in the organizational *people*. Learning, in that same context, is usually defined as changing the person's behavioral patterns to coincide more closely with the needs of the organization, thereby increasing organizational effectiveness (Gattiker 1990).

The major concern is the observed change in the trainee's on-the-job behavior because we never can tell what has happened to the person's thinking processes. Those processes are obviously not accessible. Occasionally, training fails because the organizational climate outside the technical group does not support behavioral change. For example, it is well known that many suggestion schemes fail because employees quickly recognize that their suggestions are rarely, if ever, acted upon (Bass 1987, p. 185). When training succeeds, the results may not be that obvious but when it fails, it can cause you to lose a lot. The costs of training are clear and immediately available, but the benefits are difficult to measure, are often long term if when they do occur, and may benefit others if the trainee leaves the organization. However, in some research that attempted to measure the benefits, the researchers found that those benefits to the organization, when they occurred, outweighed the costs by a ratio of 6 to 1 (Wexley 1986, p. 283). That is quite a return on an investment. In that example, the benefits were documented through interviews, monthly productivity reports and other file information. Of course, the validity of these results depends on the accuracy of the measurements used and the observed behavioral changes of the trainees.

Because of the apparent value to the total organization of the technical function, it is usually the professionals, the knowledge workers, who are trained. This applies primarily to management training. If an economic theory is used, general training to increase the knowledge worker's value to him/herself is not the responsibility of the organization. However, skills training applicable to the job is and should be a constant funding operation by the organization (Pfeffer 1994, p. 21).

When training goals are outlined by individuals who cannot observe and/or manage the trainees after they have been exposed to training, the expected behaviors may not relate to those valued by the functional organization. All structures operate at both the formal and informal levels. The formal structure can reflect an ordered, predetermined system to the outside groups. There are often documented processes that can be shown to outsiders as evidence of logic and consistency. But it is the informal structure that often gets things accomplished. For example, if one were to strictly follow the formal organization chart, a person in engineering would have to go through the chief engineer, who would then contact the purchasing manager in order to request that manager to obtain a delivery schedule for an important component from the appropriate purchasing clerk. Of course, that doesn't happen. The engineer will either pick up a phone and call the clerk or go over to see him or her directly. The informal organization has its own rules, policies and procedures. This is not a novel situation.

> The dilemma, however, is that the training programs were designed and developed in order to support the formal system, the: "accepted way" of doing things. The reality was quite different. . . . (Bentley 1992, p. 180)

Therefore, the questions are not: Why train the technical staff, or what should they be trained for, or what will the benefits be, or even should there be any training at all? Training must be an ongoing activity, directed at producing forecasted behaviors in order to maintain and improve corporate outputs and reduce organizational technical obsolescence. As one last prescription—the training content should be a cooperative activity among three people: (1) yourself, as the manager supplying the training goals, (2) the expert training personnel, delivering applicable and effective training in accordance with your directions, and (3) the trainees, who recognize that the results of training will be directly useful to them as part of the organization.

7.0 REVIEW AND PRACTICAL TIPS

One of the better-known classes of universal theories concerning structures is that of classical management theory. *People* was the major component that affected that structural design and it still does today. Classical theory has the logical and consistent approach of the production-oriented industries in which it was developed. Fayol managed a large coal-mining company and Taylor, a steel mill. Therefore, these designs still have applications in industries that are relatively slow to change and whose strategy involves satisfying fairly constant market needs.

On the other hand, contingency theory is relatively recent, more flexible, and oriented primarily toward responding to more of the variables in human behavior. It recommends no "one best way," because it recognizes that all organizations and situations are different, but it does provide generalized guidelines for "if A happens, then B is probably appropriate." Probably *does not mean* absolutely. This is flexible theory building. It attempts to evaluate the particular situation in terms of some of the psychological aspects of the people who work in the organization and the requirements of the environment. The informal organization is no longer an enemy to be crushed or manipulated as in human relations theory. That informal organization is a major consideration that interacts either positively or negatively with the formal organization. There are many ways to organize, depending upon the environment, the people and the times.

If the universal theories can be considered analogous to the deterministic physics of Newton, contingency theory can be considered analogous to the uncertainty of the atomic physics of Einstein. As we learn more about organization design, we find (just as the physicists do when learning about the physical world) more complexity and many alternatives as we get closer to our own organizations, rather than the relatively simple and rigid pyramidal structure of classical theory or the manipulative structure of human relations theory that are supposedly applicable to all organizations.

This progression into complexity seems almost inevitable when changes in the environment and in organizational people are considered. Technical organizations are the producers of innovation, and that is the support of future organizational growth. Organizational participants are no longer interchangeable workers; they are the human capital that produces innovation and consequent growth. The formal structure must adapt to this change with increased complexity. The informal structure that interacts with it is at least equally complex. For example, we now have research that shows that people in different departments really do think differently. The design engineers are not being difficult (according to the contingency models of Lawrence and Lorsch), they really have different ways of looking at things than the sales engineers. The structure must be responsive to these differences if the organization is to grow.

Functional structures can be either discipline or product oriented, and each design has its advantages and disadvantages. Designing functional structures includes different criteria than designing project structures. One of the principal variables is that of time, with functional structures considered to be "immortal" and project structures considered to have a limited life with limited, defined goals. Conflict levels and differentiation levels (or differences in attitudes among departments) become higher when the structure must handle rapid change. Both the long-term functional and the short-term project organizations have opportunities and problems in coping with environmental change. There is, however, one general rule for all types of organization structures: no matter how well they are designed to fit the perceived present contingencies, they are never completely finished. With time, the fit between the structure and the economic and social environment decreases, making structural redesign a repetitive activity similar in many ways to the normal yearly budgeting process.

In this chapter we continued our discussion of the management model by describing many alternatives that can be used in the design of the technical organization's functional structure. That semipermanent organizational structure included both the formal and the informal repetitive interactions among people. There were tools proposed to measure "where you are" vs. "where you want to be," using ordinal scales to value alternatives. This was the prescriptive part of this chapter. The two major structures of technical organizations were then described: "functions," permanent parts of the organization, and "projects," created to solve specific problems and then to be dissolved. So far, the prescriptions for organizational design have been limited to functions. Projects will follow in the next chapter. Finally, it was proposed that one of the central issues in every function is the need for ongoing training, to support and improve the way things are done.

8.0 "ONE-SENTENCE" SUGGESTIONS

1. Designing technical structures is a never-ending task.

2. Functions should have relatively stable goals and change much less than projects do.

3. The major job of the functional manager is to train people, but that training must support functional goals.

4. Organizing by product lines promotes faster response to markets. Organizing by discipline is less expensive.

5. The informal structure is often the leader in requiring change to the formal structure. How do people interact now and how should they interact in the future for improved effectiveness?

and finally, always use the inputs of others because:

I not only use all the brains that I have, but all those that I can borrow. (Woodrow Wilson) and remember, when designing your structure, if the eraser wears out before the pencil, you're doing something wrong.

SUGGESTED ANSWERS TO CASE QUESTIONS

1. If George were running Corvis, the organization would probably have two different internal structures. One would be a very flat structure that would be organized

in a project orientation for rapidly changing products such as doll's clothing and doll houses. The other would have a relatively fixed production line for the dolls themselves and other products that did not change very rapidly. This would be similar to the classical pyramidal structure.

2. The organization would be set up on a pyramidal basis, using the assembly line as the central core. Production would be primary, there would be high centralization of authority, and direction would be top down.

3. No, they are only responding with other problems. None of them has any specific operational hypothesis that can be proposed, with the best one selected for testing in order to correct the defined problems. It seems the situation-manager interaction influences both parts. The situation is influenced by the manager and vice versa. Therefore, it is quite difficult for George to put on Albert's or Mike's "hat." Don has to be able to "listen" to what his differentiated people are saying and attempt to integrate the answers into a framework for everyone to use.

4. Don Corvis might do several things to improve communications. One quick solution during the meeting would be to request that each of his managers try a little experiment in communications. Suggest that each one tell another manager what that other manager has just said. In other words, "I'll tell you what you told me and you tell me if I'm correct. If I can do that, and you agree that I have it right, at least I understand it." Perhaps a more practical process would be to have each manager prepare a position paper outlining his suggestions, in an engineering report format, before the next meeting. By that time, everyone will have had an opportunity to review and compare the other people's positions. It seems that rapid change is coming to Corvis manufacturing, and it might be appropriate to set up a project team to handle several of the new product developments. For example, a project team might try to develop a new plastic for doll bodies or an improved method of developing and changing production lines, or might install multipurpose lines that can handle different products under computer control.

Some theories are suggested in this chapter. Would YOU use any of them? How? Do you agree with my answers? Why?

REFERENCES

Ackoff, R. L. **Creating the Corporate Future.** John Wiley & Sons, New York. 1981.

Ackoff, R. L.; Finnel, E. V.; and Gharajedaghi, J. **A Guide to Controlling Your Corporation's Future.** John Wiley & Sons, New York. 1984.

Argyris, C. "Being Human and Being Organized." **Current Perspectives in Social Psychology.** Hollander, Erwin P., and Hunt, Raymond G. (eds.). 2d ed. Oxford Univ. Press, New York. 1967. pp. 573–585.

Bass, B. M. and Drenth, P. J. D. **Advances in Organizational Psychology.** Sage, Newbury Park, CA. 1987.

Beer, M. "A Social Systems Model for Organizational Development." **Systems Theory for Organizational Development,** Cummings, Thomas G. (ed.). John Wiley & Sons, New York. 1980. pp. 73–114.

Bentley, J. C. "Facing a Future of Permanent White Water: The Challenge of Training and Development in High Technology Organizations." **Advances in Global High-Technology Management,** Gomez-Meija, and Lawless, Michael W. (eds.). JAI Press, Greenwich, CT. 1992. pp. 179–201.

Chandler, A. D. Jr. **Strategy and Structure.** MIT Press, Cambridge, MA. 1962.

Filley, A. C., and House, R. J. **Managerial Process and Organizational Behavior.** Scott, Foresman, Glenview, IL. 1969.

Galbraith, J. "Environmental and Technological Determinants of Organizational Design." **Studies in Organization Design,** Lorsch, Jay W. and Lawrence, Paul R. (eds.). Irwin/Dorsey Press, Home wood IL. 1970. pp. 113–139.

Gattiker, U. E. **Technology Management in Organizations.** Sage, Newbury Park, CA. 1990.

Kotter, J. P. **Organizational Dynamics: Diagnosis and Intervention,** Addison-Wesley, Reading, MA. 1978.

Lawrence, P. R., and Lorsch, J. W. **Organization and Environment.** Harvard Univ. Press, Cambridge, MA. 1967.

Likert, R. **New Patterns of Management.** McGraw-Hill, New York. 1961.

Meir, R. L. "Explorations in the Realm of Organization Theory, Decision Making and Steady State." **Readings in Organization Theory: A Behavioral Approach,** Hill, W. A., and Egan, D. (eds.). Allyn & Bacon, Boston, MA. 1967. pp. 471–477.

Mintzberg, H..- "Organization Design: Fashion or Fit?" **Harvard Business Review** (January–February 1981):103–116. Excerpts reprinted by permission of the *Harvard Business Review.* Copyright @ 1981 by the President and Fellows of Harvard College. All rights reserved.

———. "The Pitfalls of Strategic Planning." **California Management Review.** Fall 1993. 36(1):32–47.

Morse, J. J. "Organizational Characteristics and Individual Motivation." **Studies in Organization Design,** Lorsch, Jay W. and Lawrence, Paul R. (eds.). Irwin/Dorsey Press, Homewood, IL. 1970. pp. 84–100.

Osborn, R. N.; Hunt, J. G.; and Jauch, L. R. **Organization Theory: An Integrated Approach.** John Wiley & Sons, New York. 1980.

Pfeffer, J. "Competitive Advantage Through People." **California Management Review** (Winter 1994):9–28.

Shockley, W. "On the Statistics of Individual Variations of Productivity in Research Laboratories." **Proceedings of the I.R.E.** March 1957. 45(3):281.

Silverman, M. **The Technical Program Manager's Guide to Survival.** Wiley, New York. 1967.

Steele, L. I. W. **Managing Technology: The Strategic View.** McGraw-Hill, New York. 1989.

Taylor, F. W. **Principles of Scientific Management.** Harper and Brothers, New York. 1911.

Thompson, J. D. **Organizations in Action.** McGraw-Hill, New York. 1967.

Wexley, K. N. and Baldwin, T. T. "Management Development." **Journal of Management, 1986 Yearly Review of Management.** 12(2):227–294.

Woodward, J. **Industrial Organization: Theory and Practice.** Oxford Univ. Press, London. 1965.

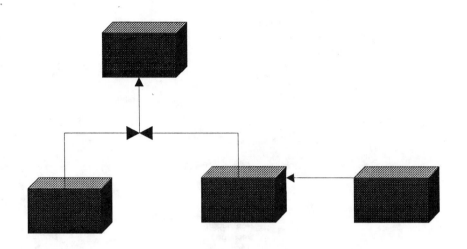

CHAPTER 9

Prescriptions for Structures: Temporary Teams

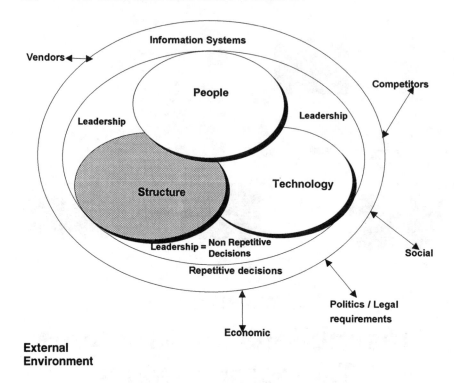

Case Study:
The Inadequate Turbine

The scene is the Amalgamated Machine Company. The company produces a broad variety of capital equipment. Power turbines are just one product line, and each turbine development project has its own project manager.

Cast
Milt Cowan: General manager, turbine division
Eric Redder: Project manager, "blue" turbine
Mary Ann Lane: Field support manager
Alan Bonson: Corrosion engineering specialist
Clyde Miller: Chief turbine engineer (Alan's boss)

Scene: Milt's office

Eric Redder was nervous. There was something he couldn't quite understand about the tone of voice that Milt used that morning when he phoned Eric and asked him to step into his office about 10:00 A.M. When Eric arrived, he found Mary Ann Lane and Milt going over some official-looking papers.

Milt: Oh hi, Eric, come in. You know Mary Ann here?

Eric: Sure. We have worked together on several projects. (Mary Ann nodded at him.)

Milt: Well, how's the "blue" turbine coming along?

Eric: Just fine. We just finished life tests on the pumps and we'll be ready to order major components for the shell and casing by the end of the week.

Milt: Well, that's great. You're getting this project going, just as we had expected, and I'm sure that this little problem that's come up won't be much of a bother. However, Mary Ann insisted that we talk about it a bit. The "blue" turbine is mighty important to this division's future because of its light weight and dependability. Those are major assets when we sell it to less developed nations for use in energy recovery activities. But you know all about that. Now, about this little problem. Maybe I'd better ask Mary Ann to explain it.

Mary Ann: Well, in field support we are very concerned about training local third world operators who don't have the basic skills that we would normally expect to find in more developed industrial areas. These operators have to learn how to maintain the turbines in the field, and therefore we have to consider some aspects of turbine maintenance that just don't come up often with our other customers in more industrialized nations. For example, we have just learned that the turbine model before yours, Eric—you remember the model Q—has a bad reputation in the field because it rusts in semitropical environments.

Eric: Well, how does that affect the "blue" turbine? Our project charter definitely states that the turbine is to be operated in an air-conditioned, environmentally controlled power stations only.

Mary Ann: We can't always control how our customers use our products. Even though the operating instructions are clear, and it may be the customer's fault, if the turbine begins to corrode, our reputation suffers. We don't want our reputation to suffer, so how can we prevent this type of problem from happening with the "blue" turbine?

Eric: Now look here, I've been working with this project for over a year now. We've already had our first design review and we're pretty well along in procurement. I don't know if anything can be done now about this type of problem, but really, why should we in project management get concerned? I always thought that the education of the customers and their operators was the responsibility of field operations. We build the turbines, write the manuals, and then turn them over to your group. Why bring me any more problems?

Milt: Of course, you're right, Eric, but could you think about this problem a bit? Maybe this is one of those things that we can resolve here in the plant rather than out at some desolate power station.

Eric: Well, I suppose that I could have our design evaluated for corrosion resistance, but I'm not sure what the effect will be, especially now that we're ready to order major components. They might all have to be changed to stainless steel and the delay and cost could be appreciable.

Milt: OK. Take a look into it. Coordinate with Mary Ann here. And, oh yes, by the way, top corporate headquarters has been watching this project because it could be so important to us. I'm sure that they wouldn't like it if costs were to increase very much.

Seeing that the meeting was over, Eric and Mary Ann left Milt's office. Eric was very unhappy. Mary Ann was sympathetic but offered no immediate advice. She did, however, give him some standards on field corrosion and offered to help in any way that she could later. Eric went back to his office and made a phone call to Alan Bonson.

Eric: Hi, Alan, how are you doing? I'm OK, but say, we have a minor problem that just came up and I'd like to get your opinion on it. Can you meet me here in about half an hour? Fine, see you then.

Later, Alan came into Eric's office, his usual twenty minutes late. After exchanging greetings, Eric covered the problems with Alan and asked for an analysis of potential corrosion problems if the turbines were operated under the new standards that Mary Ann had passed on.

Alan: Look, I'd like to help, but I've got four other project managers to satisfy. Your budget for corrosion testing is about used up, and I'm leaving on a two-week vacation tomorrow, so you'll have to hold everything until I get back. Sorry, but you should have thought of this before.

He then got up and left Eric's office. Eric was really angry. He sat for a while cooling off, then got up and walked across the building to the office of Alan's boss, Clyde Miller. Luckily, Clyde was in and there was no one else with him at the moment. That was unusual, because Clyde was handling about seven major turbine projects in addition to the "blue" turbine.

Eric: Hi. Got a minute to talk?

Clyde: Come in. Why not? You're a project manager. I'm supposed to report to you and the rest of you guys who have only their own projects to consider. I'm trying to keep you guys happy in addition to administering and training the technical prima donnas who really turn out the bright, creative products that we need to keep us in business. (Noticing that Eric was not responding to his joking, Clyde stopped, then continued in a more serious tone.) What's wrong? It looks like somebody has been raining on your parade.

Eric explained what had happened at the meeting that morning with Milt Cowan and the subsequent meeting with Alan. Clyde offered to look into the situation and report back by that afternoon. Clyde then called Mary Ann and Alan and spoke to both of them on the phone. Later that afternoon, he stopped in to see Eric.

Clyde: Hi there. I may not have all the answers but I have a few suggestions about that corrosion problem.

Eric: Well, what do I do now? And remember, suicide is not a viable management alternative because it sort of keeps you from getting any promotions in the company.

Clyde: Very funny. You have several problems. Mary Ann's specifications would require a complete redesign of the housing. It also needs new operating manuals and extensive accelerated corrosion testing of the pumps that we just finished testing and accepting. Alan says that he's sick and tired of doing all your work in a hurry, but that's no real problem; he says that about everybody. I did find out, however, that he hasn't made any definite plans for his vacation. There's always another problem, though, and here it is: according to company policy, there's no way that we can reschedule him without his approval. There might be a way, though. He's always wanted to visit our plant in San Diego because he has a sister there. Of course, I'm not telling you how to spend project money but we do have some vendors in that area and he could evaluate their ongoing quality programs for us. What do you want to do?

QUESTIONS

1. What would you define as the underlying problem here? (Remember that sometimes obvious variances are merely symptoms of underlying diseases.)
2. Are there any procedures Eric should follow to resolve the corrosion problem? What are the alternatives and how would you rate them?
3. How should he handle Milt's request to minimize costs and delays? Is this an unusual request? Why?
4. What should be done about Alan? Are there any procedures to follow? Who is responsible?
5. What relationship should Mary Ann have to this project now? Why?
6. If you were Milt, how would you have handled this problem with the "blue" turbine?

1.0 OVERVIEW AND INTRODUCTION

Chapter 8 covered the *structure* and included a more detailed description of and prescriptions for the designs of the *functional* structure. There is another major structure that every technical manager must be familiar with in addition to those of *functions*. It is *projects*. As you know now, functions are the more or less permanent groups that go on "forever." They manage their work within their group boundaries and then pass both the management responsibility and the work on to another functional group. An example would be the way in which a company produces a new "widget." The functional "sales" group would produce the specifications that it expects the "widget" to have, based on the "sales" group's knowledge of customer requirements. Those specifications would then be sent to the functional engineering group for development of the "widget" designs. These designs would then be sent to the functional manufacturing group for production. Quality control, acting as the internal customer, would finally test the finished "widgets" against the specifications (also acting as an internal functional group). If approved by quality control, the "widgets" would then be sold by the "sales" group.

These more or less permanent structures of the technical organization are the functions. Each of them has slightly different internal processes and procedures that are well known to their people. Those processes remain relatively stable over time because the tasks that each of these functions performs within the company are well established. Functional groups provide an ongoing and necessary support to the total company as predictable and specialized work centers. The work itself moves across organizational boundaries within the company and when overall control of this relatively predictable flow is maintained by upper management, these structures can work very effectively. When that control is lost or overlooked, those structures can easily fail.

This chapter continues the discussion of structure by dealing with the "short-lived" structures of "projects" and suggests how to design the administrative and social support systems that support project success. Project organizational structural designs are often the most difficult to use in many technical organizations because of differences from the typical top-down structure of engineering or scientific functional structures. During project operations, the situation is often directed "bottom up." It's not the same.

> The pyramidal structure acquires its form from the fact that as one goes up the administrative ladder, (1.) power and control increase, (2.) availability of information increases, (3.) the degree of flexibility to act increases and (4.) the scope of the decisions made and the responsibilities increase . . . the (project-matrix) organization is almost opposite. (Argyris 1979, p. 23)

Not only are the structures different, the processes are opposed. This makes the project situation quite complex when one considers that project efforts are usually directed in an opposite way (bottom up) from functional operations or efforts (top down). For example,

> (in a project effort) . . . we begin with an expected set of results and seek to find or build a collection of processes to bring it about (Result drives process). With operations the reverse is true. We begin with a process (factory, plant, refinery, line, etc.) and search for materials to feed it and markets in which to discharge the result (Process drives product) (Gilbreath 1988, p. 7)

Typically, projects are structures that cross functional boundaries to get a major task done. That task is always unique and is either so large or so important to the company that it cannot be handled within the regular, functional structures where each subtask is completed within a department and then handed on to the next group. The usual work flow is inadequate to handle the speed needed to complete projects.

Organizations that exist in a fast paced environment cannot rely on relatively slow movement of decisions through hierarchical levels to determine appropriate operational responses to environmental and market changes, to resolve technical uncertainties, or to mediate disputes between units that must collaborate in order to produce a product. . . . (Martell et al. 1992, p. 28)

In projects, there is always a manager to forecast, measure, expedite, and control the total work flow. The project structure is created to achieve a predetermined purpose with the understanding that the structure will be dissolved either when that purpose is accomplished or when it becomes evident that the accomplishment cannot be achieved within the project limitations.

1.1 More Definitions

The term "project" as used here, generally includes programs, task forces, and matrix or matrices. My experience has been that the term "project" could be used for any short-term, cross-organizational boundary, limited-goal kind of structure. Programs usually include several projects as components in achieving a major task. An example could be that of a program to build a new transport airplane. That program would have a project concerned with the wings, another with the jet engines, another with the controls, and so forth. The major criterion of matrix organizations seems to be that the participants are either hired specifically or else are drawn on a part-time basis from their permanent "functions." An example could be a matrix structure that has assigned to it on a temporary basis, as needed, engineers, quality personnel, purchasing agents, accountants, and manufacturing machinists. These people will do work for the matrix while on temporary duty; when their work is done, they will return administratively to their functional groups. The matrix manager will be the only individual who stays with the matrix organization from beginning to end. An analogy could be that of a newly created total "company" within the existing overall company. Most projects today operate as a matrix insofar as the assigned personnel are concerned. According to the following author,

It is not a true matrix when—
1. The ad hoc or project group has no budget of its own.
2. The group has no clear purpose tied to larger purposes of the organization.
3. No formal mechanism exists for making the output of the "matrix" group input to the rest of the organization.
4. The leader of the committee, task force or project has less authority than functional managers have.
5. People are not given released time from functional tasks to work on matrix tasks.

6. Successful task force or committee work does not figure in raises, bonuses or promotions. (Weisbord 1978, p. 29)

There actually are some progressive technical organizations that have adopted these criteria wholeheartedly but they are few in number, in my experience. Unfortunately, in many instances, if the above criteria were applied to the projects or task forces created by many top managements, those would not qualify. In my opinion, the failure by top management to provide the required physical and organizational wherewithal as noted above usually stems from an unwillingness to delegate effectively. They want the job done but cannot bring themselves to delegate control over the resources. Delegating effectively means that the delegator loses the authority and delegates the responsibility, yet still retains that responsibility. Authority is digital. Responsibility is infinitely analog or expandable. Delegating ineffectively occurs when the delegator tries to maintain the authority but delegate all of the responsibility.

On the other hand, there are some other companies that find the tools of project management so useful that they implement parts of the "project-matrix" techniques and structures in getting work done within their functions. I have been particularly impressed when the group manager designates a task force leader within the engineering department and provides her with a clear purpose, provides formal tools for including accomplishments into larger engineering operations, assigns specific people to that task force leader and rewards both the task force leader and all the other task force participants when the task was over. That reward occurs even if the end result didn't wholly meet the original work requirements. Sometimes, even when extraordinary efforts are expended, projects don't succeed. Mother Nature doesn't cooperate all the time. Some tasks force "failures" may provide the data that prevents a larger expenditure of funds, time, facilities, and people. There is never a "sure thing" in science and engineering. In one of the many laws attributed to the mythical "Murphy", ". . . Nature always sides with the hidden flaw." In my opinion, this is not recognized enough.

2.0 THE PROJECT MODEL: PRESCRIPTIONS

We now move on to description and prescription for project structures. In the following discussions, we will deal primarily with the most difficult project structure, that of the matrix. The matrix is most difficult to manage because people assigned to it contribute only on a part-time basis. There is insufficient socialization in many cases to support the building of an integrated team approach. The matrix manager therefore deals with a limited commitment, at best. However, describing matrix operations will minimize the requirement to keep referring to projects, programs, or task forces. This follows the central idea in any technical design; if we deal with the most difficult part first, we can handle easier parts later on. To reiterate, the matrix is a project in which the team participants contribute as required on a part-time basis. They remain in their permanent functions for administration. Matrix team participants have "two bosses": their permanent functional manager and the part-time matrix/project manager. Matrices, projects, programs, and task forces are designed with three limiting parameters. In order of importance, these are:

1. **Task** (defined by specifications or contracts)

2. **Time** (defined by a delivery schedule)

3. **Cost** (defined by a budget).

Task (defined by specifications or contracts): The nature of the task, goal, or end result that the project is supposed to accomplish is the first parameter considered in choosing a project structure. *Task,* in projects, is like *people* in my overall organizational model; it is the most independent variable. It is dictated by the customer and there always is a customer, somewhere. When a matrix/project is initiated because of an in-company task, the customer is the organizational person who finally says, "This is good" or "This is not good." In a manufacturing company, it may be the final quality inspector. When the customer is outside the organization, the sales personnel will be the data feedback to say if the output was acceptable or not. An example of an outside customer would be the computer software buyer who is the end user. If the software doesn't satisfy, it is returned. That is failure, even if the software itself met all the original specifications. In that case, the specifications were wrong because the customer is the determining factor. Project specifications are usually fixed initially but they can be changed when it becomes apparent that the customer will not accept the end result as originally defined. (Customers have been known to change their minds.) When that happens, there is a "change in scope" and that often means a complete structuring of the matrix/project. In some instances, when specifications cannot be well established in the beginning, it might be well to consider setting up a miniproject/matrix with a very short life, called a "feasibility study." This small effort is intended to define the specification. With these results in hand, management can then decide if a matrix/project should be set up or not. Usually the project specifications cannot be changed within the project without the customer's consent.

Time (defined by a delivery schedule): Most of us think of time as invariably flowing along, but that is not the way it is regarded in matrices (or projects). It can be modified by management within the project. For example, if the progress is unsatisfactory, overtime can be used or more people (if that can help; sometimes it hinders) can be assigned. This changes the "time" or when all else fails, it might change the delivery schedule. Changing time, however, involves cost, the third variable.

Cost (defined by a budget): To a functional manager, it would be usual to state that "Budgets are rarely changed." Functional operations are repetitive and can use past history as one basis for forecasting. Therefore, changing a functional budget as time goes on is usually an unlikely event. However, the unknowns are larger in projects than in functions and there is rarely any historical basis to rely cost forecasts upon; thus, most cost estimating becomes a series of negotiating sessions between the matrix manager and various functional managers for the resources and facilities to be assigned. The project manager attempts to cover all contingencies and the functional managers attempt to minimize them. Invariably, there will be some overexpenditure unless the project manager maintains a reasonable "contingency" fund. Otherwise, management has to approve cost changes as they are incurred.

In addition, because projects/matrices are dissolved when the goal is achieved, the cost measurement systems must be different from those of functional operations. Projects-matrices produce only costs. Functions produce overheads and possibly profits. When the matrix goal is a salable product, others will sell and service that product. The project team will be dissolved and then possibly reassigned to other projects. Therefore, there should only be an assignment to the project of those costs that are controllable by the project manager. For example, because the matrix/project team cannot affect the fixed portions of overheads, those costs should not be charged to the project. It's useless and even destructive. Similarly, if the project manager cannot change project participants' salaries but can control hours, those salaries should not be charged, only the hours expended. But if an appropriate responsibility for salary review and raises is placed with the project manager, charges for salaries should include the variable overheads associated with them (i.e., only direct costs). The central rule is: if you can't change it, it's not your responsibility. (This had caused me to have some fascinating discussions with an accounting group, in the past.) Assignment of operating overheads to project expenditures is not helpful.

2.1 Project-Matrix Structures

A caveat first. Project structures can be as varied in their operations as the number of project managers and the organizations that employ them. Every project structure enjoys (or suffers) changes based on the actions of the individuals who participate in it. Therefore, the descriptions and suggestions that follow do not even have the limited theoretical rigor of the previous chapter on functional organizations. There is more opinion and anecdotal data here. Consequently, the structures must be deduced from the way that the internal operations of projects are described. In other words, this is the way it has been done successfully by others. You have to decide if it will work for you.

A project structure, by definition, always includes higher uncertainty and greater complexity than a functional structure and often results in more or less institutionalized conflict. This could be an asset (if innovation is desired) or a liability (if predictability, coordination, and lowest cost are desired). Project structures support both individual flexibility and balanced, open decision making, but at a higher cost than the more orderly functional structures. These structures include an unusual characteristic: everyone in them reports to at least two supervisors; this occurs even if the technical organization has no formal structure for it. When there is no formal structure, project managers tend to exert more personal influence or informal control on project assigned personnel, while functional managers have only the formal control. The effect is still the same: two bosses, even if the formal structure doesn't show it.

A typical project or matrix organizational structure (as before, the terms project and matrix are used interchangeably here) could look like the one in Figure 9-1. The project managers for each project report to the chief project manager (and implicitly to the client or customer). When formally organized this way, each engineer reports both to his or her functional manager and to a project manager. Each finance representative and each quality control person also reports to a functional manager and a project manager. This particular structure provides for a typical project organization to consist of a project manager, an engineer, an accountant, and a quality engineer.

Each person in a matrix has at least two sets of responsibilities, which can often clash with each other. Managers head up and attempt to balance the dual chains of command (both functional and project). This dual reporting structure provides the organization with the capability for meeting the varying needs of changing technical requirements or goals and is intended to deal with the consequently higher levels of uncertainty and complexity. One human side effect is that successful project participants are usually able to handle higher levels of uncertainty than those who regularly work in a functional structure. If we agree that uncertainty increases as one goes up in the functional organization, by definition anyone who has successfully completed projects is probably trained beyond his or her present functional responsibilities.

Successful experience in operating under a matrix constitutes better preparation for an individual to run a huge diversified institution like General Electric where so many complex, conflicting interests must be balanced than the product and functional modes which have been our hallmark over the past twenty years. (Lawrence and Davis 1978, p. 132)

2.2 Functions vs. Projects (Stability vs. Change)

The uncertainty and complexity that is typical in the upper levels of large multinational organizations such as General Electric are not necessarily found only in

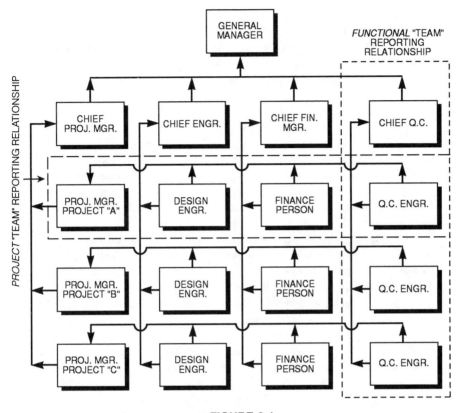

FIGURE 9-1

these larger structures. Managers in medium and smaller companies have similar problems and uncertainties to deal with.

The predictability and coordination of the in-place functional organizational structure provides few guidelines if an environmental change has not been accounted for in the structural design. The creativity needed for new product development, diversification plans, reorganization studies, and alternative investment analyses (as typical examples) is usually not part of most groups that are functionally (as opposed to project) oriented.

When a change occurs, the response of the functional structure is often inadequate to meet the challenge, and a different (project/matrix-oriented) problemsolving structure is then developed. If that project structure reaches its goals, it is dissolved because it is no longer needed. The functional group can handle the regular flow of work.

Occasionally, when the problem is expected to reoccur, but in a slightly modified way with novel and uncertain aspects, simplified project structures can become incorporated into the functional structure as a type of special unit intended to handle these problem modifications. An example could be a research and development department that is organized functionally, with internally developed short-term project organizations intended to achieve specifically assigned and limited research tasks. Assuming that an initial task, that of developing a quiet air compressor for municipal street repair work, is successfully completed by the "quiet compressor" project, the tasks of developing a line of "quiet compressors" for other applications could be assigned either to the functional development group or to a miniproject team for development set up solely for that purpose.

The size of the task is not the only major rationale for using project structures; the nature of the task itself is important. Onetime, creative tasks with high uncertainty and high situational ambiguity require project structures. Continual, logical, and relatively repetitive tasks with lower uncertainty and lower situational ambiguity are handled at less cost by the functional organizations; the existing procedures should work very well. If solving a new problem is important, will cross the engineering boundaries and is a first-time effort, it generally requires project structures. In one organization where both new and repetitive types of tasks must be handled, such as in the technical department, it is often difficult to decide whether the task can be handled normally or requires a project setup. The decision is dependent upon the situation. This is a continuation of the no "one best way" to organize, even within the department.

However, this is not intended to suggest that there are no recommended repetitive ways to set up project structures. They depend on the organizational situation and the decision criteria that management uses. Each organization can develop standard operating procedures to be used in organizing projects. The model of project organizational structure that follows and the processes used to operate it are only one alternative. The model utilizes concepts that I have proposed in many technical industrial organizations. Your structural design for projects might include this model, but only as a starting point. Because your situation is unique, you should:

- Modify the model to fit your situation.
- Test it.
- Revise the model, improving it.
- Test it again, and so on.

A project structure is intended to be a totality designed to handle tasks involving great uncertainty. However, as the unknowns decrease, structural modifications should be made until, at a logical point (and only you know where that is), the organizational structure parallels the functional design and becomes part of it. The following is a possible partial list of the project design sequence with which a project structure is intended to cope:

1. Establishing objectives (purpose, time, cost, resource limitations)

2. Defining internal task dependencies (logical development of sequential and concurrent activities)

3. Work scheduling (performance vs. time)

4. Estimating (costs expected to be incurred as a result of work scheduling)

5. Deciding on work assignments and authorization (who does what and how work shall be organizationally controlled and measured)

6. Procuring services, materials, and equipment (requisitioning and purchasing methods)

7. Attaining required resources (personnel, funds, facilities)

8. Setting up information systems (reporting and measuring data).

The organization of the structure (or repetitive human interactions) could include these four major areas: charters, work breakdown systems, operating procedures, closedown methods.

These are all covered here instead of being described in other sections of this book. Although some of this information might be more appropriate elsewhere (such as in the financial information systems of Chapter 13), because of the overall interrelationship between project structures and the operations that they are intended to accomplish, I have placed them here.

3.0 CHARTERS

Projects have to start somewhere. Someone or some small group has to perceive a new problem that is disturbing the organization and/or an opportunity that satisfies some requirement of the market or can provide a major organizational improvement. Project goals may include marketing, sales, profits, quality, or any other organizational, economic, and/or social goal. Because projects themselves may be somewhat repetitive structures in a particular company, the charter is intended to be the starting point that defines how to start a project. It is the foundation upon which all the other project operations will rest. Charters are different for different organizations, but they have to exist either explicitly (as I am suggesting here) or implicitly in the minds of the technical managers. Without charters, the organization starts from ground zero each time in an attempt to develop a method of operations. That would be wasteful and obviously unnecessary. There may not be "one best way" for all, but there are commonalities to *each* organization that can be defined for that particular organization. The charter outlines the requirements for the initial planning of any project. It is similar to a contract that outlines the relationship between two separate companies or between the company and a customer, because it defines relationships. An internal charter defines the relationship between the project and the functional organization. It doesn't contain all the operational details that each team leader within a project would need—that will be the responsibility of the team leader to define for his group—but it's a start.

On a personal note, developing a charter, a work breakdown structure, and project manual in addition to doing all the other "up front" work before the project begins is almost always time consuming and occasionally quite difficult. You have to think before you do. I have witnessed scenes in which senior management has demanded that the project manager, "Stop all this time wasting and get to work on the project!". That's a road to pain. There are enough unforeseen problems in projects to keep most of us busy dealing with them; we really don't need to deal with predictable problems at the time that they occur. If you can predict them, try to solve them when there is less time, economic or (worst of all) psychological pressure. One problem that should be quite important at the beginning is the authority relationship that the project manager will have with project participants and with the functional managers from whom those participants are drawn. It's easier to solve problems when they are merely theoretical, rather than when they occur. Conflict can be enervating, not to say dangerous when considering your future in the company.

The charter is a document that will be modified several times before becoming the basis of the project manual. It is first developed by the responsible project manager and then presented to top management for preliminary review. If the results seem to be within the general expectations of top management, it will be presented as a model to the project team leaders. They will be expected to create their own charters to cover the parts of the project that they will contribute. The final revision will take place after the project manager has consolidated the charters of the team leaders, including their respective estimates, and the entire team then reviews the total project charter. The charter contains the following information:

1. The basis: The importance of this project or its priority, comparing it to other existing projects in the company during the time that the project will exist. There is an old saying that there are no "absolutes" except death and taxes. There are no "absolute" priorities. Therefore, based on the workload, facilities loading, availability of people, funds, etc., *today,* what is the priority of this project as compared to all the other work that is presently scheduled to be done? How does it compete for scarce company resources?

2. The subject: Describes the project—what has to be done, who the "customer" will be and when goals will be achieved. For example, *the subject* in the charter may describe "a new transport aircraft that will fly so many miles without refueling, and carry such-and-such a load," but that would be merely a starting point. The team leader responsible for the aircraft fuel-handling facilities must develop a subject and a charter for his specific task as would all the other team leaders on the project. Because this charter is intended to be used only within the organization, there is no requirement for the legal terminology an external contract would require. Such a contract is beyond the scope of this book and probably that of most technical managers too.

Describe the end result. The description doesn't have to be extensive. Using the example of the new transport plane, it is not necessary at this preliminary stage to cover every last subsection design, etc. Just note the major characteristics.

Who will accept or reject the end goal? The customer has to be a real person. An amorphous "buyer" won't do. Who will the project define as the customer?

When will it happen? In my opinion, a sure way to stop a project is to set unrealistic or arbitrary specifications or goals and when they should occur. Even though this charter will have to be approved by the project team later, it is wasteful to establish goals that cannot be accomplished or cannot be completed within the time set. The result would be hostility, conflict and lack of enthusiasm. As an example, one manager set a goal for a particular accomplishment at three months even though the team leaders protested that this schedule was unrealistic and it needed at least six months. The answer was: "You always put fat in your estimates. I'm just squeezing it out." A gut feeling is not an adequate substitute for logical, technical discussions during the estimating process. It's possible, of course, that everything will go well and the three-month goal will be met. Then the team leaders will receive no reward because the manager was "right." If the team leaders exceed the three-month goal, the manager was "right" again because the team leaders were obviously inadequate, in his distorted opinion. If this situation continues, setting goals becomes a game with team leaders really adding "fat" and managers trying to find it. Because team leaders know a lot more about their own tasks, they can hide the "fat" in more ways than the manager can extract it. Why waste the time? A tough questions and answer session will uncover any discrepancies without the necessity of relying on a gut feeling.

3. The cost, the price and how determined: Cost and price are obviously not the same. The project will determine the cost. If the output is to be sold to a "customer," it is management's responsibility to determine the "price." The price may be above the cost, equal to the cost or even below the cost if management wants to open a new market. That is not really the concern of the project. It is concerned almost entirely with the cost. But how is cost defined?

Depending upon the organizational situation, the cost may not even be defined in dollars. For example, if the project manager has no control over salaries in engineering operations, those costs may be defined as the number of engineering hours in the various categories of engineering skills needed. Another alternative is to have applicable costs estimated by the responsible team leaders when the project is broken down in the next stage using the work breakdown structure. At this preliminary point, the cost might be quickly built up using any similar tasks done in the past and then relying on the "expert" opinion (thoroughly documented, of course) of the team leaders as to the differences between this project tasks and those used as the historical underpinning.

4. Standards: Definition of acceptance criteria. How will we know when certain tests have been passed? When the project is finished? Do we have any final test needs?

5. Changes: How to handle changes. What happens if it just "misses" the final test and therefore fails to pass? What if personnel are transferred out? What if the scope of the project is modified? What is defined as a "change" and what is defined as merely a problem to be solved? And so on.

6. Specifications: This is a listing of resource people and facilities. Who will be there, when and what will they be able to use? This is necessary in any event to obtain the costs. People and facilities are not interchangeable. If this preliminary charter is changed because top management wishes to have other personnel assigned or facilities used, that will affect the cost, possibly the delivery and the end result.

7. Protection clauses: What will be the authority of the project manager to add/remove people from the project, give raises, assign overtime if necessary, etc.? What will be the dispute resolution procedures? What are the rules on communications such as letters, memos and reports, within the project, to the customer, to vendors, to other company divisions, etc.? How and when will progress meetings be run? Who will participate?

The suggestions above are intended to indicate that the development of a charter for a project is not a fixed, onetime process. It is iterative and a part of the development of the project itself and the structure with which it is expected to operate. That structure might be modified in the future as the project moves along. There have been charters with modification numbers in three digits because of changes. The first time that you, as a project manager, go through the charter process to design your project structure, you will probably have to do it alone, because *no one else* can really do it. Some apparent questions pop up immediately:

How can activities be estimated without the concurrence of project participants, who have not yet been recruited?

How do you know the price or cost at this point, before the detailed
planning needed for the work breakdown structure?

Creativity is difficult enough to forecast, but without the concurrence of the
individual who is expected to perform that creative act, it is almost a waste of
time. For example, it's very difficult to schedule inventions without gaining the
concurrence of the inventor but it can be done later. That concurrence and those
of the other managers assigned to your project can be obtained during the work
breakdown system development that follows. At this preliminary stage, gaining
accuracy may not be as important as providing a speedy response to the person
requesting the project. When the designated project manager makes clear to the
requester that a "ballpark" plan will be presented almost immediately, it should
be obvious that the plan is not final and has not been approved by the project
team leaders that will be assigned if the initial "ballpark" plan is accepted. If
the requester understands this and the immediate feedback charter is tentatively
approved, then time, people and money can be expended on getting a more
accurate charter based in inputs of the project team leaders. At that time, concurrence with the preliminary charter should be discussed with the team leaders and
the people concerned. Usually, those other team leaders will suggest modifications to the initial charter before they agree with the preliminary forecasted tasks,
costs, and accomplishments.

Developing a charter can probably be accomplished in a three-step process.

Step 1: The project manager develops the charter alone and presents the results
as only a generalized "first pass" to the manager who requested that a project
be established. This first pass is a rough estimate merely to decide if it is
worthwhile to spend the time, money and equipment to get a more realistic charter.
At this point, there have been no complete estimates from any participants, no
quotations obtained from potential vendors and no scheduling of required facilities. It's merely a response that implicitly says: Is this what you generally had
in mind?

Step 2: If the manager who requested that the project be established accepts
this rough charter, the project team leaders are selected and given copies of the
charter. They are requested to meet within a few days with the project manager
after completing their own charters to evaluate and compare them. At this meeting, the charters are assembled and negotiations among the team leaders and
between those leaders and the project manager are expected to result in a final
charter. This revised charter is presented to management. If approved—

Step 3: The team leaders pass out copies of the charter to every technical participant over whom they have jurisdiction. These leaders then meet with those participants in order to gain their inputs and eventual approvals. With no changes, the project may begin. If there are changes, the project team goes back to Step 2 to work out any differences. Any differences are then presented by the project manager to the manager requesting the project for an approval. Now the project can begin to develop the work breakdown structure.

This process of resolving differences or conflicts through negotiation among project team leaders is minimal training of people who have to operate as part of a cooperative team. Teamwork is not the same as individual creativity. Not better, not worse, just different. In most functional technical groups it is individual performance that is reviewed and rewarded. There is an implicit assumption that positive individual performance will somehow impact the total organization positively. When individual team participants carry this assumption into performance at the team level, there is either no effect or a negative effect at that level (Martell et al. 1992, p. 45). Individual performance at the expense of team achievements is detrimental. Individual performance in support of team achievements is positive. There's a different motivational situation in working in project teams. There are new rules to follow and these may not be the same as those followed in functions. Using project charter-discussion meetings to clarify the expectations of both team leaders and participants and to establish the project work environments as different from those of functions is perhaps one of the more valuable outcomes of these meetings. It is a rather pragmatic approach in which

> Project managers will seek to create their own policies and rules, to create their own procedures and methods, rather than comply with those established by others. Project-oriented managers, because of their twin drives of expediency and pragmatism, will value only methods that contribute, that work. They will shun the rest as burdens or obstacles. They should be judged by their resourcefulness and adaptation, in contrast to operational (i.e., functional) managers of the past, who were punished when they exhibited these traits but rewarded for strict compliance. (Gilbreath 1988, p. 13)

The next management task is the development of a work breakdown structure.

Work breakdown structure: The work breakdown structure is a graphical representation of the work packages that each team leader agrees to complete. See Figure 9-2. It includes the initial planning and control mechanisms intended to forecast project progress, measure actual achievement, and point out potential problem areas. The complexity of these control mechanisms equals the complexity of the project that it is expected to help control. As the project becomes less uncertain and ambiguous, the complexity of the work breakdown structures and

CONDITION 1: FIRST SHIP

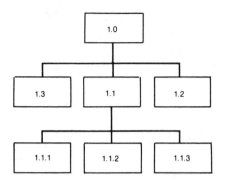

```
1.0    Spaceship
1.1    Nose
1.2    Body
1.3    Tail
1.1.1  Life Support
1.1.2  Steering
1.1.3  Nose Housing
Etc.
```

CONDITION 2: SIMILAR SHIPS

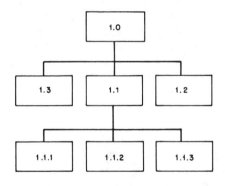

```
1.0    Spaceship
       Proj. Mgr.
1.1    Engineering
1.2    Finance
1.3    Quality Control
1.1.1  Mechanical
       Engineer
1.1.2  Electrical
       Engineer
1.1.3  Civil Engineer
Etc.
```

CONDITION 3: PRODUCTION

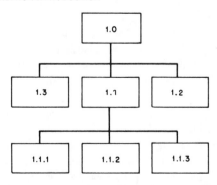

```
1.0    The Company
1.1    Production
1.2    Engineering
1.3    Quality Control
1.1.1  Spaceship
1.1.2  Turbines
1.1.3  Generators
Etc.
```

FIGURE 9-2: Typical Work Breakdown Structures

control mechanisms decrease (and the costs of operation should then decrease) until the most straightforward method or the standard functional level of effort budgeting method emerges. The work breakdown structures, noted here, are therefore only examples in a family of planning and measuring structures.

A brief review of estimating processes used in developing typical work breakdown structures follows.

> Historically a number of estimating methods have been used; terms that have been coined to describe them are subjective estimating, comparative estimating and synthetic estimating. Subjective estimating relies on the skill and experience of the estimator and does not place much reliance on data accumulated from past projects. Comparative estimating is made by considering the project or tasks in the light of previous projects and with due allowance for differences in scope and technology, extrapolating the historic data to reflect the new proposal. The synthetic estimating method consists of breaking the project into a series of very small and discrete blocks of work which can then use historically proven relationships to generate an estimate. Synthetic estimating in its true sense needs a very detailed knowledge of the tasks and is most often applied to relatively small operations such as planning the production of a component where all the operations can be estimated using speeds, feeds and human factors data. Broader synthetics can also be derived for larger scale estimating such as numbers of drawings per kilogram of product and numbers of man hours per drawing. (Webb 1994, p. 265)

The initial work breakdown structure is usually modified and simplified as the uncertainty in attaining project goals change. To illustrate, let us assume that the project in question is concerned with building an interplanetary spaceship under three different conditions of situational uncertainty (see Figure 9-2). The first or most uncertain situation (condition 1) shall be that the ship is the first one we have designed and built. This situation has the highest amount of uncertainty and ambiguity. The second or less uncertain situation (condition 2) occurs after the first ship has been built and others of similar configuration but slightly different destinations in space have to be built: medium uncertainty and ambiguity. The last situation (condition 3) occurs when many ships have been built and they then become part of the organizational product lines: lowest uncertainty and ambiguity.

Condition 1 is primarily product oriented. It provides maximum control over product components, but it has the highest labor and management costs. It is the most differentiated and therefore incurs the highest cost for project integration and coordination. With the first ship delivered, condition 2 can be used for the design of the work breakdown for the next ship or group. This condition is task or functionally oriented but still structured in project mode. There is less differentiation. Participants' skills are used across all parts of the project, thereby incurring lower integration costs, which are major concerns of the project manager as the chief integrator. Risks are more definable because the work breakdown is becoming more person and skill oriented. When the spaceships become part of the organization's production lines, condition 3 should apply. At that time, the work breakdown is almost the same as a functional budget. It has the lowest integration costs (and therefore needs less project management) because it is the least differentiated method of control. In this last case, the risks are no higher than for any other product, and the uncertainty for all products is similar. Now the spaceships are just another product.

3.1 Changes as Projects Move Toward Functions

Although the work breakdown structure is primarily an administrative control and budgeting tool, it could also reflect changes in the structural design as a project changes. Projects move, with time, from high uncertainty to lower levels of uncertainty as they approach completion. The work breakdown is supposed to be changed with this change in structure, becoming more simplified as uncertainty decreases. Table 9-1 is indicative of this general change in work breakdown methods as projects move into production phases. It reflects the three typical conditions of work breakdown noted before and the equivalent changes toward a functional structure in the organization.

3.2 Project Procedures

Project procedures can be categorized for descriptive purposes into three general areas, although, of course, the three areas are not separated in the actual operation of projects.

Table 9-1. Changes from Project to Functions

Time and Purpose	Structure	Performance		
		Repetitive Decisions	Cost	Management Tasks
Long range and innovation	Project	Work Breakdown	Very costly	Highest differentiation and highest integration
Middle range and market response	Functional and project	Budget and work breakdown	Middle	Less integration
Survival or slow growth	Functional	Budget	Lowest	Coordination

1. Those that deal with people,

2. Those that deal with costs,

3. Those that deal with accomplishment

3.2.1 Those That Deal with People: Dual Reporting

Project team members have two sometimes conflicting reporting channels. They report both to their functional managers for administrative operations and training and to the project manager for technical competency and team inputs. Each team member, therefore, wears a minimum of two hats. They are representatives of their functional "home" and contributors to the project success. This situation has been described by many researchers (Clark and Wheelwright 1992) as Dual reporting. Dual reporting is a procedure that is intended to reduce the negative results of the built-in conflict patterns when technical personnel are assigned to projects but remain in functional groups administratively. Projects have limited lives and this often causes role conflict. "Which manager do I respond to, my functional boss or the project manager?" Or perhaps even more typically, "When I am assigned to several projects, how do I respond to the conflicting requirements of several project managers?" Dual reporting is one tool used to resolve this type of conflict. The management process is based on these definitions of structural responsibilities.

1. Functional manager: Prime responsibility is to train the person assigned to the project and to ensure that his or her administrative needs (i.e., vacations, time sheets, attendance, etc.) are satisfied. The major input is coordination of individual technical expertise for the project.

2. Project manager: Prime responsibility is to provide financial resources, operating direction, and support for the people assigned to the project. The major input is coordination and management of the various project teams.

Therefore, under dual reporting, project personnel report to the project manager for day-to-day operating guidance and to their functional manager for training and administrative support. The performance review process of personnel assigned to projects under the dual reporting procedure should therefore be relatively straightforward. At appropriate time intervals (e.g., six months, one year, the person's scheduled progress review, or whenever someone is moved off the project) the functional manager to whom the person reports administratively prepares a work loading sheet for that person.

It shows the names of the project managers to whom the person reported during that time period and the amount of time spent on each project during the total time interval. For example, if the review is a six-month review, there are approximately 1,000 work hours to be accounted for. The functional manager produces a schedule for each person in the group, showing where they were assigned during that time period. The schedule sheet shows the names of the project managers and how the 1,000 hours that the person expended during those six months were distributed among those managers. One copy of the sheet for each person is then sent to the appropriate project manager. The project manager meets with each person to review his/her performance and tells that person the score, which is a function of performance during that prior time period.

For example, George worked for project manager A for 100 hours, for project manager B for 800 hours, and for his functional manager for 100 hours. Assuming that the potential scores can be between 1 (absolutely terrible) and 10 (absolutely wonderful), if George gets a 3 from project manager A, a 9 from project manager B, and an 8 from his boss (since he worked in the functional group during that last time period in addition to working on various projects), his weighted score, which is reported to his functional boss, is:

$$
\begin{array}{ll}
3 \times 100 = 300 & \text{(for project manager A)} \\
9 \times 800 = 7{,}200 & \text{(for project manager B)} \\
8 \times 100 = \underline{800} & \text{(for functional manager)} \\
 8{,}300 &
\end{array}
$$

The weighted score is $(7{,}200 + 300 + 800)$ divided by 1,000 hours) $= 8.3$. The scores are summarized by the functional manager. The final score of the person is related both to the personal evaluation (performance?) and the time spent (importance?) on a particular project. Because this mechanism includes time spent in a functional area, it is an overall evaluation. This example is indicative of the way dual reporting can work with several project managers and a functional manager. Each project manager should rate independently of all others. Therefore, all the project manager sees is a sheet with an employee loading and a space for his or her own evaluation. Because each rater gets a separate sheet, it is obvious to that rater how much weight the evaluation will have vs. all the other evaluations that are to be done.

The advantages are:

1. There is higher coordination of inputs received by the functional manager in determining the adequacy of functional training received for the needs of project managers. If the person is scored low, it probably means that he or she is uniformly regarded as inadequate and needs training.

2. Dissatisfaction is less likely to be smoothed over and high performance is less likely to be overlooked, as all three people, the person, the project manager, and the functional manager, are involved in the evaluation process.

3. Functional managers are still responsible for the continual, long-term growth process of the person.

4. Project managers are responsible for evaluating the person's on-the-job performance.

5. The person is more likely to observe a direct connection between performance on project teams and compensation: more connection equals more effective manipulation of rewards that are connected to motivation.

The disadvantage are:

1. Administrative costs go up to operate this evaluation process.

2. It is possible that no person will ever receive an extremely good or an extremely bad score. That is unlikely, because a logical response from the functional manager could be: "Why didn't you tell me about this before the review process? I could have (either) commended the person for his excellent work sooner, assuming that he had scored well, or put him into training, assuming that he had scored poorly, if I had known sooner."

Dual reporting supports coordination between project and functional areas and control of participants when they are assigned to different projects. It is probably one of the more important control mechanisms for the project manager, as people are the most independent variable. Control over expenditure of funds can be next because it is usually emphasized by senior management control over accomplishment is third.

3.2.2 Those That Deal with Costs: Financial Networks

The work breakdown method is the mechanism that assists in:

1. Forecasting how the work will be done,

2. Defining the interfaces among various work packages,

3. Fixing responsibility for finishing these various work packages by assigning them to various project operating managers.

Financial networks are set up, using hierarchical numbering systems (see Figure 9-2, showing the conditions 1, 2, and 3 numbering system) that indicate which package is related to which. When the person tells the project manager that a task is to begin, the project manager can inform the financial control center (accounting department) that a particular charge number is now open and can be charged. Time and material charges can only be accumulated against those numbers in the work breakdown that are open. When the particular task that has been assigned to the particular number has been completed, that number should be closed. This prevents errors such as charges being made after a task is completed.

When the estimated costs forecasted in a particular charge number are used up, that number is automatically closed by the accounting department. Then the alternatives are 1. the work stops, 2. the project manager adds more funds and reopens that charge number or 3. the person responsible for accomplishing the particular task tries to charge another charge number. In the last example, when there are no open numbers in the project, the only place that people can place their time charges is in the functional overhead accounts (unless they can find another project available with an accommodatingly open number). Assuming the work continues, with only two alternatives for charges, project or functions, there is an automatic disclosure of progress and/or problems, as there is a plan for each of these alternatives. Projects use work breakdowns, and functions use budgets. When a charge is completed, something has been accomplished, like the reaching of some tangible milestone. A system that allows charges without responsibility for deliverables is like offering a full checkbook with unlimited funds and all the checks signed. However, when the project manager can open or close any part of the hierarchical work breakdown method, she or he plays a major part in an effective financial control. With the ability to open or close charge numbers, the project manager is in control. The advantages of this type of financial control are:

1. Expenditures can be matched against critical work packages of the project.

2. Underexpenditures of particular packages that are closed upon completion of tasks can be reallocated within the project by the project manager. (This happens rarely, of course.)

3. Overexpenditure of funds that affect the total project *cannot* occur without the prior approval of the project manager (i.e., no surprises), because each charge number contains only a limited amount of funds or time. The limitation on spending is ensured by the limited resources available under that particular charge number. If additional funds are needed, they can be added only by the project manager.

4. Any disputes between the project and the functional managers quickly surface if project personnel try to charge functional overhead accounts because these charges have not been budgeted for. This sometimes can occur if the project is not "ready" to accept the person's planned charges; for example, when there are delays in other parts of the project. Usually however, this system raises this kind of problem before it becomes a major area and then it can be resolved through negotiations among the project manager, functional managers and senior management.

The disadvantages of this type of financial control are

1. This control cannot stand alone, but must be coordinated with a system to measure progress. Expenditures by themselves do not indicate anything; only when they are matched against progress (or lack of it) is this tool useful.

2. The administrative costs in opening and closing project work packages are not high, but they are not free either; all controls cost something.

Going a step further, any system that merely matches financial or time expenditures against plan is inadequate for effective project control because the main task of the project is to achieve the overall end goal on time and within budget. Reporting that the expenditures are on target with the plan does not indicate that the end point will be (1) on target, (2) higher than target, or (3) lower than target. Hardly an acceptable state of affairs! The project manager is concerned with only one goal: the estimate at completion. And that is what he or she watches. It is easy to calculate (although sometimes difficult to get because people may be hesitant to cooperate). The formula is:

actual + estimate to complete (ETC) = estimate at completion (EAC)

The actual is reported by the person expending the time or funds through the regular financial or accounting routines. The (ETC) for each component of the work breakdown method that is open for charges is reported by that same person with the same frequency that the actual is reported to the accounting department. By adding both numbers to obtain the (EAC) and plotting the sum as time goes on, the project manager can compare it with the original EAC and determine when the variance between the original EAC and the present EAC requires corrective action. (See Figure 9-3.) The ETC focuses attention on the end goal. There are other concepts such as estimate of percent completion. This seems to

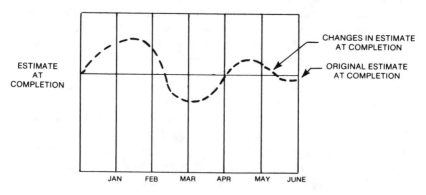

FIGURE 9-3: Estimate at Completion over Time

me to be less effective than the ETC because I always got the impression that the percent completion emphasized what had been done rather than what was yet to be done. But, it's a matter of preference.

When the chart of the repetitive calculations of the EAC overtime is flat, the person is probably not estimating the remaining tasks in front of him or her but is simply reporting arithmetic differences. It's virtually impossible to predict the future with no error, and that's what a flat EAC curve indicates. When this occurs, it requires looking into by the project manager. The shape of the EAC curve is quite revealing. If it has a sinusoidal curve of diminishing amplitude with a midpoint around the original EAC, it probably is as good as it will ever be. It means that the person is analyzing the future work to be done and is correcting as the work progresses.

3.2.3 Those That Deal with Accomplishment: The Design Review.

If it were possible to plot uncertainty as the ordinate (of course if we could measure it, it wouldn't be uncertainty) vs. project elapsed time as the abscissa, it has been my experience that the curve would generally be in the shape of a decreasing S. Many of the factors that cannot be forecasted with any certainty at the beginning of a project become fairly well defined during the initial phases. If no one else can, the project manager must "guesstimate" where the major problems will probably occur during the project life. On the other hand, plotting a curve of cumulative expenditures of cost vs. time in a similar fashion would probably result in another S-curve, but reversed (see Figure 9-4).

The starting phases of most projects are always less costly than the middle and ending phases. It is almost impossible to spend a lot of money or time in the beginning phases, if for no other reason than it takes a certain amount of time to recruit and apply personnel, issue purchase requisitions and turn those requisitions into purchase orders that eventually result in invoices to be paid from project funds.

Assuming that there is a similar vertical scale for uncertainty and cumulative

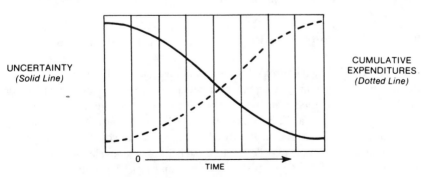

FIGURE 9-4: Uncertainty and Expenditures

expenditures vs. the same horizontal scale of time, it becomes apparent that after some point uncertainty will quickly decrease and cumulative expenditures quickly increase. That is the optimum time to review the project for the first time (usually at about 20 to 30% completion into the project). Sometimes review of project expenditures, prior to the first design review, are called a feasibility study. The progress and learning achieved in this preliminary phase or feasibility study can be used to redefine project goals, time schedules and budgets. The design review process is fairly completely detailed elsewhere (Jacobs 1979), and I suggest that you become familiar with it. When the design review is used as a control on sequential reaching of project goals, it can be matched with the financial controls of opening and closing various work packages.

The advantages of design review are:

1. There is an ability to evaluate progress against goals at a predetermined interval.

2. Financial exposure is limited to the time before the design review, not the whole project.

3. It is possible to match cost and achievement.

The disadvantages are:

1. The cost of bringing all the project work packages to a halt during the design review process, restructuring them if necessary, and starting them all over again is high.

2. If the uncertainty curve does not have a typically decreasing S-shape, because everything is dependent upon some crucial end test, the design may not be particularly helpful. (Suppose you won't know if something works until the first field test, which cannot be scheduled until the project is almost over? An example is the first atomic bomb test of 1945.)

3.3 Politics, Negotiations, and Other Important Things

Most of us are trained to think logically and consistently during our academic studies. In our beginning careers as "knowledge workers," we are rewarded for maintaining this mental framework. But when we are promoted and have to manage other people, a completely logical approach is probably the surest way to failure. We learned that in the discussions about the component *people* in previous chapters. People may be creative and behave in innovate, nonpredictable ways that may not fit the consistent systems and procedures that we devise. That's a great advantage (especially when we're wrong) and a problem (during the few times that we're right). This can be unnerving when we manage the stable environment of the functional organization. It can be downright confusing in project operations.

Project managers must deal with the same other groups that functional managers do but they have an added degree of complexity. That complexity is typically due to two factors.

1. The overall nature of most projects, when they can affect all parts of the parent company, with the heightened uncertainty associated with trying to achieve unique project goals,

2. Most importantly, getting along with both the functional mangers who supply the part-time contributors to the project and with senior management who fail to understand the special problems of project operations.

It's not easy, but there is anecdotal data that could assist. According to one author, asking questions is a very appropriate start to resolving problems before they occur. For example, dealing with the overall nature component,

1. Do I have all the needed views represented on the team?
2. Do I know the customer?
(**my comment**: *very, very important—This has helped me in many situations.*)
3. How well do I understand the application?
4. How certain are my product . . . specifications?
5. How solid are my cost-volume projections?
6. How realistic is the timing . . . ?
7. How solid is our projection of competitive reaction?
8. How well characterized is the technology?
9. Is the review process comprehensive? (Steele 1989, p. 142)

And then there is politics (a bad word?), socialization (a better one?) or cultivating the appropriate manners (that's a good one) to get along. In an ongoing technical organization, projects can be an unusual (and annoying?) situation for most functional managers. Many of them don't like to delegate control of their personnel, funds or facilities. In some cases, senior management is reluctant to even establish a project organization for these reasons. Projects are less predictable than functional operations and there is always some tension between projects and functions. However, there are other companies that prefer using projects to accomplish tasks because senior management feels that it is a broadening experience for both the project and team leaders. That experience, they say, better equips these managers for future progression into top corporate echelons. But even in that happy situation, which, unfortunately, occurs not often enough, the internal culture of a company can vary greatly among various functional groups.

And that culture may not support project operations very well. Therefore, it behooves the project manager to get along. In other words,

> The success of project management may be less sensitive to the specifics of the organizational system than it is to the cultural ambiance that exists in the organization. (Cleland and Kocaoglu 1981, p. 51)

With experience and knowledge of the internal culture of the organization, the experienced project manager learns how to get along.

> Over a period of time, the individual responds to experiences and learns how to accommodate to them. This mode of accommodation or adaptation is a part of the individual's personality. Socialization is a process by which a society inculcates values, norms, ideals, and beliefs to its members, and may be understood in terms of learning theory. . . . However, there are different ways in which individuals learn to interact with such authority. . . .
>
> Functional departments are oriented toward the acquisition, maintenance, and development of specialized resources—people and equipment, for instance. Project teams are oriented toward the production of output. As a result, relationships, authority patterns, and evaluation all become highly complex and usually negotiated—because there are no clear and simple answers. The multiple dimensions of organization foster an ongoing tension—which can be highly creative, if appropriately balanced—between the underlying resource orientation and the desired output orientation. The matrix form encourages a far more useful view of organizational realities, because it legitimates the sort of bargaining and negotiation that is essential when many goals—for instance, both output and resource goals—are to be met at once. It also encourages a general managerial viewpoint well down in the organization, where members see these alternate dimensions interacting and begin quite early in their careers to understand the trade-offs required, and then, as managers seek new answers. (Tosi 1990, p. 284)

However, the neophyte can have a very difficult learning program as the move is made from competent technical contributor to persuasive project/matrix management. Although the idea of "politics" is often viewed negatively, in reality it can be defined as "The use of means not sanctioned by the organization to attain sanctioned ends or the use of sanctioned means to obtain unsanctioned needs." (Pinto 1994, p. 33.) This same researcher further suggests (p. 35) the following:

> Understand and acknowledge the political nature of most organizations.
> Learn to cultivate appropriate political tactics.
> Understand and accept "What's in it for me?".

The difficult path is treading between behavior that is aggressive and predatory, interpreted as intended to obtain power (the ability to get things done by yourself, either by directing others or controlling resources) and refusing to recognize that all behaviors do not fit within predefined company directives.

3.4 Handling Changes: The Impact Statement

No project has ever been completed without changes. The causes could originate anywhere from company management ("Sorry about that, Pete, but your best engineer has to be transferred to another project that's in a crisis. I'm sure you understand."), from the market in general ("Sales says that competition has produced the same product with half the tolerances you are estimating, and at a lower cost."), or even from the eventual customer ("Sorry, that's not the color we really wanted for our spaceship, so just change it."). Changes can even be caused by the project itself. ("The design review indicates that 30% more time and funds are needed to meet the final specifications."). No matter where changes come from, there has to be some mechanism built into projects to account for them. That mechanism is the *impact statement.*

The sequence for handling changes is similar to the sequence for setting up the original project:

1. Determine the new or revised estimate to complete.

2. Complete a new charter if necessary.

3. Reconstruct the work breakdown.

4. Write the impact statement, which explains the change and its impact upon the project.

What is now required to reach the project goals and how will that be done?

The form is not as relevant as the content. Something that is major has changed, and if the forecasted estimate at completion is affected to the extent where the project will not meet its targets, the project manager's responsibility is to present those data as quickly and as accurately as he or she can. When there are insufficient resources to correct conditions within the existing budget, the time to issue the impact statement is now.

The project manager is no different from any other manager in a decision-making situation; he or she also has to absorb uncertainty in making decisions, but each of us can only absorb a limited amount. That limited amount is generally defined as the contingency resources available within the project. When the change is great enough to exceed this, a new estimate at completion is needed. When the amount expected to be spent is greater than that in the existing budget, regardless of the reasons, it is vital that the next upper level of management, with greater amounts of resources, be advised and that either those resources be made available or the scope of the project be reduced to use those that are available. I have found that impact statements are a very valuable defense against sudden changes imposed from any source, but especially from your own management. When the cost of changes is known, those changes may not be imposed after all.

3.5 The Trigger Clause

Projects don't go on forever, and when there is a change requested, there has to be a time limit placed for a response. I have found that it's much easier to negotiate the extent of the "trigger" clause at the beginning of the project when the first charter is being created. For example, if an impact statement is issued to the next upper level of management (uncertainty and resources increase as one goes up the corporate ladder), the receiver of the impact statement is allowed (*fill in this blank*) days to respond. The clause I always liked was, "If there is no response in 10 days, 'Silence means assent' and the requested funds, resources, and personnel will be allocated immediately to the project." A slightly more aggressive clause states, "If there is no response in 10 (or whatever) days, 'Silence means reject' and the project will immediately be closed down." Occasionally, senior management would protest. in that case, the questions was, "How long do you need to make a decision?" Bringing this up in the beginning of a project when the charter is first outlined should start some interesting discussions. Bringing it up later when the first major change occurs is not recommended. That can be difficult.

4.0 PROJECT CLOSEDOWN: DOING IT RIGHT

Eventually, however, the project will be completed. Planning for completion should begin before the expected completion date.

Closing a project requires almost as much planning and skill as starting one. The closedown has its own special needs. There are generally three main areas to be considered:

1. The client-user-customer inter relationship known as the outside

2. The functional needs internal to the organization—the inside

3. The project itself—the project summary.

1. The outside: A list of open items is drawn up by the project team, listing the tasks to be completed in order to reach the project goals. This list is proposed to the client-user-customer as all the things that the project manager feels still have to be done. If accepted, a new work breakdown is developed to accomplish the tasks on the list. In effect, the scope of the work has been defined and a new project (for closedown) is set up. By checking the amount of funds available in the cumulative expenditure curve (difference in the y axis of the curve) and the amount of time to complete the open items (difference in the x axis of the curve), we know how much money and time are still available to complete the project. If the customer does not accept the list and demands other inputs, then we may have to issue an impact statement requesting more of both.

2. The inside: This is a subjective list of tasks that should be completed to take care of the administrative aspects of the project. My list, which follows, is not the only possible one, but it does provide suggestions for you to use when you draw up your own list applying to your particular situation.

(a) Personnel: Review who you, as the project manager, will need, for how long, and when they will be reassigned back to their functional areas. When they are reassigned, write any dual evaluation reports needed, because you might not be working with them when the next six-month or yearend review is due. Just do this for the elapsed time from the last review.

(b) Test reports: Complete the documentation. The tests were done, but where are the reports?

(c) Capital assets: Did you buy anything for the project? What are you going to do with it when the project is completed?

(d) Inventory: Did you have anything left over? What will you do with it? One project manager used to notify the customer that there was X amount of material left over. If not advised otherwise in ten days, he would scrap it and send the value received to the customer. It's an idea.

(e) Documentation: Have the blueprints been updated? What about spare parts lists, manuals, and other documents? Remember, nobody ever seems to need that stuff until you have just been transferred to the next bigger project.

(f) What else?

3. The project summary: This is a brief outline of the project history: what it was expected to do, the changes that occurred, and the results achieved. Because all the other project documentation is available, such as minutes of meetings and drawings, the summary is not supposed to be voluminous. A simplified two- or three-page outline summarizing all the "wonderful" achievements would be just fine. Remember how helpful it would have been, when you were starting the project, to have a summary of similar projects to review. Now's the chance to leave one for the next project manager.

5.0 DEVELOPING THE "HOW TO" OF PROJECTS: THE PROJECT OPERATIONS METHODS

The development of functional organizational structure is glacially slow when compared to that of the project organizational structure. That is because environmental uncertainty has a much more gradual impact on the total organization than on projects. Rather than take the time to evaluate the design elements and assign weights to them, as recommended for functional organizational structures, we are forced by rapid change in projects to use a more radical and faster method of design. This allows fast response to uncertainty. The project plan should consist of the following three steps.

Step 1. Forecasting: The project manager (with possibly a small planning staff) is responsible for the initial documentation of the project charter, which includes a list of tasks and responsibilities and the recommended work breakdown structure. After approval of this by senior management, the selection of the project personnel (actually only the first line of management under the project manager) then takes place through negotiations between the project and the functional managers concerned (e.g., controller or financial persons, chief engineer for engineers). When that has occurred, the first meeting of the project takes place under the direction of the project manager, and the whole group evaluates a preliminary list of tasks that the project manager has developed as part of an initial program plan. Then the appropriate people on the project accept the responsibility for each task.

When a person accepts tasks, he or she initials the preliminary task list, thereby accepting the responsibility for developing the budget to match this preliminary list. If the group disagrees with the list, the first order of business is to develop a list with which all its members agree. A revised work breakdown is then unanimously developed and costed, and a revised charter written. An organization chart for the project is drawn up and people initial their own jobs on the chart as an indication of preliminary acceptance of the tasks assigned, the work breakdown, and the charter.

Step 2. Measuring: At this point, a cumulative spending curve can be drawn by the project manager. This will be the basis for determining the design review schedule and the schedule of dates (including general subjects) of future project review meetings. As uncertainty should be decreasing with elapsed time, but not on a linear basis, this is the opportunity to request a schedule of accounting reports, which will also decrease in frequency of publication as the project moves along. Weekly reports on project costs until the project is 25% finished are vital, but they are almost useless when the project is 75% finished. The dual-reporting mechanisms, the work breakdown release dates, and the estimate at completion reports should then be in place.

Step 3. Initiation: With all of this complete, a project organization chart is drawn up showing names and responsibilities of people, as team leaders on the project, who have some responsibility for the expenditure of project funds. Each participating person then initials the chart as an indication of agreement. After the project manager and senior management sign the chart, that chart becomes a notification to the company controller that project funds have been approved and can be spent by the project manager as parts of the project work breakdown are opened or closed. It is both a responsibility and a funding authorization. When impact statements are issued, this chart must be reapproved.

6.0 TOP MANAGEMENT: THEORETICAL PROBLEMS

There will be crises, there always are and there are tools that the project manager can use to handle them. One is to have a "contingency" fund that is not associated with any task in the work breakdown structure. These resources would be used by the project manager to finance unforecasted problems that do not justify the sending of an impact statement. Another is to design the size of the work packages to fit the stage of the project. Because uncertainty is very high at the beginning of a project and expenditures are fairly low, it would be wise to have smaller work packages (both in time and funds) at the beginning stages of the project in order to obtain rapid feedback during those crucial beginnings. However, occasionally major problems occur and this is the time when senior management might interfere thereby causing difficulties. As one researcher puts it,

> It seems during periods of crisis, top management should think that the organization needs a firmer hand and reinstitutes the authoritarian structure. "There is no room for organizational toys and tinkering, the matrix is done in." Thus the matrix is the readily available scapegoat for other organizational problems such as poor planning and inadequate control. (Stuckenbruck 1981, p. 81)

7.0 REVIEW AND PRACTICAL TIPS

This chapter covered prescriptions for organizing and managing short-term project structures. These are the organizations that deal with major problems or opportunities that have to be handled across normal functional boundaries. The deliberate decision making of the functions, with an insistence upon following predetermined organizational standards, cannot respond with the fast-changing, innovative processes that are required. But projects can. In many cases, this fact is lost upon upper management. Sometimes senior managers are promoted because, as functional managers, the problems they faced were within their past experience and therefore solvable. Over the years they had no "failures." A failure in those terms is defined as not meeting a predetermined goal. Conversely, not meeting goals is almost the norm in projects.

It sometimes seems as though senior managers expect that changes in external criteria such as customer changes and unforeseen technical complexities should not materially affect the project's progress. They may feel that these changes occurred either because of inadequately defined project budgets or poor management practices. They are less sensitive to sudden, unplanned shifts of personnel, arbitrary diminishing of project resources and minimal delegation of authority to the project manager. This places the project manager in a difficult position. It is the rare project manager who achieves promotion into senior management— since he has had "failures." This contradicts the common sense view that because project managers can solve new and unusual problems, they are the ones that are most promotable. His/her future in the organization can even be at risk when there are successes because of the endemic conflict between every project and functional organization. In many examples, senior management equates conflict with inadequacy instead of the healthy growing pains that it really represents. I've seen too many situations where senior management equates "conflict" with "bad" things, rather than a fast way to resolve legitimate differences. Senior management likes it when things are "quiet." Between projects and functions there will always be conflicting claims for people, finances, and facilities.

Finally, every project manager eventually realizes that the project cannot achieve the tight integration of the functional organization. The functional organization has the time to integrate and acclimate individuals into its social milieu; projects do not. Functions have the ability to review and formulate new budgets every year; projects do not, they are limited to a review of what's left to do. Functional managers have a great advantage when it comes to exerting authority. As the permanent head of their group, they can set relatively stable goals for people, supervise their training and reward them appropriately. They are the de jure (legal) and de facto (actual) managers. Project managers have less authority because even if authority is delegated properly by top management (and this is unusual), there is a relatively short time period to set goals. Project managers don't have the time the responsibility, or the project funds to properly train people and their ability to reward is limited.

Project managers are usually the de jure (formal, but socially powerless) managers; the team leaders are the de facto (informal but real) managers. Most project structures are very flat with few levels between the top and the bottom. The team leaders report to the project manager and, in fact, the project is similar to a relay team. In a very simplified example, the project manager may organize and start the project "race" but the baton is typically quickly handed to the engineering team leader. When that team leader is finished, the baton is probably handed to the manufacturing team leader. Then the baton is handed to the sales team leader, and so on. The project manager, of course, is vitally interested as the "race" progresses and checks the progress of each team leader as the work is done. There is particular interest shown at each handoff from team leader to team leader to ensure that the prior work was completed satisfactorily and that the receiving team understands exactly what it has to do.

This example is simplified, but it illustrates the great need for forward planning of both the project goals and the relationships among team leaders. Project work environments, therefore, are more socially driven than are most functions. During the creation of the project charter, the work breakdowns, the meetings, and the written communications, the project manager is always challenged to maintain an undefinable, but nevertheless required, team spirit. In functions, calm deliberate decisions that fit within predetermined company standards can be the norm. Although that may be desired in technical groups, projects are invariably more hectic, possibly more creative and inevitably more "failure" prone than functions.

In Chapter 10, we cover the third central component of our organizational model: technology. You will see that technology is the least independent of the three components, people, structure and technology. It is directly affected by whoever uses it (people) and how the organization determined how it shall be used (structure).

Technology will be defined in many ways. The technology of production, operations and management are all dealt with. Technology is an important variable, and one that cannot be overlooked if your technical organization is to adapt to its situation. In general, technical management is not a simple study and, because the effect of technology upon technical organizations grows more important every day, we should understand it thoroughly.

8.0 "ONE-SENTENCE" SUGGESTIONS

1. There is an answer to the classic question: Why don't we have time to do it right the first time but we always have time to do it over? The answer is:

    It takes a patient and intelligent manager to understand that excellent "up front" work prevents trouble later on. Unfortunately, because we never know what trouble we have avoided there are no disasters to talk about. (Except success, of course. But "they" expected that, didn't they?)

2. If you are a project manager, pay particular attention to defining the "details". For example, what is your authority? Do you control the expenditure of funds, etc.? (Check the sample charter in this chapter. If those questions are not answered satisfactorily, you will never be able to get control of your project.

You are then a coordinator, not a manager. Find out who the real manager is, then—your boss, his/her boss? Who?

3. Linear reporting of results against a forecast is useless to a project manager. You want a fast reporting schedule in the beginning when forecasts are being made. Later when manufacturing takes over (and the funds flow like water), you can settle for monthly reports. It's peculiar but because you can't do too much about it as the project ages, what good are reports at that time? There has been more money made or lost in the beginning on the drafting table than in any manufacturing group.

4. It is possible to estimate costs on a "new" project with no inputs form anyone else within 25–30% of the final actual costs. Just use any similar historical project as a basis. Most projects within a company are similar.

5. Design review meetings are a very powerful tool to find out what's happening. Be sure to schedule them when you begin to lay out the original project schedule.

and finally:

Success in business is 1% inspiration and 99% perspiration. (Thomas Edison) and if there is a choice between doing lots of planning and just "get started", do lots of planning. Remember that when it hits the fan, it will not be uniformly distributed.

TWENTY QUESTIONS THAT
COULD SAVE YOUR PROJECT

by Dr. Arnold M. Ruskin[1]

Abstract: Successful project managers must understand and meet performance objectives, plan, control schedules and costs, develop and guide personnel, and use resources productively. This article helps project managers assess how well they perform these vital functions.

Tight schedules and budgets are often used as reasons not to plan or control a project adequately. "How" you might ask, "can I take time or spend effort to

[1] Portion reprinted, with permission, from *IEEE Trans. Engrg. Mgmt.*, vol. EM-29 no. 3, August 1982, pp. 101–102.

plan and control my project when I am already strapped for schedule and budget to do the work itself?" This kind of thinking can lead to trouble.

To help you discover if your project is on a potential disaster course, consider the following questions. Check the one answer to each question that best describes your situation; then find your score at the end of this article.

1. What's the biggest challenge in your project?
   A. To satisfy your client (or boss).
   B. To be as good as our client's (or bosses') alternative choices.
   C. To keep our costs low.
2. I know what my client's (or bosses') alternatives are." Translated, this means:
   A. "I know *all* the choices my client (or boss) has for each part of my project."
   B. "I know all the groups that do exactly what my project is doing."
   C. "I know most of the groups who could do the main part of my project."
3. What is your best estimate of the daily cost to your client (or boss) of a delay in meeting project objectives?
   A. $_____
   B. Less than/more than the daily cost of the project.
   C. I have no idea.
4. What kind of a project plan do you have?
   A. A pretty formal one that my task leaders and I worked on together, covering the entire course of the project.
   B. An informal plan that I worked up, covering the next six weeks or so.
   C. Nothing written down. I know where I want the project to be a few months from now, and I talk about it to my task leaders.
5. If someone asks your task leaders to describe your overall project objectives and the strategy to attain them, what would they say?
   A. They'd say to look in the project plan. We worked on it together, it's all spelled out there.
   B. I'm not sure what they'd say. The objectives and strategy are in our project plan, but we haven't talked over the details.
   C. One answer, depending upon the case at hand: Study the _____, Design a _____, Build a _____. etc.
6. If intermediate performance, schedule, or budget objectives are not being met, what happens?
   A. I have contingency plans written out.
   B. I don't encourage that kind of negative thinking, but I have a pretty good idea how I would revise our approach to get back on track.
   C. I'd just have to sit down with my task leaders to figure out what our next move should be.

7. Are there any potentially big uncertainties whose effects you have yet to evaluate?
   A. No.
   B. Yes, but we know what they are.
   C. I'm not sure, but we already have more problems than we can handle.
8. How often do you slip milestones/overrun the budget?
   Answer: _____ percent of the time. How does this compare with industry averages?
   A. Favorably.
   B. A little more often.
   C. I can't really make meaningful comparisons. There's a lot of variation depending upon how busy we are overall.
9. How often do you compare actual accomplishments with the schedule and actual expenditures with the budget?
   A. At frequent, regular intervals.
   B. Whenever I can.
   C. Whenever I hear that there's a problem.
10. About that schedule and budget, do you have them all worked out for the entire project?
    A. Yes.
    B. We have started working on them, but we won't have them finished until the project is well underway.
    C. We won't get a project schedule or budget until after we have completed most of the work.
11. When did you last revise your project schedule and budget?
    A. Within the last couple of weeks. We update them every time we approach a middle sized or larger milestone.
    B. A couple of months ago. We are due for new updates any day now.
    C. I can't remember. We're moving so fast now that I have had to put my time on other things.
12. How useful are your schedule and budgetary progress reports?
    A. Very. We can compare them item by item with our plans to see where we stand.
    B. So-so. My administrative helper can figure them out if I ask; they're not so well organized or easily deciphered. They are subject to manipulation, too.
    C. Not very. We receive them as a formality, but we never look at them.
13. How often do you compare resource availability with the project's needs for resources (personnel, equipment, money)?
    A. Rather regularly, to make sure that we'll have what is needed to finish the project.
    B. As we enter each new phase of the project.

   C. Whenever the project's in trouble.
14. Check one:
   A. I know specifically which parts of the budget are behind schedule/under budget, which are on schedule and on budget, and which are over schedule/over budget (and why).
   B. I know generally how we're doing on the schedule and budget overall.
   C. I don't know where we stand on the schedule or budget.
15. When responding to a client's (or the boss's) request for a change in scope, I review the entire project plan, schedule, and budget and determine what impacts the change would have on all three components.
   A. Always.
   B. Sometimes.
   C. Never.
16. How would you describe your project team members?
   A. Most are bright, ambitious self starters. I have to work pretty hard to stay ahead of them.
   B. A few of my people are full of ideas. The rest plug along and do their jobs.
   C. My people are cooperative and hard working.
17. How often do you evaluate your task leaders?
   A. I try to sit down and talk about performance and problems at the beginning and midway through every project phase.
   B. At the end of the project, when I turn in appraisals to their permanent supervisors.
   C. Whenever it seems that someone isn't performing well.
18. How do you deal with a project team member who is not doing his job?
   A. First I talk to him, either formally or informally, depending upon the situation. We usually agree on same measure of performance to achieve in a specified period of time. If he continues to fall short of the standards that we've agreed upon, then I negotiate with his permanent supervisor for a replacement.
   B. I usually let things go for a while, hoping that the team member will do better. If he doesn't, I have a talk with him to point out where he should improve. As long as he seems willing to try to improve, I don't try to replace him.
   C. Usually the situation deteriorates to the point that the team member either seeks reassignment or resigns.
19. Imagine that you were offered the management of a project twice as large as your present project. Now imagine who your task leaders would be.
   A. They would include all my present leaders. I would multiply their effectiveness by obtaining other personnel to work under them.
   B. They would include some of my present task leaders and some new talent that I would hire to replace others.

    C. They would be practically a whole new team of people who are more capable than my present task leaders.

20. Write down the average percentage profit on your projects done for clients or the average rate of return on your projects done internally.
Answer: % _____. How does this figure compare with the same type of figure for your industry?
    A. Favorably.
    B. It's quite a bit lower.
    C. I don't know the figures.

Give yourself 5 points for each A answer, 3 points for each B, and 1 point for each C and tally your score. If it's 90 or higher, other project managers can learn from your example. If it's between 75 and 90, you have things under control and can probably handle a more demanding project when this one is concluded. If your score is between 50 and 75, watch out for potential trouble. You're already in trouble (and you probably know it) if your score is less than 50. Less than 30? Then get help in a hurry.

The questions cover five areas that every successful project manager must pay attention to: understanding and meeting performance objectives; planning; schedule and cost control; personnel; and productivity. Once you've got your overall score, read the brief analyses below to pin-point areas that need more of your attention.

All in all, the best tip for managing a project is to manage as if you were demonstrating that you are competent to be assigned project management responsibility. This approach will prod you into paying attention to the essentials of good project management. Then you will both do a good job on your present project and earn the right to manage more demanding efforts.

PERFORMANCE OBJECTIVES
QUESTIONS 1–3

A low score in this section means that you don't really know if you are meeting the client's (or the boss's) project objectives. This blindness can not only threaten your present project but almost make it nigh impossible to be assigned another project responsibility by the same client (or boss).

One trap that you can fall into is to believe that your client (or boss) has only one way to solve his problem—the way represented by your project—and that your competitors are only those groups who do *exactly* what you do. In doing so, you may overlook another approach which is more effective, more timely, or less expensive. This oversight could cause you to fail to meet the client's (or boss's) real needs.

PLANNING
QUESTIONS 4–7

If you scored less than 13 on these four questions, you probably consider a project plan a needless exercise. You prepare one to please the client (or the boss), then put it aside. You shouldn't. It is a road map meant to guide your project into the unknown future. You can travel without a map, but you won't get to your destination as fast or as efficiently as those that have one. And it's equally important for your task leaders to be part of the planning effort so that you have the benefit of their insights, obtain their commitments, and are sure that everyone is working toward the same goal.

SCHEDULE AND COST CONTROL
QUESTIONS 8–15

A surprising number of project managers who seem to be surviving could do better if they had more control over day-to-day schedule and budget matters. You may one of them if you scored low here.

Schedules, budgets, and timely reports of work accomplishment and resource expenditures and commitments are all controls at your disposal. But even the most detailed won't help unless you understand them and use them regularly to check on your project's progress.

Whether you have a project plan or not, you expect certain results. If you have to wait until a major project segment is concluded to know how you are doing, it may be too late to do anything. For most projects, schedule and expenditure information is stale and nearly useless a week or so after the reporting period.

Project status reports (you get them weekly, don't you?) should be your tools, not a pro forma exercise to meet someone else's requirement. If they are not, sit down with your project staff and accounting staff and tell them what information you need to make them more useful to you.

Time represents a precious asset that must be used efficiently or lost. Very few project managers are careless about accounting for money per se. Yet unproductive uses of time are as wasteful as squandered expenditures. Moreover, it almost always costs more to play "catch up" than it costs to work steadily. Steward both project time and money carefully in order to complete your project within the planned budget and on schedule.

PERSONNEL
QUESTIONS 16–19

Every project and project manager ultimately depends on people for success. You do not depend on yours if you scored below 15 on these questions. Few people can do well on their project assignments unless they know their own strengths and deficiencies and are *consistently* encouraged to do better. There are many techniques for evaluating people. The best require that you do so routinely and that you put the evaluation down on paper for later comparison.

Removing staff members who aren't contributing to your project is just as important. If you don't discipline an erring staff member, others will become careless about their own performance. Appropriate discipline, recognized by the entire project team, will do wonders for everyone's morale.

Question 19 is the acid test of your ability to develop task leaders and their ability to handle people and their other responsibilities. If you do not rate high on this question, the odds are that you will not be able to accept a larger responsibility, particularly if it is significantly larger. In your own interest as well as theirs, make sure that your task leaders get the necessary training.

PRODUCTIVITY
QUESTION 20

Profits and internal rate of return are measures of successful projects. Productivity is the key to achieving such success. An increase in project productivity of, say 5 percent, may increase the profit by as much as 50 percent. That should be a big incentive for you to encourage greater productivity.

SUGGESTED ANSWERS TO CASE QUESTIONS

1. The underlying problem here is that Eric has reacted to an informal request for information as if it were a management-directed change in scope.

   This *could* be a change in scope of the program in the future; therefore, Eric should lay out the alternative recommendations that he could make in an engineering report format. That format is quite straightforward:
   - Description (to and from, date, identification)
   - Recommendations
   - Any other information.

   We can assume that Eric (as is typical of most project managers) is generally competent when it comes to his project. There is no reason he can't come up with several suggestions including "ballpark" costs and benefits. For example, he could probably tell Milt in general terms what it would cost, how long would it take, and what the probable (risk?) effect on the existing project would be if:

(a) The casing were redesigned to be corrosion proof.
(b) A fail-safe mechanism were built into the turbine to preclude its operation unless it were covered by an appropriate shelter. External sensors could be designed in.
(c) The existing design were sold by the company for a limited time, and then only to highly industrialized customers who would permit the company's service personnel to inspect periodically. At a reasonable later time, a new turbine for more severe external conditions could be redesigned.

There are other alternatives that I'm sure you have considered at this point. The idea is that there are probably many alternatives, and before valuable time and money is spent, these should be documented along with recommendations for action. However, Eric should not stop work on his turbine without getting directions (and additional resources) from Milt.

2. No, except for written suggestions as to how the problem could be solved, if an appropriate addition were made by Milt to the project, funding, schedules, and specifications. All alternatives provided by Eric should have subjective "ordinal" values in order to give Milt some ideas as to how Eric would rate them.
3. Management always requires minimization of cost and delivery delays. It's nothing to be concerned about. A request for minimizing cost and delays is, if not explicitly stated, an implicit part of every impact statement. In this case, Eric should provide a preliminary cost-benefit analysis that forecasts the potential cost to implement any of his recommendations vs. the time that he is allowed to start the implementation. Milt may delay a decision. If so, he should be told by Eric what that delay could mean in terms of additional rework, design, etc. To reiterate, those alternatives should be spelled out for Milt with an "ordinal" number assigned to each alternative. The higher the number (from 1 to 10), the more acceptable the answer that Eric would recommend.
4. There has been no decision and no resources provided by Milt to commence the corrosion design work at this point; thus, nothing should be done about Alan. If he is unavailable when the work is to be done, either alternative sources must be found in the company (or in outside design and test labs) or the work will be delayed until he returns. If Milt does decide that the corrosion work is to go forward immediately, I suggest that Eric deal with that problem by possibly including a trip to San Diego for Alan as part of his impact statement, but this should be pointed out to Milt. In this example, the person responsible for the decision to spend additional funds is Milt. Eric is responsible for seeing that Alan or an outside source completes the job within budget and to schedule,

with Clyde supervising the actual work done. The trip to San Diego might even be contingent upon the work outcome. In project operations, responsibility is rarely a single, clearly defined sequence. As noted, other alternatives would be to wait until Alan returns from his vacation or to subcontract the corrosion testing to an outside firm.

5. Because Mary Ann now has information that could be important to the success of the project, if I were Eric, I would request that she be added to the team, lay out any additional tasks for her to preclude this problem and other field service problems from occurring and have her inputs disseminated throughout the turbine project.

6. If I were Milt, I would not have gotten Eric involved until I had satisfied my own level of uncertainty. For example, how bad is this problem in the field? What can our potential losses be? What alternatives do I have for a change in the existing "blue" turbine project? What would corporate headquarters want? Eric has not been given enough information to come back with a specific recommendation. As noted above, as he has no idea which alternative was suggested, it is advisable for him to come back with several from which Milt can choose. If he thinks of the one that will satisfy the upper levels of management, that is fine, but he might not, and then his work will be wasted. If Milt had done his "homework," it would have saved everyone a lot of time. This is a case of poor absorption of uncertainty.

Do you agree? What other answers do you think would apply?

REFERENCES

Argyris, C. "Today's Problems with Tomorrow's Organizations." **Matrix Organization & Project Management,** Hill, R. E., and White, B. J. (eds.). Div. of Research, Graduate School of Business Administration, University of Michigan Business Papers, #64, Ann Arbor, MI. 1979. pp. 5–31.

Clark, K. B., and Wheelwright, S. C. "Organizing and Leading Heavyweight Development Teams." **Revolutionizing; Product Development; Quantum Leaps in Speed, Efficiency and Quality.** New York Free Press, New York. 1992. Chapter 8.

Cleland, D. I., and Kocaoglu, Dundar, F. **Engineering Management.** McGraw-Hill, New York. 1981.

Gilbreath, R. D. "Project Management." **Project Management Handbook,** Cleland, D. I., and King, W. R. (eds.). 2d ed. Van Nostrand Reinhold, New York. 1988. pp. 3–15.

Jacobs, R. M. "The Technique of Design Review." **Proceedings of the Product Liability Prevention Conference.** American Society for Quality Control. 1979. PLP79 Proceedings.

Lawrence, P. R., and Davis, S. M. "Problems of Matrix Organizations." **Harvard Business Review** (May–June 1978). Reprinted by permission of the *Harvard Business Review*, Copyright © 1978 by the President and Fellows of Harvard College. All rights reserved.

Martell, K.; Carroll, S.; and Gupta, A. K. "What Executive Human Resource Management Practices Are Most Effective When Innovation Requirements Are High?" **Advances In Global High-Technology Management,** Gomez- Meija and Lawless, Michael (eds.). JAI Press, Greenwich, CT. 1992. pp. 3–30.

Pinto, J. K. "Successful Project Management: Do You Know Your Politics?" **PMNETwork** (July 1994):33–35.

Ruskin, A. M. "Twenty Questions That Could Save Your Project." **IEEE Transactions on Engineering Management.** August 1982. EM-29(3):101–102.

Steele, L. W. **Managing Technology: The Strategic View.** McGraw-Hill, New York. 1989.

Stuckenbruck, L. C. **The Implementation of Project Management: The Professional Handbook.** Addison-Wesley, Reading, MA. 1981.

Thamhain, H. J. **Engineering Management.** John Wiley & Sons, New York. 1992.

Tosi, Henry I. **Organizational Behavior and Management: A Contingency Approach.** PWS Kent, Boston, MA. 1990.

Webb, A. **Managing Innovative Projects.** Chapman & Hall, London. 1994.

Weisbord, Marvin R. "Organizational Diagnosis." **A Workbook of Theory and Practice.** Addison-Wesley, Reading, MA. 1978.

Woodward, J. **Industrial Organization: Theory and Practice.** Oxford Univ. Press, London. 1965.

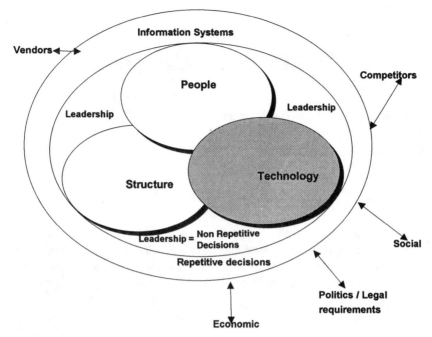

External
Environment

CHAPTER 10

Technology: How We Do Things, Techniques and Systems

Case Study:
The Pregnant Computer

Cast
George Jackson: Design team leader
Michael Jacoby: Chief engineer
Mary Hughes: Fabric designer
Sam Disko: Data processing manager

Scene: The John Textile Works
 It was late on a beautiful spring afternoon in South Carolina and Mike Jacoby decided to take a five-minute break to smell the roses. Just as he stood up to lean close to the open window in his office, Sam Disko walked in unannounced, as he usually did.

Sam: Hi, Mike. Got a minute? (He proceeded to go ahead without waiting for Mike's answer.) I'd like to talk to you about some things I've noticed about computer usage in the company in general, and specifically right here in the engineering department.
 Mike: OK, Sam. I've got a meeting in half an hour, but pull up a chair and let's talk. What's the problem?
 Sam: Well, you know that my department is responsible for procurement and operation of all computers in the company. We make sure that when they are "born" in this company they will work the right way. We service all the departments, but my fellows tell me that they are having a lot of trouble with your engineers because your people don't want to follow the right processes.
 Mike: Tell me about it.
 Sam: Well, you know how we design a software system. After a request for our work is approved by you, I send in a system analyst to work with your people to do an information flow diagram and a rough layout of what the programs are supposed to do. When your people sign off on the layout, I give it to our programmers, and when they get finished, they run trial data and present the results to your people.
 Mike: OK, what's the problem?
 Sam: I've never been able to get a system signed off right away by your people. They always say that we misunderstood, or that's not what they really wanted, or they wanted something a little different. They never really tell me what

they expected to get. I've been here over five years and I've never gotten a straight answer yet and I'm getting tired of it. Oh, another thing, after a lot of work we finally get the computer systems to run, but only after a lot of downtime and meetings. It's wasteful and inefficient.

Mike: Well, let me look into it and I'll get back to you tomorrow.

After Sam left, Mike phoned George Jackson, then Mary Hughes. He briefly told each of them what Sam had said and asked them to discuss it with him the next day at lunch. The scenario at lunch the next day was:

Mike: Well, that's the story. What's to be done about it?

George: Mike, that guy Disko and all of his people are just a big pain in the neck to me. For example, all I wanted four months ago was a simple modification to our design program so that the graphics display would be able to check our piping diagrams for the process plant. I wanted to be sure that we didn't have any more fiascoes like we did last year when a designer put two pipes in the same place because she didn't have a complete assembly drawing to look at first. When Disko's people finally came over, the last piping design was almost finished, and by the time they seemed to understand what we wanted, they had wasted four days of my best designer's time in these endless meetings that they seem to like to have. We finally got the job done without them but it meant redrawing and that's a lot of work.

Mary: Mike, my problems with them are a little different. I am never sure how to design the fabrics to meet the special requirements from marketing, as fabric design is still pretty much of an art, not a science. So because I wasn't sure what I was looking for, I couldn't exactly tell Sam's systems analyst what I wanted. Finally, I signed the approval form just to get rid of him. I then took a course in computer programming for several weeks and started to program my own terminal. When Sam found out about it, he complained because he said that my attempts at programming weren't up to his standards. I think he called it a simulation program or something like that and said we were interfering with his mainframe.

I then asked him (you must remember, because you approved the requisition) for a standalone personal computer and when I finally got it, I found that any programs I wrote for it had to be approved by Sam's group. They had programmed my new personal computer before I got it to accept only those programs that they unlocked from the mainframe. They said that this was necessary to prevent unauthorized use of the computers, and they wanted an approval form signed every time I tried to use that personal computer. They called it the computer delivery process, just like it was having a child in the hospital or something. They said that they had to be sure that my programs wouldn't interfere with theirs. I finally just stopped using the darn thing and went back to designing by using the old system: my eye for color matching and a desk calculator to compute the fabric variables. It's too much bother to get those guys to do anything. They use more paper forms than anybody else I know and I just don't have the time to fill them out if we want these new fabrics out in a fast delivery cycle. So I just do my designs the old way, by trial and error and then down into the experimental weaving room to try them out.

Mike: OK. Everybody has problems and we'll have to deal with them in a while, but right now Sam is complaining about your groups. What can be done about it?

QUESTIONS

1. Does the company policy for centralizing computer procurement and operations accomplish its intent? What is that intent? Is this the major problem or is it something else?
2. What would you suggest if you were Mike?
3. How would you optimize the use of computers in the firm and still prevent their unauthorized use? What are unauthorized uses and why should they (or should they not) be prevented?
4. How should George's group be structured, as a function or a project? If Mary were to head new fabric design, how should that group be structured? Why?

1.0 OVERVIEW AND INTRODUCTION

Earlier in this book, we gave the general definition of technology as the processes by which organizations change inputs into outputs. It is a major component of our general management model. There are other more detail oriented definitions,

We need to define technology as encompassing three fields of activity: product, process, and information technologies. These three different substantive fields of work have different traditions and different educational foundations. They are each recognized as fields of specialization, but are rarely thought of as an integrated

entity that constitutes technology. (p. xx) . . . technology is "the system by which
a society satisfies its needs and desires." (p. 8) (Steele 1989)

The general management model doesn't differentiate into three fields as suggested by the above author because product, process and information are intri-
cately interwoven in an organization, and attempting to describe and then prescribe separate parts would not be useful. Every product is the result of some
process that in turn is measured by the flow of information about that process.
Therefore, the component *technology,* as defined in the model's terms, is the
many ways that individuals (i.e., people) interact predictably (i.e., structure)
with one another to achieve organizational goals. We deal here with a more
limited field of activity: that of the technical organization.

Technology has achieved an importance equal to people and structure only
within relatively recent times. Its interactions with the other two components
used to be as a junior rather than an equal partner. Changing inputs into outputs
(i.e., the most general definition of technology) was almost a secondary consideration, as management was primarily supposed to spend its time controlling the
machines and the structure. Technology was expected to result from these other
two components. Until the beginning of this century, it had a very limited role.
It consisted only of well defined production methods, and except for minor
changes in products that required retooling, the technology remained fairly constant. Then, as well as now, technology was useful for making the product more
quickly, more cheaply, and with fewer inputs of human effort and raw materials,
but it was not a central issue because markets were expanding and customers
were satisfied. It was not a major component until the advent of the assembly
line of Henry Ford in the early part of this century. Before that, products were
made in small groups and the technology controls did not exist except in the
minds of expert plant mechanics. The more modern definition, that of all the
processes of changing inputs into outputs, is relatively new and includes many
more complex aspects of management decision making.

In the past, this practice of limiting technology improvements to minor changes
in production or products did not affect the people component greatly. The job
skills were basically the same for years. However, with the advent of mass
production, the situation was radically changed. Technology began to be change
workers from craftsmen into machine feeders and maintainers: in effect, adjuncts
to those machines. A familiar but overemphasized example of the use of people
in this type of technology was on the automobile assembly line, and at the time
this change was not necessarily negative for that people component. Henry Ford
raised the daily wage of assembly line workers to the then fantastic wage of
$5.00 per day. This wage was, of course, based on the worker achieving the
amount of production that was established by time and study standards. If the
worker did not maintain that pace, the result was not $5.00 a day, it was
unemployment. Much of our industrial strength in the past was built on this

novel technology. During that period, the people did not affect technology as much as technology affected the people. People served the machines.

The other major component, structure, also seemed to be relatively static. The "one best way" of the classical management theorists early in this century, Taylor, Fayol, Gulick and Mooney, defined a rigid hierarchical structure. Their contemporary, Mary Follett, suggested that the structure should be modified according to the needs of the particular situation but she was unusual and her ideas did not become widely accepted. Then, if a particular organization did not succeed well, it was thought that the company had not met the standard of the "one best way." If the structure was well designed in accordance with the "one best way," the technology could be assigned to the production or manufacturing engineers, who would do the job of eliminating unnecessary motions and simplifying the work. That would reduce costs, resulting in increased profits as sales went up. Unfortunately, that never completely applied. If sales revenue did not go up because of changes in customer demand, they could also decrease with costs, then profits would remain the same or even decrease. Profits may grow as costs are decreased only when the difference between prices and revenues increases. (I recall being exposed to this "unlimited market" idea in an introductory course in time and motion systems. We were supposed to find the "best way" to organize the work and were told that the structure and the technology would then be optimum. It didn't make any sense to me then but I didn't say anything because I had to pass the course. That professor was sure that this process was right.) Now, we know that if prices are forced down by customer resistance as costs decrease, profits may stay the same or even decrease.

There have been changes since those times. Decreases in product life cycles, with quickly changing market needs, now require positive company responses involving more than the production processes through which the organization transforms inputs into outputs. Production methods and new product development have been joined by management decision-making processes affecting people and structure.

In addition to changes in manufacturing technology, there have been other changes such as those in techniques to measure (and control) the organization's internal processes. Typically, these would include the computer-based, information processing methods and modern control tools, such as quality assurance techniques.

Therefore, the transformation of inputs into outputs involves more than tangible processes. Accordingly, in recent times technology has become interactive with and as important as people and structure. In prior chapters, we covered some of the thinking and decision-making processes available to the manager in a very preliminary discussion of technology. In this chapter, we will continue by following a by-now familiar path of description of applicable research findings about technology, followed by prescriptions and specific recommendations. Contingency suggestions (as a part of situational theory) will be important in that path.

As noted before, in the early part of this century when technology was defined primarily as the unchanging methods of production, it was supposed to have a lesser effect on the company's economic success. The search was for the "one best way" to manage people and the structure. Research has changed all of that and expanded the technology variable from the narrow definition of production processes to include all the information handling methods and the managerial controls such as those typified by quality assurance.

Therefore, technology, as redefined, is no longer contained entirely within a relatively unified section of manufacturing departments. Even though manufacturing systems are still central issues in technology, there now is an expanded cluster of working and management operations, tools, and materials. It now includes both the measurable organizational and manufacturing processes and nonmeasurable managerial actions. These are quite complex, because they interlock with and affect each other.

In order to understand this, we will use a familiar but artificial analysis method. We will untangle the human and production interaction by assuming all other things to be equal and investigate only one variable at a time. In the following, the important variable is technology.

2.0 TECHNOLOGY: INPUTS INTO OUTPUTS

In previous chapters we said that classical management theory considered people to be relatively interchangeable. The pyramidal, hierarchical organizational structure that was developed as the best (and only) way to manage this commodity was intended to clarify and help simplify the work that people did. Therefore, although the design of the way people interacted in the structure was a company problem, the solution to that problem was well marked and clear, as the resulting structure was expected to generally reflect the society of that time. Because there are organizations that still use the technology (and structure) of the early classical theorists, a brief review of the historical bases might be useful to you. Knowing where something came from helps in understanding it, as it may exist today.

2.1 Some Historical Roots

In the late eighteenth century, the industrial revolution that took place in a social and economic environment rewarded technological applications in manufacturing very handsomely. Owner-managers who could gather labor and materials into one location (using a new invention called the factory) and transform those resources in a *predetermined, repetitive* fashion were considered to be technologists of the highest order. They produced valued products in a continuous stream. These products were then sold to an accepting and hungry market, resulting in major profits for the entrepreneur. Cottage industries, in which small producers used the labor of their families and close friends and manufactured products in their homes, could not compete with the efficiency of these centrally organized factories.

Moreover, the political scene completely supported free or laissez-faire trade of manufactured goods and developed a fascinated interest in the financial benefits that accrued to these entrepreneurs (Bronowski 1965). The development of accurate machine tools and cheap energy sources to drive them defined the primary mass production technology, which suited the resulting patriarchal organizational structure very well. Productivity increased when work was simplified by being divided up and machines were used to speed up that simplified, divided work.

A classical example of those times was the productivity of pin makers, which could be increased more than 200 times through task specialization and effective tooling. The master pin maker who straightened the wire, cut it off, headed it, sharpened the point, and put it in a paper was outclassed by technological change. Dividing the job into approximately twenty operations performed by several workers increased the worker's relative skill through job simplification and minimized the nonproductive planning tasks of preparing or setting up. It also allowed skills to be learned quickly and to be repeated. The concept of specialization, implemented through a change in technology, undoubtedly contributed to the success of those technologically adept pin-making companies and the demise of other companies that did not change their technology (Smith 1977).

Then, the improvement of technology was a cost-driven activity. The idea was first to develop the production tasks and methodology and then to devise a disciplined manufacturing structure to manage it. The goals were clearly order, logic, and acceptance of authority. The primary objective was optimal performance within rigid constraints of a pyramidal type of hierarchy, with communications only in vertical directions. Orders were sent down and information was sent upward through the structure. People's jobs were clarified, spans of control were limited, and responsibility and authority were clearly placed together. Resolving of any uncertainty was the task only of the owner-manager, and the structure was intended to minimize costs in using available technology to carry out that owner-manager's wishes completely. This system was believed to be ideally applicable to all factories and organizations.

However, several years into this century, there were occasional protests and suggestions from some researchers that were intended to modify this approach. But these researchers were not considered to understand the "correct approach" for designing company structures, considering the technology. Because the technology of mass production was relatively new and hierarchical-designed structures worked well with it, one wasn't supposed to argue with success. The design worked then and sometimes still does, but recent research indicates that it has a limited application. It works best with mass production technologies such as those with relatively fixed, well-understood manufacturing processes.

The prime example of the technological (or classical) approach to the design of organizational structures was that of Taylor (1911). That design exemplified the efficiency, control, and pragmatism of the production machine applied to the design of the organization. Science was supposedly applied rigidly in the factory. Authority was centralized and vested in planning specialists. The new technology of mass production was optimized, using task specialization and relatively complex production methods. The individual worker's tasks were made simpler, but the planning and coordination (or the technology) that the manager had to do became more complex. The worker no longer had any scheduling or methods flexibility; that was all placed in the manager's hands. These new management responsibilities were difficult, but positive results were achievable with a machine-like, disciplined approach. The planning tools used to direct this technology were limited, but they were all that was required.

(I can still recall seeing many manufacturing operations for which planning the best machine feeds, speeds, and loading and unloading operations for a machine operator was supposed to be done with a stopwatch and a small table of standards. It was the day of the time study man, the efficiency expert, methods engineers, piece rate setters, and job classification specialists. Taylorism flourished because the technology of the time suited it well.)

The typical organizational structure supposedly included the best technology of production, but there had never been any testing or research on this design, because there had never been any large-scale evaluation of organizational effectiveness. The obvious answer when an organization succeeded was that it was well organized. Failure had an equally obvious answer—the company was not well organized. It almost seemed to be a matter of faith or of ideals, with little empirical backup data. These ideas were the mainstays of management texts for many years.

Taylor's work meshed nicely with the description of bureaucracy proposed by Weber (1947). That description covered the specialization of skills and tasks, specific limits on discretion embodied in a known set of rules, professional impersonal behavior, and a complex hierarchical structure concerned with administrative control. It is the description of the "professional." The Taylor-Weber model of the organization in which the "professional" (i.e. patriarchal) manager operated was based in the intuitive logic and personal experience of managers as they attempted to develop order in managing technical organizations. They believed that the machine model of the organization was surely the best, because the machine-driven technology had obviously produced the best social and economic arrangements, for them as well as for their customers.

However, in more modern times, when this machine model was tested by researchers who observed the success patterns of companies that used it and compared them with the pattern of nonusers, the researchers found that the nonusers sometimes also succeeded very well. Moreover, some companies with well-documented internal procedures/systems and clear, unambiguous organization charts were even declining in the face of successful competition from companies that were very loosely organized. Both kinds of companies, of course, had the same technology. Is it possible that something was missing?

2.2 Technology and Structure: Improving Modern Organizations

One important study of recent times that we covered very briefly before (Woodward 1965) attempted to determine what the real relationship was (if any) between "structure" and "technology." That study used a simplified definition of technology: the methods of production—including not only manufacturing but also management techniques. Those techniques had changed radically from the ideal of mass production that came out of the industrial revolution.

The research findings relied on data obtained from approximately 100 manufacturing firms in southern England. The original purpose of the research was to determine if there was any relationship between organizational structural design as the independent variable and business success as the dependent variable. Measurements were made of the organization's structure, such as the chief executive's and the first level supervisor's span of control, levels of management, the ratio of indirect to direct workers, and the ratio of staff to line personnel and these measurements were compared against business success.

Initially, the researchers found nothing. There was little or no correlation even when comparisons were made on size of the firm, production control techniques, and organizational change. No empirical structural relationship emerged to predict business success. Firms classified as being less successful also had all kinds of organizational structures. Empirical research had not supported the logical machine model or any other model at this point.

The research team decided to re-evaluate the data it had gathered after thinking through some of the inherent design concepts that classical theory possessed. Taylor and his contemporaries were engineers who practiced successfully in heavy manufacturing industries. They had generalized on that experience as if it were applicable to all administrative organizations. The similarities between the background of the people who had developed classical theory and the technology that they had used had been overlooked. The research team felt that this was a major point to be considered.

The members of the team then redefined their independent variable of organizational design in three separate technological methods of *production*. When they reclassified their data into these three categories, they found that economic success occurred when there was a closer match between the different kinds of organizational structures and particular production technologies. The companies that were economically successful had structures that were very similar to each other within specific categories of technology; those that were less successful did not. The three categories were:

1. Unit or small batch production: The manufacture of articles to customers' individual requirements. This is the oldest and simplest form of production. Because each article is produced to order and is modified to suit some customer's needs, the processes used are sequential to each other. The unique requirements of the customer's specifications define the machinery used and the nature of the product. In a factory environment, the manufacture of tools and dies, custom machinery, specialized test equipment, and very large industrial turbines are typical of this unit or small batch technology. The worker is not a machine adjunct. His or her skills are central to the manufacturing process. She plans and controls her own work.

2. Large batch and mass production: The technology with which Taylor and his associates were familiar. It produces standardized products in sequential manufacturing with little deviation, except with extensive setups and reworking of production equipment. The customer gets a product that is essentially the same as all those that are produced. Some examples are bathroom fixtures, moderately priced clothing, and most types of automotive vehicles. The worker is both a machine adjunct (i.e., assembly line worker) and a worker controller (i.e., maintainer).

3. Process production: The flow manufacturing process of the refinery, the pharmaceutical plant, and chemicals. The factory can be considered to be one large manufacturing tool and the product can be varied, but only within very narrow limits. A refinery, for example, can turn out various kinds of gasoline but it can't be converted into producing something other than petroleum-based products, such as maple syrup. The worker is highly skilled, but those skills are primarily used to control the process. The technology of the process has been defined by the plant designers, not the plant operators.

The three-way classification included twenty-five firms in process technology, thirty-one firms in mass production, and twenty-four in unit production (Woodward 1965, pp. 52–65). Woodward found the results shown in Table 10-1. Organizations with structural designs that varied greatly from the above median figures for their category of technology were less successful than organizations whose structures were close to the medians.

Ratio of direct to indirect workers decreases as more technologically complex production processes are used. (A refinery uses more complex technologies of production than a plant in large batch and mass manufacturing, and that plant in turn is more complex than a tool and die shop.) Process industries employ

Table 10-1. Technology Variables for Unit, Mass, and Process Production

	Unit	Mass	Process
Ratio of direct to indirect workers	9:1	4:1	1:1
Chief executive: levels of management	3	4	6
Costs in wages: %	35	30	12.5
Ratio of manager to total personnel	1:28	1:16	1:7
Ratio of staff to industrial workers	1:8	1:6	1:2
Span of control: first-level supervisors	1:24	1:50	1:15
Span of control: chief executive	4	7	10

many maintenance people and relatively few production people. A trip through an operating refinery can be an unusual experience for the uninitiated. There seems to be a great jungle of hissing, stationary machinery and insulated pipes with no people visible, except for the occasional worker adjusting a valve or tightening a pipe connection. It is only when you open the door of the control room that you find the workers (and there are still very few) watching dials and adjusting controls.

Levels of management and span of control of the chief executive increase with technological complexity. This indicates that the unit technology company has an organizational structural pyramid that is flat, whereas the process industry has a narrow, high structure. Unit production executives are more closely involved with the day-to-day operations typical of a customer-oriented shop environment. Process production executives manage more by committee and are more concerned with working group activities that are intended to minimize the production downtime of the particular process.

Ratio of managers to total personnel and percentage of costs in wages show the effects of increasing specialization as technological complexity increases. The generalist worker who sets up and controls his own operations becomes less prevalent and the number of managers coordinating the lower skills of workers increases.

Span of control of first-level supervisors is the only internal variable that appears not to follow the rules. Small spans are an indication of the breakdown of the working force into small groups in both unit and process operations. According to Woodward, the relationships between the first line supervisor and the highly skilled generalists of unit production or the highly specialized workers of process production are more intimate and informal than in the mass production or large batch technology firms studied.

The implications for successful design seem to be:

Unit production: If your company produces special or unique products to meet the specific requirements of a customer, it should have a flat organizational structure (i.e., few levels between workers and top management and a wide span of control), few line managers, and a small staff. A high proportion of costs will be in wages for very skilled generalists who independently plan and execute their own work (e.g., the systems development and programming section of your computer operations).

Mass and large batch production: This is the organizational model familiar to Taylor and is very close to the bureaucratic model of Weber. Production processes should be routine, efficient, and mechanical, with workers tending machines that are paced by engineering standards (Gibson et al. 1976). Work control and supervision are separated from the worker, and any deviation in standard working methods is not acceptable. The first line supervisor is primarily a disciplinarian in this model and a problem solver in the other two models. The organizational structure is the familiar equilateral pyramidal shape of the classical theory, with spans of control and intermediate levels of management somewhere between those in unit and in process production.

Process operations: If your company produces continuous printing of any kind, chemicals, pharmaceuticals, or fabrics in facilities that are operated on a multishift basis, it should have a tall organizational structure (i.e., many levels between the workers and top management, with a narrow span of control), a high proportion of line managers to workers (this may mean fewer workers but the same number of managers as in a comparable size unit production company), and a fairly large number of indirect maintenance and staff personnel.

This and other replicating research (Zwerman 1970) provided an important link between the technology of production and the structure of the organization. Taylor and the other pioneers of organization design were right when they suggested that their personal successes could be extrapolated to other companies. But they were not aware that the technical and management organization structures applicable to steel mills and other large batch or mass production manufacturing companies might not fit all production technologies, especially those in unit and process industries.

Woodward's research showed that there is no direct connection between operating success and organization design unless the *specific manufacturing technology* (as an intervening variable) matches the structure correctly. When the structure matches the technology of production, the correlation between structure and economic success becomes clear. The technology component is no longer a built-in, static kind of process; now it is an active component or variable that must be considered in the organization's design. Management discipline, logic, or intuition probably is necessary but it is not enough.

However, the definition of technology that we use includes more than the methods of production. It includes *all* the processes of changing inputs to outputs, and that involves innovation and product improvements/changes. Do these parts of technology also affect the way the organizational structure is designed? There is research that provides some potential answers.

2.3 Technology Defined as Innovation

Burns and Stalker (1961) defined technology slightly differently and produced results that explained some other structural designs of the firm. These researchers were concerned with companies in which the products had considerable variety and could be changed during development processes. Product innovation was the independent variable and organization structure the dependent variable. Their data were based on the reported perceptions of management personnel and on the communications patterns within the total organization.

Burns and Stalker found two general types of structures. One kind was *mechanistic* and applied during stable conditions. The other structure was *organic* and applied during conditions of change and innovation. The typical mechanistic structure included rigid descriptions of functional specializations, precise definitions of duties, and a well-defined system of command through which orders flowed downward and information flowed upward. This is the epitome of the middle (mass production) model in Woodward's categories and the classical theory of Taylor.

Organic structures had less formal job definitions with internal communications that flowed up, down, and horizontally to wherever they seemed to be needed by the participants. These structures were more collegial or consultative, less directive, and very adaptable to changing environmental conditions. They were, however, more expensive to operate because of the additional coordination costs that were required. The two technological extremes of Woodward, unit and process productions, seem to fit this organic mold.

These alternatives of organic and mechanistic models help to explain the metamorphosis of the loosely organized, smaller technical firm using flexible unit production modes with a great number of product innovations into a tightly defined, more rigid larger company using mass production modes with fewer innovations and product changes. The high costs required with rapid change and innovation and unit production methods are reduced through operations being channeled into more fully defined organization charts, responsibility diagrams, and production standards. It's a different technological world in which the firm seemed to have changed its organizational structure as it continued its success.

2.4 Rate of Product Change

This research seems to have provided different kinds of answers to the questions in the technology and structure interrelationship, but are they really different or just part of a larger theory? If we think about the total organizational structure as being a monolith or a strictly uniform design for each company, we definitely have a conflict between theoretical results. But, as you know, technical organizations can have both flexible (i.e., advanced research and development) and rigid (i.e., mass production) departments within the same overall corporate structure. Technology has affected how the people will work as well as how the structure is designed.

We reviewed some of the ideas of Lawrence and Lorsch (1967) in previous chapters. Those findings can be applied again when considering the different component of technology. Lawrence and Lorsch were also concerned with how technology affected departments *within* the organization. To review briefly, their data involved observations and the results of questionnaires administered to the personnel of firms in three kinds of industries: plastics, food, and container manufacturing. These represented fast- (plastics), medium- (food), and slow- (container manufacturing) changing technological, scientific, and economic environmental conditions. The three industries did not exactly fit across Woodward's ideas as all three of them used a type of process technology (Lawrence and Lorsch 1967, p. 191).

However, the researchers discovered differences in responses for the same kinds of technical groups between companies when they asked "What kind of organization does it take to deal with various economic and market conditions?" and "What are the differences in cognitive and emotional orientation among managers in different functional departments . . . or groups?" (Lawrence and Lorsch 1967, p. 11). They selected three functional departments of the company: science, technical production, and marketing groups. Even within the single continual production-oriented process technology of Woodward's classification, they found that effective organizations that had lower differentiation and minimal integration needs between internal groups were typically concerned with high volume, prompt delivery, consistent product quality, and the production of a relatively unchanged product at a minimal cost.

The most successful company in this case was the one involved in high production of relatively fixed product designs of cardboard containers. It had very little product change and was almost *mechanistic* in its operations. Those with higher differentiation needed special integration functions, and although they also were concerned with the same criteria of volume, delivery and quality, they were more involved in rapid product change and innovation. Typically, it was the plastics manufacturer that was very *organic*. A company that was *mechanistic,* according to Burns and Stalker, was typically very structured in its use of technology. A company that was *organic* was not.

Differentiation was defined as the presence of behavioral subsystems that were consistent within their own groups but very different from the behaviors of other people in other groups. *Integration* was defined as the qualities of the state of collaboration that exists among departments required to achieve unity of effort by the demands of the environment. In these cases, the technology was the rate of change in products that each company manufactured to satisfy the requirements of the environment. See Table 10-2.

The container industry's customers wanted minimum innovations and proven containers that could be processed at high speeds on automatic packaging lines. They did not want technological or product innovations that could make their existing packaging processes obsolete. Improvements were secondary to continuing production, as is, and lowest prices were important. In the food industry, there was competitive pressure for some innovation and change in the foods produced. Selling prices didn't make a great deal of difference to the market and those prices had little internal relationship to volume or costs. Therefore, moderate change could be paid for through the increased costs that this rate of change would cause. But the most product change occurred in the plastics industry, which provided specialty plastics tailored for specific customers. Although production was process oriented, with relatively few workers needed to monitor automatic and semiautomatic processing equipment, the product life cycle was very short. The development of plastic materials, moreover, was more of an art than a science, as there was a poor understanding of cause and effect relationships. The customer environment definitely wanted high innovation, was willing to pay for it, and operated under major conditions of uncertainty.

3.0 TRYING IT IN OUR OWN SITUATIONS

We can use these research findings by applying a little self-analysis to our own technical situations. Because differentiation was defined partly by the questionnaire reports of the organizational participants, it had to reflect the subjective perceptions of those participants. Therefore, if this research applies in some way to your own organizational situation and if high uncertainty is perceived in the environment by your technical organizational participants, perhaps your structure should have high differentiation and the best organizational design would be the organic model of Burns and Stalker. Conversely, if low uncertainty is perceived, the structure should have low differentiation, with a mechanistic or classical theory model.

Table 10-2. Differentiation and Integration in Three Industries

	Container	Food	Plastics
Degree of Differentiation	Low	Medium	High
Major Integrative Device	Direct managerial contact	Individual integration	Coordinating department

In these examples, changes in technology such as production methods, product innovation and adaptation, and the information processing responses of organizational participants all impact both the structure and the people. The classical management theorists were correct in their limited design concepts of the bureaucratic, mechanistic, and relatively static organizational designs. They erred, however, in suggesting that these concepts were universally applicable, because they could not account for variations in success with companies that were not organized "correctly." The error probably came about because the classical management theorists were familiar with only one technology.

A similar error of supposed universal application occurs in some modern laissez-faire oriented management where there is an implicit suggestion that all structures should provide free and open environments that support people in developing their job potentials. In the first case, it is virtually impossible for organizations to use the same general rule that there must be a match among the components of people, structure, and technology. No two internal groups operate *exactly* the same way. The engineering group is always different from the manufacturing group to some extent, even when there is supposedly low differentiation. Problems occur when senior management doesn't recognize this and expects the internal company departmental structures to operate all in the same way. There is no universal design.

Another example of senior management trying to limit structural design by not considering technology would be the dangerous decision to limit the structural freedom of the small research and development (R&D) group with the intent of having it match the mechanistic, disciplined, logical manufacturing group that mass produces the new and wonderful "widget." That "widget" was developed in the heretofore free and open organically structured R&D group. With a change to a standardized mechanistic structure, there might not be any more "new and wonderful, improved widgets."

Conversely, it might seem that the manufacturing structure does not fit the free and open forms of the high tech organization. It must, in this example, build a tightly organized, closed organization to produce "widgets" systematically. That means hiring different kinds of people, setting up different organizational structures and policies, and using different technologies to produce with and make decisions. And because organizations change with time, the small, entrepreneurial company may have to become departmentalized into many different groups as the technology changes. Structural success in the past may have little relationship to success in the future.

4.0 SUMMARY OF TECHNOLOGY AS INNOVATION

High uncertainty, innovation, and unit or process production technology seem to require organic, very flexible structures with highly differentiated people.

Those firms in the middle range or mass production technology seem to require mechanistic structures and more collaborative (and unimaginative?) people who can coordinate both with each other and with other groups within the firm. These summary statements could begin to unravel some of the tangles in the interaction of technology with the organization. But let's see if these recommendations have been supported by other research findings.

5.0 TESTS IN ORGANIC FIRMS: THE RESULTS

Some research was done by Walton (1979) on using open, organic structures with worker-defined working conditions. He gathered data from about three dozen technologically varied companies, generally in process manufacturing technology, and found no clear long-term indications of success in applying organic structures. His major definition of organic structures included:

Self-managing work teams: They developed their own goals and norms.

Whole tasks: The reverse of task specialization. Simple operations organized into meaningful wholes for operators.

Flexibility in work assignments: Movement from one set of tasks to another; systematic rotation of tasks.

This technology resulted in almost uniformly early gains in organizational performance. For example:

1. Higher production efficiencies were reported. These were derived from less waste of materials, less downtime, or more efficient methods.

2. Quality improvements were significant.

3. A reduction in overhead was common, due to a leaner supervisory and staff structure, with less paperwork.

4. In several cases, a more rapid development of skills produced promotable people at a more rapid rate, increasing the number of operators who were promoted to foremen outside of their own department.

5. Turnover and absentee rates were generally lower.

In fact, Walton found that almost every organization that employed this structural design reported initial gains. However, with the passage of time, other patterns began to emerge. Some plants returned to conventional mechanistic patterns within a short time, others regressed from the ideal organic structure toward these mechanistic patterns after a few years of successful evolution, and a very few persisted permanently, although in a limited fashion. As an explanation for this, Walton (1979) pointed out that " . . . new demands may also tax the system's ability to perform and survive, producing a return to more conventional patterns."

Closer supervision is a familiar response. It involved typical questions like: "What are those guys in advanced R&D doing? Dammit, we needed this new widget on the market last month. I want a written report every day on their progress!" The increased speed of innovation (or technology) seemed to force an inappropriate return to mechanistic close controls. Creativity seemed to be acceptable but only when it met a predetermined schedule! (*A personal note:* How can that happen? Isn't that an oxymoron?)

The other new demand occurred when companies came under severe or long-term competitive pressures. Under these conditions, upper management began emphasizing cost reduction and near-term results, insisting upon discipline and compliance with their programs. Authoritarian decisions and "do it now" commands don't support group efforts and innovative decision making. It's important to remember that innovation can only follow if the firm survives to support that innovative effort.

This research indicates that the organic and loosely constrained organizational structure works well only when technology and external economics are supportive and when there are few extreme pressures to contend with. A lifeboat is no place to hold a participative, free discussion on future organizational planning. Neither is it the place for highly differentiated behavior patterns, because rowing a lifeboat together requires a great amount of cooperation (or integration). A technologically innovative and organically developed decision to use either gas turbines or nuclear energy to propel the lifeboat is also inappropriate, since you only have oars and the delivery schedule on these other items is fairly long, even if the vendor knew where in the ocean he had to deliver them. Therefore, the naval tradition on lifeboats is justifiably very authoritarian, with task-oriented, closely supervised groups. Most technical organizations that are in economic trouble and wish to survive, therefore, tend to adopt a mechanistic structure that almost instinctively demands integrative or collaborative behaviors from people. Again, there is no "one best way" to organize, because time in terms both of survival and of growth or development must be considered. Even if the structure matches all the manufacturing requirements of technology (Woodward), the environmental uncertainty (Lawrence and Lorsch), and the innovative, organic structures that support progress in product technology (Burns and Stalker), the technical department must consider both the organic elements to produce creative product improvements and the mechanistic elements to produce (quality, high-volume, low-cost) products. Organic structures and highly differentiated behaviors work well when there are sufficient resources within the organization to protect it from day-to-day problems (Walton 1979). And we can also find a need for the mechanistic, closed, well-defined organization structure with its well-trained, cooperative professional people for the broad middle range of technology. To reiterate, it also still seems to be the best design to use when the organization is in immediate trouble.

5.1 Suggestions: Manufacturing Technology and Structure

Table 10-3 summarizes the relationship between technology and structure. Why should we be concerned about developing an organic, relatively differentiated structure if we can predict that this free and responsive design requires changing into a mechanistic, rules-oriented organization if an economic or technological downturn occurs? The reason is that growth depends on organic structures, and perhaps some of these can be protected within the organization even if a mechanistic fire-fighting approach is needed for short-term survival. Two longer-term potential answers to this question also apply to technical organizations.

1. The organic, highly differentiated structure promotes innovation, high innovative productivity, and prompt flexibility in responding to changing market demands. In short, it promotes business success *in a relatively benign economic environment.*

2. As a general rule, both our economic environments and the product process technologies are moving away from mechanical, high-labor content into electronic, high-knowledge content for products. Two major reasons are the increased skills of the knowledge workers and the gradual taking over of the

Table 10-3. Classification and Description of Technological Structures

Classification	Description
Unit and small batch, organic	Few procedures. Short, squat structure. Few in staff. Few levels of management. Workers are generalists
Mass and large batch, mechanistic	Rigid procedures. Triangular structure. Intermediate staff. Large span of control of first line supervisors. Decrease of percentage of costs in wages. Classical theory model. Workers are semiskilled.
Process, organic	Process controlled procedures. Tall and thin structure. Many levels of management. Small groups. Greater proportion of staff to workers. Workers very skilled and specialized.

drudgery in work by automation and computers. Knowledge workers should not be wasted in drudgery. It's too costly and very ineffective.

The industrial revolution drastically changed people's life styles. Presently we seem to be heading into an equally drastic kind of new industrial revolution. Our products, our work methods, and our industrial systems are changing. The proportion of our work force in manufacturing is dropping, but the work it does is still quite critical to business success. Machine serving labor, which is measurable, has been replaced by knowledge and information work that is very difficult to measure. The latter type of work typically requires organic structures, because the person is, in effect, determining what to do and how hard to work at it.

Is it possible that we are involved in as revolutionary a change in technology as that which occurred in the early part of the industrial revolution? Then the worker moved from the farm and the cottage industry into a central, more efficient work places called a factory. Now the worker seems to be moving from the factory into the laboratory, the computer area, or, if the worker has a computer at home, the living room. (Back to the cottage industry?) The same kind of industrial discipline would not support the innovation and creativity required to meet new and changed market demands. We may be facing an information and technology revolution that is changing the way we work and live as drastically as the industrial revolution did in its day.

Although speculation and prediction can be important guides to use if we want to modify our organizations (and to improve our own position within that organization), the existing environment confronts us with today's problems to be solved before we can be concerned with tomorrow. If we are in mass production (or threatened with severe or negative changes in the economic environment), the mechanistic, logical, systems-controlled, relatively undifferentiated organizational structures are optimal. Those structures in the short run still account for much of the employed working force, even if they probably are not the way we will be organized in the future.

Where the worker defines the job, mechanistic concepts no longer support innovative productivity.

Information and knowledge workers, the innovative human capital, require organic structures, because the technology that they use is either unit production ("We need an "emergency" team to get this computer program debugged by tomorrow morning.") or process operations. ("What kind of automatic setup and run programs do we need to keep that machining line operating even if we're manufacturing different products on them all the time?") The relationship between technology and structure is complex, because productivity then depends on non-programmable behaviors of the people. That is very difficult (if not impossible) to measure. With organizational economic success, the demand for more organic and less mechanistic structures *within properly defined organizational subunits* increases. How will this occur? Recent advancements in defining the internal controls and operations of technical organizations have been tentatively set up with the design of internationally accepted quality assurance processes (ISO 9000), covered later in this chapter.

5.2 Review: Organizational Models and Technology

In much of the management literature, the mechanistic model is held up as an example of modern know-how and productivity. The technology is not considered to present the major problem; the nature of the problem-solving tasks does. Effective programming of tasks and elimination of uncertainty are supposed to help in solving these problems. The mechanistic model of the production line (mass production or large batch) is the result and the prime example of the attempts to rigidly program tasks. However, from the worker's viewpoint, it has had a mostly negative effect. This is how it is described in the rigidly controlled, technologically advanced, manufacturing environment of an auto plant:

> A typical assembly line worker can learn his job in 30 minutes. He gains no skill that might qualify him for a better job. He has little incentive, investigators claim, to acquire the instincts of workmanship, to take pride in his job, or to feel a sense of purpose. The point of all this is that my experience working in an auto plant has convinced me that alienation of workers as a personality characteristic is in the eyes of the social scene beholder. (Widick 1976, p. 71)

This writer seems to believe that alienation comes with the job, not the psychology of the worker. The interpretation is in the eye of the beholder. The job is the controlling factor. The "beholder" (in our example, you) could feel this way if you believe that these two conditions exist:

1. All workers react this way to repetitive jobs and have these same feelings.

2. These feelings, in some workers, really can affect the organization's productivity.

It may be a bit surprising to find that the first condition does not always apply. (This author is writing about a food processing plant, not auto work, but the idea is the same)

> I would characterize this work as highly repetitious, with little room for autonomy or growth and no position to rise to, but processing. Here there is also a lack of that critical job satisfier, the freedom to walk around and schmooze. Yet because workers, like the rest of us, are not of one mold, one worker said, "I like packaging because I do not want to think about the work anyway." (Schrank 1978, p. 233)

Some people apparently may not like their work for reasons unrelated to the organizational model of the job. We all carry different mental models of the world in our heads. Not all of us are the creative, innovative, I-can-hardly-wait-to-get-to-my-job kind of people, (although some of us are, of course), but no matter whether you are creative or not, it now appears that you, as the worker, are the major variable in increasing productivity.

Several years ago, the Brookings Institute economist Denison (1978) estimated that the new technology accounted for 38% of the factors affecting productivity. In other words, if we assume that technology, with its 38% effect on productivity, is kept constant, the worker controls more than 62% of the increases or decreases in productivity! (I believe that these estimates still hold although I could find no substantiation in the literature for my experiences.) If we then assume that the worker also controls the technology (which she or he does in technical organizations), it would appear that almost all the productivity changes are controlled by the worker. Therefore, it follows that the people in technical organizations are the determiners of organizational success. Of course, this line of reasoning has taken some liberties by comparing actual production with technical workers as if they were the same. If they are not so now, they will be in the immediate future. Production workers of the future will probably be programming a computer that will then control the plant.

6.0 TECHNOLOGY—COMPUTERS: CHANGING THE ORGANIZATION?

Technology changes both the formal and the informal organization, impacting the behaviors of the people. In a mechanistic classical organization, most of the response of workers has been negative: "Dissatisfaction with work seems to be a function of technology. The greatest dissatisfaction is reported on jobs with short job cycles and relatively little challenge in industries, especially the automotive industry" (Fairfield 1974, p. 39).

Although other research (Hulin and Blood 1968) shows no relationship between this dissatisfaction and productivity, there is a relationship between self-interest and productivity. Therefore, if the technology can increase that self-interest, then dissatisfaction consequently decreases and productivity will increase.

Selected according to seniority, they (the workers) studied the slimmed down V6 (engine), its automated machinery, the high quality standards that would apply, and how workers would be given responsibility for assuring that quality. . . . In a stroll down the new line, Mr. Wilson (assembly line worker) and Mr. Hayes (maintenance electrician) pointed out numerous changes their three work groups had made in the equipment. . . . Under the traditional concept, they would have told us. . . . "This is the way it is; you live with it. . . . I'm no longer a job rat under this concept. . . . I feel like I'm part of the product (Hayes 1981).

This description doesn't seem to be one of an auto assembly line, but it is. Because the worker in this very modern assembly line now appears to control many of the technological elements of productivity, there is a change in the psychological environment of that assembly line that seems to increase his or her control and self-interest in keeping the line moving. It is a change from a labor commodity resource into a type of semiartisan. It has also enlarged the ranks of management, as the worker has become a manager of his or her own area of responsibility.

Although mechanistic structures are economically most successful for mass production and large batch manufacturing, they must be changed for people if there is a change in technical methodology to remove deadening and repetitive tasks. Installing computer-controlled or automatic production machinery accomplishes this to some extent. The nonhuman kinds of work have been assumed by a machine. In some of our larger manufacturing companies, the installation of production robots has begun. The organization then becomes more mechanistic in actual production technology (that's good) and more organic in its affect on the human structure (that's good, too). We begin to move toward multiapproach, much-improved, and highly differentiated organizational designs typical of process and unit industries.

> Technically and economically, process production is at the opposite end of the scale from unit production, but from a social organization point of view, this inner system was surprisingly similar to the social system of the special order production firms studied. There was the same identification of formal and informal systems. (Woodward 1965, p. 161)

The important relationship between worker behaviors in the more human-oriented unit or process production technologies and that of organizational success was that in both cases the workers controlled their work inputs. To a large extent, they determined productivity and seemed to be able to manage their own jobs very well. If we can assume that there is also a possible future relationship that applies between these variables for mass production technologies, those machine-like organizations should be concentrating the necessary discipline and logic of production on the shop floor, where it belongs, through increased uses of computers and robots. They should then be able to provide more decision-making capabilities in worker behaviors, allowing those workers to handle unforeseen contingencies more freely themselves.

With that computer-robot technology in place, workers will be able to work more "process-like," thereby exerting psychological freedom in their control over the work flow. The mechanistic organization would be that existing only on the manufacturing shop floor involved with, say, the actual metal cutting, but the human (i.e., managing) organization would be organized in a organic, process mode. It would be like a two-layered structure, the bottom layer mechanistic and the top organic. The top layer would be closer to the refinery-type organization workers and the unit tool and die workers who manage the work flow and control it. They are definitely more than machine adjuncts.

Organizing to satisfy the needs of technology could be achieved by emphasizing more automation or computerization accompanying an organizational structural redesign into this two-layered, very "differentiated" style. It is possible that the general categorization into unit, mass, and process technologies in industry could be moving toward a consolidation into one overall design with automated manufacturing technology (organized mechanistically) that can generally operate in any of the three modes overlaid (or differentiated) with an organic management structure.

This seems to be happening on some machine shop floors. Introduction of computer-controlled, general-purpose equipment allows accurate machining of different products on the same machine, because the computer programs control setups and operations. Those programs, of course, were developed in the organic group of the factory's manufacturing engineers. Although we still have some distance to go before we can expect to see general-purpose manufacturing companies that can turn out anything from steel to gasoline or even any kind of machined product, there will be decreasing differences in overall organizational structures, first among companies in the same industry, such as all steel mills, and then between various industries such as steel mills and refineries. But that is for the more distant future.

Even though successful organizations within the same industries will become generally more similar in the future, they will still be somewhat differentiated internally, as they are today. People are different and each manager affects the organization as well as it affects him or her. The internal differences perceived by the participants (i.e., engineering is different from sales) can determine how well the firm will succeed. Therefore, although I forecast that due to improvements in technology, steel mills and refineries will become more organizationally alike, they will still be very different inside because of department differentiation.

In this case, technology (or the capacity to process information quickly and efficiently) supports differentiated social behaviors within each department. Computerization can therefore promote more organic kinds of structures for human beings inside while supporting more mechanistic structures on the shop floor and more consistent external interactions with the market or external environment. (It seems to be a good thing all around.)

However, before using this general prescription for organizational change into similar "outsides" but different "insides," which is intended to help the human side of the group, we should be concerned with the effect on the quantity and quality of the product output. Would more open organic structures, using the technology suggested above, help or hinder those outputs?

7.0 MECHANISTIC INTO ORGANIC STRUCTURES

The relationship of machine-oriented, organizational designs that manufacture large quantities of satisfactory products for a resulting excellent company profit has always seemed to be fairly clear. Now, since the research of Woodward, Lawrence and Lorsch, and Burns and Stalker it seems that this relationship is no longer very clear, and many organizations are not optimally structured to fit their technology. Is it possible that they have been operating at less than optimal effectiveness? Why have they been able to survive this long? Has the relatively recent decline in some country's national productivity been the result of poor use of technology? Let's analyze what has really been happening to our so-called mass production operations.

It seems that mass production has never really worked as well as it should have. There have been problems (Skinner 1979, p. 214). Some of the obvious ones were:

1. Poor worker reaction and cooperation in serving as machine adjuncts.

2. Many management-dictated changes in product mix, which negated the advantages of single setups and long runs.

3. Many changes in product design, which affected the way that they were made.

The most important of these was the first problem, poor worker reaction. When organizational behaviors were in conflict with the requirements of the formal organizational structure, the informal structure fought with, rather than supported, the formal structure. This might have indicated that the human discipline required in mechanistic structures in mass production was lacking and conflict was therefore a constant problem. According to some recent data on the metal-working industry that uses this structure quite a bit, it seems that we actually don't have real mass production. Therefore, a rigid mechanistic structure could be inappropriate and the poor worker reaction could indicate a need for a looser, more organic, structure. For example,

Complex metal products . . . occupy a central part in metal working. They come in myriads of types, shapes, and sizes and are made in untold millions. But contrary to popular belief, most of them are not mass produced (i.e., over 100,000 units per year). Moreover . . . over one half of the mechanical parts are made in small batches, less than fifty pieces at a time, and in terms of value, mass production accounts for less than 30% of the total. Small and medium batch manufacturing is the predominant mode of production and is characterized by the employment of general purposes opposed to special purpose machinery . . . cutting time is on the average less than 30% of the time that the work piece is in the machine. The remainder is spent on loading, unloading, idle motions, measuring, operator rest, etc. . . .

If anything, the situation has become more difficult because the number of different product types is increasing, doubling every twenty years, so the batches grow even smaller in size. (Barash 1980, p. 38)

And if you consider the total time (not only the time at the manufacturing station) that the work is typically on the factory floor, for more than 95% of that in-process time the part is waiting, tying up costly floor space and working capital (Kops 1980, p. 110). Our metal processing facilities, which are typically supposed to be mechanistically organized as if they were using mass production technology, are really using semiunit production technology. Therefore, according to the research that we have reviewed, they should be organized in a more organic fashion! Surely this would be a radical change for most metal processing companies, and I can predict that it will not occur quickly. Most organizations change their structure slowly. (The only fast changes that I have seen are those caused by rapidly dropping sales. That always seems to tighten the mechanistic design rather than loosening it.)

My feeling is that this reluctance to move quickly into organic structures is based on the behavioral patterns of the senior management personnel in mechanistically oriented companies. They are familiar with more deliberate, mechanistic (functional?) thinking and seem to be uncomfortable with supporting change intended to allow more decision making at lower organizational levels (bias against projects?) in a wholesale manner. I remember once being told that delegation and decentralization were alright within limits but that "there's no way that we are going to let everyone run their own job since that would be anarchy." This appears to be a common frame of reference typical of a patriarchal and mechanical organizational structure.

Perceptions and images, personal and collective, shape the behavior of individuals and groups when changes in their systems are proposed" (Kops 1980, p. 101). Generally, organic structures seem to be the better design for technical organizations when changing technology is a major factor.

7.1 The Real Mechanistic Structure

Do better mechanistic structures matching mass production technologies exist? We know that technology affects product quality, quantity (i.e., productivity), and organizational structure. Within recent times, quality comparisons between

American and foreign products have often been decided in favor of the imports. Japanese cars and cameras, for example, have typically been considered to be the best in their class. This has on occasion been attributed to the Japanese management system, training, or many other (typically intrinsic social) factors, depending upon the academic background of the researchers. Although these factors may be important, I suggest that there could be another approach, one that more effectively relates the technology of production to the organizational structure. When the structure supports only quality and discipline, it is really mechanistic and the worker services the equipment well by becoming, truly, a machine adjunct. Innovation is not part of that structure.

According to a recent report, the modern Japanese factory is not really a prototype of the factory of the future, as many of us might believe. "Instead, it is something more difficult for us to copy: it is the factory of today running as it should be." (Quotes and concepts in this paragraph are from Hayes 1981, pp. 57–60.) Automation consists primarily of adequate materials-handling equipment used in conjunction with standard processing equipment. There are monitoring systems that allow Japanese workers to oversee the operation of more machines than their U.S. counterparts, and production schedules are based on capacity measures derived from actual performance data (not, as often seen in other countries, from theoretical or obsolete standards). There is no expediting, and no overloading is allowed. Work is measured out to the plant in careful doses instead of being, as one manager put it, "dumped on the floor so the foreman can figure out what to do with it."

If these are some of the reasons for success, it appears that for the last twenty years or so, we have had the wrong goals for our mass production organizations. Those mass production organizations have selected other goals, such as mass distribution, advertising, financial controls, and minor new product development as more important than the day-to-day emphasis on the basic production of reliable, low-cost, defect-free products to a predetermined schedule. They seem to have selected failure rather than success as one of those goals. In international markets, others understand that difference.

7.2 More: The Computer and Technology

We have reviewed how productivity represented by structure interacts with the technology as much as any other factor. Although we have cited several definitions of technology, none has explicitly covered the effects of changes in processing information as much as they have covered the changes in production methods affecting products and organization design. Yet, there probably have been more changes in processing information than anywhere else. The abacus still works, but the technology of the computer is no more like it than the foot-driven lathe of the industrial revolution is like the modern, high-speed, automatically controlled chucker.

The technology of processing information is now as vital as the technology of processing products or creating innovation. Improvements in production processing or product innovation are important, but if it were possible to separate out the effects of all changes on the organization, we would probably find that changes in information processing technology have become one of the most important. These changes have supported a lot of the moves from mechanistic structures into organic designs by relieving people of many of the short-cycle, repetitive work patterns that they do so poorly and computers do so well.

Many of the definitions of technology used here have included the computer or its predecessors, because there always has to be some method of repetitively processing information supporting the particular organizational structure. But the combination of information processing controls and automatic manufacturing equipment that results in process technology is still not commonplace. In my opinion, this lag contributes to less than optimal technical production organizations. We use ineffective human "computers" to perform the repetitive coordinating and information-processing tasks. For example, consider the shop floor:

> In a country where 60% of the average foreman's time is spent expediting, looking for material, and in general, fire fighting because of poor scheduling . . . there is an opportunity to improve productivity that is largely untapped, and for that matter, poorly understood. . . . Before computers, scheduling in a manufacturing company was simply out of the question. (Wight 1980, pp. 93–94)

It seems that before computers, the informal, unstructured organization would often keep things going by moving material lots around, finding lost work, and changing schedules constantly. Even technologies that were intended to support mass production industries needed internal miniorganic organizational structures to keep going. The informal organization served that purpose. The inefficiencies that continually cropped up on the shop floor were then overcome by human corrective action that was not directed by the organizational hierarchy. These actions may seem to be relatively minor—finding a part here, setting up a machine there—but the aggregate cost of using this very valuable human effort is quite high. It is now possible that the valuable, creative human being can be released to assume the nonrepetitive tasks she or he is best suited for, and the nontiring computer can take over the minor repetitive tasks or the drudgery of keeping track of things. Even though the possibility has existed for some time now, it has not been acted upon as rapidly as it should have been because, in my opinion, based on observations in many companies,

1. Top management has an underlying fear that it might not be able to understand this new technology. Therefore, any movement in this direction has been kept very slow and deliberate.

2. The concept of having several different kinds of organizational structures, which this technology of computerization would support, could be a bit disturbing to traditionally oriented top managers (because of fear of the unknown).

For example, it is unusual for top managers to take the time to attend training sessions on computers (or in my experience, anything else, for that matter). Therefore, when the advanced engineering department (organic structure) wants to use computers for innovative design alternatives and suggests that each designer have his own microcomputer or a connection from his home to the company's central processing unit, but the manufacturing and production departments (mechanistic structure) want to use them only for tightly disciplined and centrally controlled production processing, top management no longer can set a company-wide personnel and computer policy. Each group has to be treated differently, and many conservative top managers have not had (and may never accept) the training to handle this change.

This type of change places a much greater administrative load on the top operating management. When the change is extended to the organization, even the most basic types of uniform rules such as vacation, working hours, and compensation areas may have to be evaluated on close to a one-to-one basis.

For example, computer operations (running the programs) require tight discipline very close to that of a well-managed mechanistic organization. They require a close collaboration between the mechanistic human structure and machine processes. On the other hand, the actual systems design and consequent programming of the computer is closer to the creative efforts of the organic human structure. This is no longer a change between two different departments; it is now a change between two sections of a seemingly similar computer-based organization. Management efforts and the costs of administration are increased. However, whether the structure is organized mechanistically, as in mass production technological manufacturing (similar to the organization of some Japanese factories), or organically to help produce the design alternatives from which innovations and new products will come, the necessity for computerization will continue to grow in all organizations. The cost of this change will occur whether a company installs computers or not.

Consider the two alternatives of computer installation vs. noninstallation for the simpler mechanistic structure. In addition to equipment purchases, installing computers typically involves the expenditure of funds for training of people to perform highly skilled, machine-supporting, self-managed maintenance tasks in the shop office, and for managers to develop rigid rules that ensure production optimization on the shop floor. The mechanistic organization will have no more changing lines to suit temporary shortages, a lack of extra parts, or any of the many small, but heretofore necessary, jobs to keep the plant operating. The computer is able to plan the purchasing, production, and shipment of a higher quality, uniform product; if something goes wrong, the cause is very apparent to the supporting human organization. On the other hand, not installing computers means spending almost as much, if not more, money and time, but in a less obvious way. Decreasing quality levels, increasing labor costs, and other effects that result from the mismatch of technology and structure and people slowly cut into markets and profits. Eventually, the competitive edge and perhaps the whole market is lost.

The cost equivalency is more difficult to visualize with an organic organization. Here the role of the computer is to increase the effectiveness of the human operator. When that operator is the organization's human capital (see Introduction) and he or she has always controlled productivity, it is more difficult to determine when, or if, that productivity has been increased. In that situation, the motivation plan is the first task to be completed if any resulting organization structure redesign is to achieve results, since that plan is as differentiated as the people to whom it applies.

In this case, the computer is a tool of the person, not of the production process. That person will increase throughput only as a function of self-interest, which the correct motivation program will promote; when motivation is high, computers do increase output in organic organizations dramatically. In both mechanistic and organic structures, one conclusion seems inevitable: computerization increases costs but it also provides great benefits either way. When there is a deliberate investment, it can pay off in greater future returns. When there is no deliberate investment, one is made by the market anyhow, since there is an increased cost to compete that can eventually drain the vitality of the firm, causing its probable demise.

7.3 Computer Applications in Managing

Using computers in manufacturing technology is an obvious application for a mechanistic organization. Other uses involve application to some of the repetitious activities that do not result directly in products. In some special applications in manufacturing technology, the computer sequences equipment and work stations, stores all the programs for different parts, analyzes system status, takes corrective action if needed, and even makes operational decisions. These are only the obvious beginnings of a multitude of management tasks which heretofore had to be performed by people.

Among these tasks are computer-aided design/computer-aided manufacturing (CAD/CAM) for automated drafting, design, manufacturing sequencing, documentation, and engineering development. The primary job is to assist the user in creating and/or revising engineering drawings. Secondary tasks typically include preparation of accurate bills of materials, flow diagrams, manufacturing process descriptions, purchase orders, machine tool instructions, instrumentation layouts, and the multiple, and often exasperatingly detailed, administrative jobs that require great accuracy. (Did you ever wonder who wrote the assembly instruction for that not-so-easy-to-assemble airplane model that you just bought, or those instructions that didn't tell you which way the handle bars were supposed to be put on the new bicycle until you were finished with the job? Someone didn't work out the small details correctly.)

Another computer application is the word processor (WP) computer that can record documentation, manuals, and internal memos; allow revisions almost instantaneously; suggest writing style changes; correct spelling; and absorb all the deadeningly repetitive work of the person who had to generate a bill of materials, for example. In these tasks as in most others, all the computers have essentially the same design:

1. Logic unit for data manipulation.

2. Interactive display for visual and keyboard interaction with human beings.

3. Data storage and retrieval unit, usually hard disc, compact disc, or tape.

4. Hard copy output device to provide something permanent on paper.

5. The software or the program that controls and directs the machine to manipulate the data that you, the operator, put into it.

Using computers, in effect, could make each operator like the highly trained worker of the process technology mode. The machine (or the process) is programmed to produce the product (the flow of information) and the skilled computer operator maintains the process machinery on a day-to-day basis. Similarly, computer systems designers and programmers themselves resemble the artisan workers of the unit technology mode. Each product (system or program) is developed to meet the customer's (user's) needs and is unique with respect to all the other systems or programs produced. The organic structure with independent manager workers in both unit and process technologies is again the optimal with the advent of computerization.

I have described and minimally prescribed how technology affects the technical organization, but potentially larger effects are expected in the future, and these have not been well covered in the literature. They include changes in resources, both inanimate (such as materials) and animate (such as the work force).

8.0 CHANGES IN RESOURCES

It seems reasonable to assume that the world's resources are limited and that we will run out of specific raw materials sometime in the future. Historically, we have developed new sources of energy and materials as the old sources have been used or the costs of extraction have risen to the point of economic unfeasibility. The proportion of product costs represented by materials and energy has always been low compared with other costs, such as labor, overhead, and administration. Therefore, most of the economic pressures for technological innovation have been concentrated on labor and time saving, rather than materials and energy saving. On those rare occasions when a particular material becomes scarce (and higher priced), technological advances have been able to support ingenious product redesigns and substitutions of more common materials. However, when there is a general concern about the finite limit of all of our nonrenewable resources and not just a few materials, the importance of conservation and recycling of materials rises to rival the historical importance of cost reduction through labor savings with automatic machinery.

Computer-assisted design of new products can balance all the cost components and perform multiple tradeoffs within seconds, doing the job that a technical designer would require months to do. Weight of materials, production usages, and energy consumption for production can also become part of the design data banks, prompting a re-examination of every machine part for machinability, processing costs, and relative availability before inclusion into the product design. Directions for production and standards for material and energy use become tighter and more effective, which decreases costs. These additional funds can then be used to support the gradual (with all deliberate speed?) reorganization of the structure to fit this new technology, which requires an organization designed for continual processing.

8.1 The Work Force

8.1.1 Hazards

The work force itself is changing in many ways because of increased demands of industry for better educated, more flexibly trained workers. These workers are becoming more valuable and therefore in the future, they will greater affect on how work is done. These are some future affects upon them.

The application of technology to reducing hazardous and unpleasant working conditions can be related to changes in the social environment (i.e., the law) and the shortages of skilled people who will work under these conditions. Changed social attitudes, management techniques, militant unions, and federal and state laws, have dramatically affected production methods and the appearance of the factory floor. At one time the factory was considered to be a monster that "alienated" workers. This attitude (which is still seen in some of the writings of social scientists, who probably never have worked in a modern one) was born in the early nineteenth century with the shift from "artisanry to . . . the more time disciplined, task oriented, large scale organization of machines, workers, management, and materials we have come to equate with 'The Factory' of industrial society" (Miller 1980, p. 100).

The advent of classical management theory in the early part of this century, with its insistence on discipline, directed planning, and hierarchical structures, did little to dispel these attitudes. Human relations theory tended to treat the differences among people as dysfunctional and attempted to fit the human situation to the work through management attention to workers' informal organizational structures. Militant unions demanded better and safer working conditions and governmental regulations on pollution control and environmental protection have continued the pressures to use technology to reduce hazardous and unpleasant working conditions. But foundries will always have fumes, coal mines will be dirty, paint spraying is dangerous, and the production of nuclear power has its own set of problems.

In many cases, the application of technology to the reduction of these hazardous working conditions results in unexpected changes to the organizational structure. Automating a process to eliminate the need for continuing human intervention can also eliminate the need for heating the work place in winter and cooling it in summer. Dangerous punch presses that now feed themselves and distribute their products to other machine tools no longer require hourly workers to operate them. The work environment (and the structure) has changed to support fewer, but more highly skilled, maintenance engineers who program and maintain this new "factory". The factory has changed, and the nature and quality of the work to run it have changed accordingly.

8.1.2 Scarcity

There is a steadily increasing preference of workers in industrialized countries for employment in the service industries instead of manufacturing . . . in recent years, the percentage of the work force in manufacturing has begun to decline from 30% of the work force in 1947 to 24.9% in 1968. The U.S. Bureau of Labor statistics projects that in 1980 it will decline further to 22.4% and a Rand corporation forecast projects that by the year 2000 only 2% of the labor force will be employed in manufacturing. . . . This migration of the labor force puts a direct social pressure on the manufacturing industry to continue to produce and remain productive as it gradually loses its labor force, which is slowly but surely moving out of manufacturing into the more attractive jobs in the service industries. (Merchant 1980, p. 75)

Perhaps an appropriate question would be whether technology has to increase because of fewer workers that choose manufacturing or whether there are fewer workers because technology is displacing much of the need for manual labor. Is there really "a steadily increasing preference of workers" or do workers prefer to work in service industries because there are fewer manufacturing jobs available? Regardless of the direction of cause and effect, there seems to be a direct relationship between the two. There are fewer workers qualified to operate the new technology and a potential scarcity of those that will be qualified in the future.

Present indications are that this scarcity will become more critical as the requirements for higher skills in workers increase (Tesar 1980). And when *changes* in technology require changes in the organization, operators will not as easily be replaced with others who have more skills. Highly skilled people will then have to be developed internally, by being trained to accept the changed work patterns and independently organized responsibilities of more organic structures.

9.0 CONTROLLING COSTS OF TECHNOLOGY THROUGH RE-ENGINEERING (DOWNSIZING?)

These organic structures can therefore be best supported when you, as the manager of a functional group, recognize the challenge and develop ongoing training programs for the knowledge workers. Training, however, is not always an unequivocal aspect of management. Whether this training will eventually result in future positive returns to the total organization is another matter. There are economic forces that cause some companies to emphasize downsizing. This has a negative effect upon the whole technical organization. Obviously, the loss of highly trained personnel is negative, but also those who remain no longer have the same commitment to their jobs, knowing that they might be next. Thus, downsizing can cause losses as well as gains. Those who are doing the downsizing in the organization may not recognize that they lose major long-range benefits— those who are being downsized take valuable experience and knowledge with them as they move into other companies. Training these people is costly since training is a capital investment.

Reengineering and downsizing are not exactly the same in their intent according to Michael Hammer, the author of the seminal work on the topic (Hammer and Champy 1992). Reengineering is

> . . . the radical redesign of a company's business processes, reinventing the way the business operates in order to meet the demands of a modern economy. It is about rethinking work, not eliminating jobs, and it does indeed succeed, as demonstrated at companies as diverse as AT&T, American Express, Sun Life Assurance, Ford and Proctor & Gamble. (Hammer 1994)

Reengineering has been attractive to many organizations because of the promises of improved effectivity and eventual profits. Downsizing, whether intended or not, is often a result. (In some cases, it is the "quick fix" that some top managers look for.) In some ways, the total thrust of this book is similar to reengineering (not downsizing). The technical groups can reinvent the ways that they operate. They can accomplish changes in the "people" through better recruiting (by setting the standards for the human resources search) and training (to increase technical competency), in "structure" (by designing how to interact in both functions and projects), and finally in "technology" (by being receptive to new ideas and maintaining these idea inputs through attending external conferences, meeting with customers, and perusing technical journals). Reengineering is not novel to the alert, technologically competent technical group.

10.0 CONTROLLING TECHNOLOGY (AND OURSELVES)—TQM

The acquisition of "technology" and the design of effective "structures" with committed "people" seems to be almost an end in itself. It is not. It is a never-ending process that requires guidance against predefined standards. These standards have to be understood and adhered to by all concerned; otherwise, there can be no long-term coordination between organizational groups. Personal independence is always limited in some ways. Legal and social rules tell us how to behave with others. Organizational rules, also defined as quality assurance, limit and direct how we work in the company. The laws that govern us limit us by defining what we *should not do*, e.g. criminal activities—bad actions that hurt others and also by defining what we *should do*, e.g. good actions such as obey traffic signals, and use standard parts when we design new products.

In my opinion, the quality assurance function is concerned with the second concept. It is the "quality assurance" group that accepts, maintains, and monitors processes. (It's not the same as Quality Control. That function measures actual outcomes to determine if those outcomes meet predetermined standards. Quality Assurance monitors the processes because the theory is that if the technology is processed according to correct and predetermined standards, the outcomes shouldn't really have to be checked. Anyhow, that's the idea.)

But because of its monitoring responsibilities, quality assurance (QA) now divides the general definition of technology (i.e., the way inputs are converted into outputs) into two subdivisions:

1. The kind of technology (Shall we design for cast iron or mild steel?)

2. The way the technology is applied (What is the correct machining process for the material that is selected?)

Although the technical groups are concerned with both concepts, the QA functions are primarily concerned with, and monitor, the second one. Perhaps it is a bit revolutionary but in very general terms, it is even possible to extend the monitoring task to include the people as well as the structure and the physical processes of technology. As an example that concerns the organization's people, QA could monitor how management follows process controls such as those typified in Deming's 14 points. These are:

1. Plan for the long-term future, not for the next month or year.
2. Never be complacent concerning the quality of your product.
3. Establish statistical control over your production process and require your suppliers to do so as well.
4. Deal with the fewest number of suppliers—the best ones, of course.
5. Find out whether your problems are confined to particular parts of the production process or stem from the overall process itself.
6. Train workers for the job that you are asking them to perform.
7. Raise the quality of your line supervisors.
8. Drive out fear.
9. Encourage departments to work closely together rather than to concentrate on departmental or divisional restrictions.
10. Do not be sucked into adopting strictly numerical goals.
11. Require your workers to do quality work, not just to be at their stations from 9 to 5.
12. Train your employees to understand statistical methods.
13. Train your employees in new skills as the need arises.
14. Make top managers responsible for implementing these principles. (Deming 1982)

It is possible that all of these ideas (although laudatory) will not be universally adopted as most managers that I have encountered seem to feel uncomfortable unless they can monitor subordinates rather than coach them. (If you are adventurous, this might be a topic of discussion at the next management planning meeting.) However, in more practical and specific terms, QA is involved in prevention, rather than correction. Typical areas of control concerned with the way that the organization operates could be:

a. Were the customer's requirements/specifications transmitted by the sales department to the engineer department correctly?
b. Was the design satisfactory?
c. Was the product manufactured correctly?
d. Was the product packed and transported satisfactorily?
e. Was the accompanying documentation available and to the right standard?
f. Was the customer really satisfied? (Fox 1993, p. 89)

The recent emergence of Total Quality Management (TQM) is an extension of the more limited QA function that is intended to include the entire organization. It has been offered as an important and novel approach to management. According to the following author:

Total quality management programs emphasize the importance of top management acting as the main drives of TQM. . . . Top management support is thought necessary to assure that the right priorities are set and that commitment to the principles of TQM exist throughout the organization. . . . Quality circles and

improvement groups are major vehicles used to allow employees to make suggestions and change work procedures. (Lawler 1994, p. 69)

In TQM the central issue seems to be that no organizational activity or process
is ever good enough. There must be a focus by *all organizational people* on the
search for and the implementation of continuous and ongoing improvement.
There is an emphasis on driving our errors, decreasing costs and improving
all company operations. This could be another manifestation of the idea of
reengineering: that is, reevaluating all the technologies the company uses to
support increased organizational effectivity (Cartin 1993). This description of
TQM can now be supported by some unambiguous recommendations. The following are some additional questions that might not all apply to your situation but
certainly can be adapted to fit.

How reliable is the product?
How durable is the product from both economic and technical standpoints?
How well is the product serviced? (Speed and competency of repair)
How well does the product meet the aesthetic needs of the customer? (How
does it feel, sound, look, taste, smell?)
How does the customer perceive the quality of the product. . . .?
(Zairi and Leonard 1994, p. 15)

These questions and others are inherent in a recent international family of
quality standards beginning with ISO-9000. See Table 10-4. (Corrigan 1994)
Further discussions of the contents of these ISO specifications is beyond the
scope of this book. However, they can be obtained from any organization devoted
to the study of quality, such as the American Society of Quality Control. These
and other specifications, such as those that have in the past been applied to
military procurement (MIL-Q-5923), are guides that can assist in implementing
the never-ending processes of TQM. In order to successfully begin TQM, Stamatis (1994) suggests that a project approach will work best. That is, following
a simplistic approach:

Table 10-4. ISO Standards

Title	Description
ISO 9000	A guideline for the selection and use of quality management . . . quality systems
ISO 9001	A model for assurance of quality systems for design, development, production installation and service.
ISO 9002	A model for assurance of quality systems for production and installation
ISO 9003	A model for assurance of . . . for final inspection and test
ISO 9004	Guidelines for quality management and quality system elements

1. Define the project
2. Plan the project
3. Implement the plan
4. Complete the project (Stamatis 1994, p. 53)

Of course, because TQM involves the whole company and is intended to support constant improvement, after the initial implementation it could be a small, but vital, functional group.

11.0 REVIEW AND PRACTICAL TIPS

The existing mass production model of the "factory," with manufacturing technology using semiautomatic machinery, large batch processes, rigid performance standards for labor, and a rigid hierarchical management structure will become less viable in the future. Organizations that have structures more closely modeled on either a unit or process technology will be able to grow by taking advantage of the economic, social, and environmental changes *as they occur*. These are more *organic* structures that will be able to continuously adapt to change.

With respect to technical organizations, although it is true that all may have similar general goals involving the technology of innovation, it is equally true that the management process that directs, assists, and controls the achievement of these goals will be changed internally. Technology, such as computer-assisted design, will affect supervision as much as it will affect the "factory." Both will become more *organic*. There will be less "monitoring" and more "coaching."

However, measuring and controlling the application of technology is the job of the QA function and, when it uses TQM, that function involves the entire organization. Quality (and technology) improvement never stops unless the company is willing to become obsolescent. Technical knowledge becomes obsolete if it is not continually updated.

Chapter 11 deals with the thoughts and behaviors of the manager. It is concerned with leadership and ethical decisions, the "glue" that holds the organization together. I will, as usual, describe before prescribing.

12.0 "ONE-SENTENCE" SUGGESTIONS

1. Technical structures must reflect the technology used. Organic organizations are optimum for projects organizations and engineering groups concerned with creativity and new product development. Mechanistic structures are optimum for manufacturing and design standards groups. They may be in the same company but they have different procedures.

2. Projects have flat structure (only one or two levels). Functions may have many depending upon the manufacturing technology (see Woodward).

3. Re-engineering does not mean downsizing. They are separate activities.

4. Total quality management is not a new idea. It means to define what you intend to do, then document it and do it that way. Quality cannot be inspected into a product.

5. Quality assurance is a help, not a hindrance to technical groups. It is similar to an internal "customer." If you do things correctly, you'll never notice it.

and finally: *Effective managers focus on one job at a time, tackling the most crucial one first.* (Anon.) And remember, in spite of all your efforts, "Murphy" never takes any time off. Problems will always happen.

SUGGESTED ANSWERS TO CASE QUESTIONS

1. No, it does not accomplish its intent, which seems to be to provide the best computer service to different groups within the company. The design group in this organization seems to be almost mechanical in operation, and therefore it could have its needs serviced by Sam Disko's people. That servicing should be on all overhead systems design in which the computer is programmed to support the designers in not allowing piping tolerance errors, and, in general, it should perform the administration work needed. On the other hand, Mary's group should have its own microcomputers and Sam should be able to provide members of her group with training so that they can do their own programs, perform their own simulations, and, in general, use the equipment as just another design-assisted tool.

   The underlying problems are major ones. They seem to be poor structural design considering the rate of change of technology.

2. If I were Mike, I would suggest that Sam assign several of his programmers to the design group on a project basis. By reporting to the users, those programmers would be part of the design team, not outsiders who have only a limited functional interest in them. I would suggest that Sam start training sessions for Mary's group, since her group would be doing its own programming tasks.

3. I would consider them differently in each group, depending upon the internal needs of that group and its goals. (Can you give specifics that apply to your own organization?) Unauthorized uses of computers are those that are prohibited by law (e.g., fraud or unauthorized use of proprietary data or access to restricted

financial data) or company policy (e.g. no playing of computer games during work hours). Legally restricted kinds of data can be protected through the use of *passwords* (i.e., the computer programs for these areas can be accessed only if the user has the correct authorization codes). Computer games can be allowed by programs that release them before and after working hours.

4. George's group seems to be almost process oriented and therefore would probably have a very tall structure, while Mary's is unit oriented and would probably be very flat.

Do you agree?

Additional discussion question: How would you use computers in the technology of manufacturing processing in unit mass or process production? Include examples.

REFERENCES

Barash, M. M. "Computer Integrated Manufacturing Systems." **Toward a Factory of the Future,** Kops, L. (ed.). (PED-vol. 1). ASME Winter Annual Meeting, NY, November 16–21, 1980. pp. 37–50.

Bronowski, J. **Science and Human Values.** Harper & Row, New York. 1965. p. 67.

Burns, T., and Stalker, G. **The Management of Innovation.** Tavistock Publ., London. 1961.

Cartin, T. J. "Quality Markets Management." **Principles and Practices of TQM.** A.S.Q.C. Quality Press, Milwaukee, WI. 1993.

Corrigan, J. P. "Is ISO 9000 the Path to TQM?" **Quality Progress** (May 1994). 33–34.

Deming, W. E. "Improvement of Quality and Productivity Through Action by Management." **National Productivity Review.** Winter 1982. 1:12–22.

Denison, E. E. "U.S. Productivity Problems Stir Up Attention on The Federal Scene." **Professional Engineer.** September 1978. 20–21.

Fairfield, R. P. **Humanizing the Workplace.** Prometheus, Buffalo, NY. 1974.

Fayol, H. **General and Industrial Management,** Storrs, C. (trans.). Pitman, New York. 1949.

Follett, M. P. **Dynamic Administration,** Metcalf, Henry C. and Urwick, Lionel (eds.), Harper, New York. 1942.

Fox, M. J. **Quality Assurance Management.** Chapman & Hall, London. 1993.

Gibson, James L., Ivancevitch, John M. and Donnelly, James H. Organizations: behavior, structure, processes. Business Publications. Dallas, Texas, 1976.

Gulick, L. "Notes on the Theory of Organization." **Papers on the Science of Administration,** Gulick, L. and Urwick, L. F. (eds.). Institute of Public Administration, NY, 1937.

Hammer, M. "Hammer Defends Re-Engineering." **Economist** (November 5, 1994). 70.

Hammer, M., and Champy, J. **Reengineering the Corporation: A Manifesto for Business Revolution.** Harper Business, New York. 1992.

Hayes, D. "Workers and Bosses Get Ahead by Getting Along." **New York Times** (July 5, 1981); F4–F5.

Hayes, R. H. "Why Japanese Factories Work." **Harvard Business Review.** July–August 1981. 59(4); 56–66, p. 256.

Hulin, C. I., and Blood, M. R. "Job Enlargement, Individual Differences, and Worker Responses." **Psychological Bulletin.** 1968. 69(1); 41–55.

Kops, L. "The Factory of the Future—Technology of Management." **Toward a Factory of the Future,** Kops, L. (ed.). PED vol. 1. ASME Winter Annual Meeting, NY, November 16–21, 1980.

Lawler, E. E. "Total Quality Management and Employee Involvement: Are They Compatible?" **Academy of Management Executive.** 1994. 8(1): 68–76.

Lawrence, Paul R., and Lorsch, Jay W. **Organization and Environment.** Harvard Univ. Dept of Research, Cambridge, MA. 1967.

Merchant, M. E. The Factory of the Future, Technological Aspects. **Toward a Factory of the Future.** Kops, L. (ed.). PED vol. 1. ASME Winter Annual Meeting, NY, November 16–21, 1980. pp. 71–82.

Miller, R. R. "The Transformation of the Factory of the Future." **Toward a Factory of the Future.** Kops, L. (ed.). PED vol. 1. ASME Winter Annual Meeting, NY, November 16–21, 1980. pp. 99–108.

Mooney, J. D. and Reilly, Alan C. **Onward Industry.** Harper & Row, New York, 1931.

Schrank, R. **Ten Thousand Working Days.** MIT Press, Cambridge, MA. 1978.

Skinner, W. "The Impact of Changing Technology on the Working Environment." **Work in America: The Decade Ahead,** Kerr, Clark, and Rosow, Jerome M. (eds.). Van Nostrand, New York. 1979. pp. 204–230.

Smith, A. **The Wealth of Nations.** Modern Library, New York. 1977. (Originally published 1776.)

Stamatis, D. H. "Total Quality Management and Project Management." **Project Management Journal.** September 1994. 25(3): 48–54.

Steele, L. I. W. **Managing Technology: The Strategic View.** McGraw-Hill, New York. 1989.

Taylor, F. W. **Principles of Scientific Management.** Harper & Bros., New York. 1911.

Tesar, D. "Mechanical Engineering R&D." **Mechanical Engineering,** February 1980; 37.

Walton, R. E. "Work Innovation in the United States." **Harvard Business Review.** (July–August 1979); 88–98.

Weber, M. **The Theory of Social and Economic Organization.** Parsons, T. (ed.). Free Press, Glencoe, IL. 1947.

Widick, B. J. **Auto Work and Its Discontents.** MIT Press, Cambridge, MA. 1976.

Wight, O. W. "Tools For Profit." **Datamation** (October 1980); 93–96.

Woodward, J. **Industrial Organization Theory and Practice.** Oxford Univ. Press, London. 1965.

Zairi, M., and Leonard, P. **Practical Benchmarking: The Complete Guide.** Chapman & Hall, London. 1994.

Zwerman, W. L. **New Perspectives on Organizational Theory.** Creenwood Publishing, Westport, CT. 1970.

CHAPTER 11

Leadership and Ethics
Holding It All Together

Case Study:

The Expert Group Leader

International Chemicals, Inc., was a major supplier of industrial chemicals and for many years had operated an internal applied research and development (R&D) group that was closely associated with the marketing branch. Whenever a customer had a particular problem, that customer would call the local salesman, who in turn would contact the marketing group. This group would then get the problem defined and bring it back to corporate headquarters for the engineers, scientists, and technicians to solve.

Cast

George Jessup: Chief metallurgist
Mike Jensen: Group leader, marketing
Melanie Michaels: Metallurgist, technician
Bob Andrews: Vice President, research and development

It was early on a Monday morning when George Jessup received a phone call from Bob Andrews to come by his office when he had a chance. Later that day, George walked into Bob's office.

Bob: Oh, hi, George. Glad you could come by as quickly as you did. I know that you've been busy running life tests on that new gas-metal interaction that the utilities were interested in to increase boiler life. The problem seems to be centered around your new technician, Melanie. She's been in charge of some test that the fellows in marketing wanted to get done quickly and according to what Mike Jensen tells me, she's absolutely refused to do anything for them. I'd like you to look into it and tell me what's going on.

George: I really don't have to do much looking to tell you about it because she told me what happened just this morning. It seems that Mike had this rush job for a steel mill out in Midwest and he, as usual, just came into our department and asked the first person he saw to work on it for him. Usually, we try to respond because we know how important it is, but Melanie is new to this firm so she tried to get him to define the problem a little more before she went to work on it. She asked him to write a functional specification so that she would know what to look for and she asked him which of the other ten "hot" jobs she was working on he wanted her to drop. That seemed to me to be a reasonable question.

Well, you know how Mike is. He doesn't respond too well to things like that so he came over to complain to me, as her boss. I, of course, told him that she was absolutely correct. She hadn't refused to do anything; all she wanted was some direction and priorities. I've been meaning to discuss it with you for some time, but we really should get some structure into what we're doing. If Mike can't even take the time to define the problem or give it a priority, how can we know either what he wants or how fast he wants it? We certainly try, but it's impossible to meet all the demands all the time. So far, he hasn't come back with an answer.

Bob: Yes, it sure sounds different when you tell it, but we've always been able to operate on any demand from marketing. Why have things changed, and what is to be done about it?

George: I don't really know, Bob, but I sure do agree that there have been changes. Why don't we discuss it at the next staff meeting?

Later that week, at the regular staff meeting, all four of the principals to this discussion were present.

Bob: Mike, let's find out what this problem really is. It sounds a whole lot bigger than just having your feelings hurt.

Mike: Well, it seems to be a combination of things. Marketing used to be able to get an inquiry from the field, run it through our experts here at the home office, and have an answer, and a satisfied customer, in a few short weeks. Now it seems that we need all kinds of paperwork, none of the experts wants to work overtime, and it's almost impossible to do anything in less than six months. Our competition is eating into our markets because they're lean and mean. What's happened to us?

George: Why don't we look at our present situation instead of talking about how it used to be, Mike? Let's analyze the problem by using the example of that requirement for the steel mill that you gave to Melanie. Melanie, what did you see happening then?

Melanie: As I saw it, there were ten really hot projects that were taking up all my time. I have been working more than fifty hours a week for the past two months and I've been trying to satisfy all the requirements placed on me, but they seem to be increasing, not decreasing, and the overload is getting to be too much. I can understand that everybody wants his or her job out first, but that is getting to be impossible. I need more structure in the jobs assigned to me, I've got to have some authority across division lines because my projects cross out of this

R&D department into other divisions of the company, and I believe that there should be some coordination and training meetings with those other divisions in order to acquaint them with our needs for prompt response.

As far as Mike's request is concerned, I believe that the questions I asked were necessary to prevent a lot of wasted work if my own definitions were wrong or if I misunderstood the problem. I haven't always been able to follow exactly what marketing wanted and this is one way to be sure that the problem is clearly defined. Also, because I cannot control the other people who have to work with me on the project, I wanted some priority to use in coordinating the requests for their time that I have to submit to their bosses. I look silly when I have to change schedules every week.

QUESTIONS

1. When Mike started with the company, the R&D group was quite small. How have the leadership requirements changed as the company has grown? What has happened to the R&D group?
2. What kinds of recommendations would you make in terms of developing answers for:
   (a) Repetitive problems of definition of the problem
   (b) Project leader's authority
   (c) Intergroup cooperation?
3. How would you set up the process of implementing your answers to Question 2 if you were Bob? To whom would you assign it? How should that person or group proceed to get solutions? How would you know if a satisfactory implementation had occurred?

1.0 OVERVIEW AND INTRODUCTION

In previous chapters, we developed a model of the technical organization, analyzed its three internal components, people, structure, and technology, and prescribed how you might use them (or parts of them) in your own personal management design. We now move directly into perhaps one of the most important sets of contingencies that affect you: the contingencies dealing with leadership and ethics. These two (leadership-management and ethics) are not components because they are primarily concerned with one person, you as the manager-leader. Therefore, they are considered to be almost like an individual thinking/acting process that keeps the three central components together. Obviously each of us, as a leader-manager, behaves differently; therefore, our leadership skills are applied and perceived differently by others. I use the term leadership-management initially to follow some research that attempts to define leadership as unlike managing but then I consolidate those definitions. For example,

> Leadership is different from management. . . . Nor is leadership necessarily better than management or a replacement for it. . . . Both are necessary for success in an increasingly complex and volatile business environment. (p. 103)
>
> Management is about coping with complexity. . . . Without good management, complex enterprises tend to become chaotic. . . . Good management brings a degree of order and consistency to key dimensions like the quality and profitability of products.
>
> Leadership, by contrast, is about coping with change. . . . Major changes are more and more necessary to survive and compete effectively. . . . (p. 104) (Kotter 1990)

Although this distinction may be useful for senior levels, as I see it, technical management spans both of these aspects. We design and implement processes to enable control but we also cope with ongoing change. However, when we accept that definition, it's obvious that we need both functional managers and project leaders. But to minimize confusion, we'll assume that it is reasonable to subsume leading and managing into one category, that of directing and supporting people and making nonrepetitive decisions as noted in prior chapters. This applies specifically to technical groups as the following writer says.

> . . . the time worn dichotomy between leaders and followers has less relevance in a contemporary work setting dominated by highly educated professionals. . . . Professional employees are far from sheeplike in their thought processes and behavior. Faced with marginal leadership, a typical professional will simply conform to the obligatory formalities and do what is necessary to survive until a new leader is assigned. (Doyle 1994, p. 89)

So it really doesn't matter a great deal in the context of this book. In most technical organizations, complexity and change can be almost interchangeable. Innovation requires it. Managing can be understood as dealing with the stable and structured part of the organization. This includes development of standardized organizational policies and procedures, setting up the manuals and standards that apply to product design standards, and documenting the relatively fixed rules that the company uses when determining how to apply Mother Nature. For example, the manager may be responsible for supervising professionals in outlining standards, purchasing processes using those standards, and storing sizes of product raw materials, finishes, and components.

It is also possible to consider that leadership deals with change and innovation. That means the nonrepetitive decision-making process involving that ever-changing resource: people. An example of leadership could be establishing a new project team, supervising emergency field service of failures of new products as part of quality assurance, and implementing a management by objective system for the technical group. Management and leadership are interchangeable in the technical situation. Consequently, those terms are treated as equivalents here. This is supported by the following:

> Managerial authority refers to that part of the leader's authority which has been delegated to him by the institution he works in. Leadership authority refers to that aspect of his authority derived from the recognition of his followers of his capacity to carry out the task. Managerial and leadership authority reinforce each other; both are, in turn, dependent upon other sources of authority, such as the leader's technical knowledge, his personal characteristics, his human skills, and the social tasks and responsibilities he assumes outside the institution. (Kernberg 1984, p. 48)

We will consequently follow the by-now familiar pattern of definition, description, and prescription; then attempt to develop some suggestions for leadership/management behaviors that seem to be particularly applicable.

Leader-managers exhibit similar behaviors as noted in the following research.

> Surprisingly, managers spend relatively little of their interaction time with their superiors . . . seldom more than one fifth and usually closer to one tenth . . . (p. 479)
>
> Studies of what managers do consistently find that their work is particularly oral. . . . A successful leader, therefore, must have the ability to selectively "hear," retain, and transmit vast quantities of oral information and, perhaps even more difficult, selectively utilize a vast volume of written information provided routinely by the organization. . . .
>
> . . . Political activity—in the sense of developing and maintaining a network of contacts throughout the organization and its environment . . . is a real part of managerial work. (p. 483) (McCall, 1983)

2.0 PROMOTION AND HELPFUL IDEAS

It is very interesting that the political activity of the leader-manager seems to have a direct influence on promotions within many organizations, at least, according to the following researcher.

The startling finding that there is a difference between successful and effective managers may merely confirm for many cynics and "passed over" managers something they have suspected for years. . . . Could it be that the successful managers, the politically savvy ones who are being rapidly promoted into responsible positions, may not be the effective managers, the ones with the satisfied, committed subordinates turning out quantity and quality performance in their units? (p. 127)

The traditional assumption holds that promotions are based on performance. This is what the formal personnel policies say, that is what new management trainees are told and this is what every management textbook states should happen. On the other hand, more "hardened" (or perhaps more realistic) members and observers of real organizations (not textbook organizations or those featured in the latest best sellers or videotapes) have long suspected that social and political skills are the real key to getting ahead, to being successful. Our study lends support to the latter view. (p. 131) (Luthans 1988)

Although this indicates that a lack of social and political skills can be a detriment to promotion in technical organizations, it should be noted that these skills are a necessary but quite insufficient basis for success in our situations. Obviously, the manager-leader must possess those skills, but in technical groups, the supervisor (i.e., leader or manager) cannot manage successfully without the willing cooperation of subordinates. It is those subordinates who are the "knowledge workers," creating the innovative products and services needed to support both present operations and future company growth. The organizational "bottom" is therefore as important as the organizational "top".

Although other less structured organizational groups in the company may not necessarily require the same degree of logic in their operations as the technical group does, technically qualified "knowledge workers" cooperate best with the supervisor when they recognize that the supervisor has that logical foundation. They understand that in order to process and comprehend the great quantity of information that the technical manager receives daily (see McCall 1983), the foundation is a necessity. When that foundation exists in the supervisor, social and political skills can be built upon it. If it is not there, subordinates are less willing to cooperate and they expend only enough effort to maintain their jobs. That is one definition of failure in technical organizations.

In other words, social and political skills follow technical competence. One of the aims of this book is to assist you in obtaining those skills, in addition to extending your foundation to include the management systems and tools from which your personnel system can be built. As noted before, henceforth the term "leader" is interchangeable with the term "manager" or "supervisor."

2.1 Some Sources and Definitions

The study of leadership has a long and honorable past. Some well-known authors on this subject have included Homer, Marcus Aurelius, Plato, and Caesar. Each had his own viewpoint, and although those viewpoints are not as universally applicable as their authors might suggest, many of their basic concepts still survive (Bass 1981). The "thoughtful" leader of Marcus Aurelius, the "man of action" of Caesar, the "shrewd and cunning" leader of Homer, and the "statesman-philosopher" of Plato may still be found in modern research literature. Niccolò Machiavelli's treatise on leadership, *The Prince* (1940), which was written during the Renaissance, is still quoted today as a guide for a specific type of leadership behavior. The past is still with us.

The importance of this subject should never be underestimated. I consider leadership as the force or cement that binds the total organizational components together. In one sense, it is the process that induces one or more individuals to act in accordance with the wishes of another individual. Thus, the intended end result of the process is basically to induce compliance. Leadership may or may not include "authority" in the formal sense. Authority tends to reside in the formal organizational position. Leadership may or may not. It is quite possible to have a leader in the informal organization.

How can you get that compliance? We would usually start our descriptions by emphasizing the leader. However, describing the leader while keeping all other things fixed cannot be done. By definition, the situation includes the followers, and followers are inextricably intertwined with the leader. It seems obvious that a leader with no one to lead is an impossibility, but this impossibility did not appear to be insurmountable to early researchers in leadership theory. According to them, leaders were either born or else learned specifically how to lead by observing other leaders. These researchers assumed that a good leader can lead no matter what the situation, and that the leader would always be recognizable by exhibiting obviously superior personal attributes. The dilemma of no followers or the particular situation did not exist for them; it was just assumed away.

One of the earliest theories in this historical group is called "trait theory." It explicitly states that good leaders have inherent characteristics that support successful leadership behavior every time and in any situation. According to this theory, if we are not born as inherently good leaders, we should be able to analyze the good leader's characteristics, find out which characteristics we lack as potential leaders and improve those characteristics that are missing or inadequate. This kind of thinking is universalistic and has the obvious problem—it doesn't work in all situations.

But trait theory and other universal leadership theories are still parts of the foundations of this management literature; thus, we will review them for the few parts that may be useful before going on. Then we shall move quickly into the more realistic and useful person-situation contingency relationships. Some of those contingency relationships assume a relatively stable situation and some do not.

2.2 Descriptions of Several Universal Models

Trait theory: We will discuss two basic models: trait theory and social behavior theory. As noted before, one theory that is part of the oldest and most traditional group of theories describing leadership is trait theory. A trait is the ". . . differences between the directly observable behavior or characteristics of two or more individuals on a defined dimension" (Mischel 1968, p. 5). Because those differences are subject to interpretation, it appears that the trait being observed must be some type of construct or an abstraction of the observer. There is an impressive literature list that identifies "supposedly essential characteristics of the successful leader, over a hundred, in fact, even after the elimination of obvious duplication and overlap of terms" (McGregor 1960, p. 180).

These traits are assumed to be the generalized and enduring parts of the leader's personality that cause the leadership behaviors. It is easy to understand how important the determination of these essential leadership characteristics is as a research project. If it were really possible to determine a combination of stable mental characteristics for successful leadership, it would then be possible to predict who would and who would not be a successful leader. It might even be possible to find out what leadership traits are missing in a particular person and train that person, eliminating the deficiencies.

> These ideas implicitly assume that the leader generally determines the success patterns of the group being led. Typical traits or characteristics such as choice of associates, biography, height, weight, physique, appearance, speech fluency, intelligence, scholarship, knowledge, insight, judgment, originality, adaptability, introversion-extroversion, dominance, initiative, integrity, responsibility, self-confidence, social skills, and a myriad of other attributes were measured or correlated. The correlations were not very high, but some typical findings were that "The average person who occupies a position of leadership exceeds the average members of his group in the following respects: (1) intelligence; (2) scholarship; (3) dependability in exercising responsibility; (4) activity and social participation; and (5) socioeconomic status." (Bass 1981, p. 65)

The writers of the classical school, such as Fayol, Taylor, and Urwick, extended the definitions of traits to include the thinking activities performed by the manager on the job. They felt that typical activities such as "planning, organizing, and controlling" determined a leader's success in the organization. These activities were supposed to be both basic and universal. If the leader did them well, he or she succeeded; if not, he or she failed. This approach defined behaviors (and inferred traits) and was supposed to demonstrate how every organization could be managed effectively.

As an outgrowth of this classical, relatively fixed approach to managing, but modified to suit more modern ideas, McGregor (1960) proposed his well-known theory "X" and theory "Y" leadership concepts. His concepts define the leadership traits as resulting from the relatively fixed belief patterns *of the leader*. These belief patterns were defined as two separated sets of beliefs concerning the attitudes and behaviors of subordinates: those supporting theory "X" and those supporting theory "Y."

The believer in theory "X" assumed that subordinates consisted of people who had the traits of being passive, not inclined to work, and resistant to organizational needs. The theory "X" leader's resultant behavior was consistently directive, with close supervision of subordinates. The theory "Y" leader, on the other hand, assumed that the group being led was already motivated and that each person in it would want to work as much as he or she would want to play. The worker had an internal desire for more responsibility. That leader's behavior was consultative and intended to assist the workers or the individuals in the group to achieve their organizational objectives. The theory "Y" beliefs were valued more highly for many years, as they seemed to match the human relations school of management, which was popular at the time. There were many training sessions in which managers were supposedly indoctrinated with behaviors intended to support the beliefs of theory "Y," such as the need for listening to subordinates, gaining their participation, and attempting to develop a team approach regardless of the particular situation.

There is an obvious problem with trait theory. There is no objective way to separate the traits being measured from the evaluation processes of the person doing the measuring.

> Traits are used first simply as adverbs describing behavior (e.g., he behaves anxiously), but this soon is generalized to describe the person (he is anxious) and then abstracted to "he has anxiety.". . . we quickly emerge with the tautology, "He behaves anxiously because he has the trait of anxiety." This is the danger of Trait theoretical explanations. (Mischel 1968, p. 42)

This problem, by itself, is sufficient to cast doubt on the theory's validity and consequent usefulness: one of its basic assumptions is not supported—that of universal application regardless of the situation. For example, does the person behave "anxiously" all the time, part of the time, when in traffic and late for work. If that person behaves differently at different times, then the "trait of anxiety" is not fixed in the individual. When applied to leadership theory, the simplistic definition of "anxious" behavior, as solely a function of the fixed internal state of the leader with no provisions for the stimuli of and interaction with the subordinates and the situation, is not realistic.

However, although trait theory is not a panacea, it is still used where the situation is expected to be very predictable and the subordinate behaviors are limited. The prime example would be in the military where recruits are taught to obey orders instantly and junior officers are taught that their rank is sufficient to require immediate responses in subordinates.

That's not the usual situation in technical organizations, of course; there are always changes in the situation there. When those changes interact with the subordinates' and the leader's personal attributes, there must be behavioral adaptations. Those adaptations or interactions are the basis for *social behavior* theory in the leadership literature. It is social behavior theory that suggest an approach that is universal, but it is a bit more useful because it considers the relationship between the situation and the leader's behavior, too.

Social Behavior Theory: Social behavior theory is typically illustrated in the organic-mechanistic classification we have discussed in previous chapters. If we define a leader's response to fast situational change as always being quick and flexible, as opposed to always being rigid and unchanging, we have the expected behaviors needed in organic structures vs. those of mechanistic structures (Burns and Stalker 1961). This theory maintains that leadership also takes into account the values and expectations of those being led (Likert 1961). The leader must present behaviors that are perceived by the followers to be supportive, involve them in decision making, and increase their influence in defining and completing organizational tasks (Litwin and Stringer 1968). These behaviors are expected, every time, to build group cohesiveness and the inferred motivation to produce by supporting the individual's freedom and the taking of initiatives.

Many of the ideas in this social behavior theory have been included in a managerial grid (Blake and Mouton 1964) that measure leadership behavior with two perpendicular axes: *concern for people* and *concern for production* (see Figure 11-1). As an illustration, a leader may exhibit behavior that seems to be high in concern for people and low in production concerns. This would then result in a country club type of organizational climate where everyone was relatively happy but production was low. (This climate would confirm the idea that there is little relationship between morale and productivity. In fact, there might be an inverse relationship.)

In contrast, we would have a leader who is supposedly high in concern for production and low in people concerns. This would result in a rather directive, mechanistic, production-oriented organizational climate. Blake and Mouton have given tests to each person attending their leadership classes. Those tests indicate a "score" for each of the axes. Accordingly, the optimum leader is one who is a 9,9 (both axes are measured from 0 to 9, with 9 being the highest score). That is the leader who simultaneously has major concerns for both production and people.

2.2.1 Advantages of Universal Models

These universal models have the promise of all universal models; they promise the "one best way" to manage. Trait theory is appealing because traits are

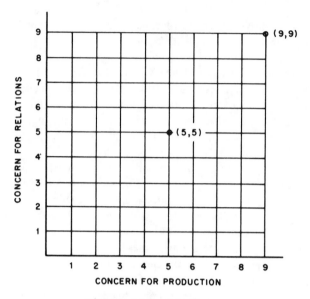

FIGURE 11-1: Grid Theory

supposedly part of the "package" of the best leader. Traits that are not in the package can probably be awakened through adequate training and experience. This theory assumes that everyone can be a good leader if they are taught how to be. In effect, a good leader is always effective, no matter where that leader is, and those who are not effective at present can be trained.

Social behavior theory is similar to trait theory in its predictions of universality. It is different when it suggests that there is an interaction between the situation and the *potentially* improvable attributes of the leader. And, of course, it also suggests that personal attributes can be modified through training to fit some predetermined and fixed situationally optimal condition. Both the situation (or the organizational climate) and the leader are predictable, and the interaction between them is defined as eventually a perfect 9,9. With this theory, the leader will invariably succeed if trained to respond correctly to that environment.

2.2.2 Disadvantages of Universal Models

The problems in trait definition severely limit the proposed universality of trait theory. Social behavior theory is also limited, because all situations are not alike and they may require a different set of "concerns" (i.e., production and relations) than the theory allows. Also the interaction of the leader's behavior with a particular or unique organizational situation is not considered. We know that all situations are not the same. Organizational situations change over time. Other changes affect their technology of production (Woodward 1965), their mode of decision making (Thompson 1967), and the perceived uncertainty (Lawrence and Lorsch 1967). All situations do not use the same ground rules, and if all leaders were trainable to fit this theoretical optimal situation, we would not have companies going bankrupt after many years of success. Their leaders wouldn't let it happen, because if they succeeded before, they would always succeed. Therefore, there are problems with these theories.

We can find discrepancies even when using some of this theory's predictions. The theory would generally predict that all leaders should have a high concern for production with a similar concern for people, and yet such a leader in a production-oriented, mechanistic situation would probably fail (i.e., a 9,9 leader in the Blake and Mouton grid theory). Directive leaders with a high concern for production and a low concern for people would be needed in this example. On the other hand, the advanced research and development group in the same company could not tolerate this production-oriented brand of leadership; for it, the theory "Y" leader or one concerned primarily with people would be best.

There are places for a drill sergeant, theory "X" type of leadership behavior and there are places for a consultative, collegial, theory "Y" leader. What is even more interesting, when the organizational situation changes, if we don't have the ability to change the leader, we may need both in the same person: someone who could be either "X" or "Y," depending on the contingencies. We can see that these theories are limited and can be applied only when the organizational variables will support them. There are other more pragmatic models, however, that make no claim to universal applications. As usual, they are more complex and therefore more applicable to real situations.

2.3 Descriptions of Situational Models

The situational models are concerned with the changing interaction between the behaviors of the group being led and the leader. The group may be affected initially by the externals of technology, information systems, and/or environmental change, but the primary concern is always with the results of those changes on the basic group-leader interaction in whatever form it takes. The interaction is divided into two parts: the formal part, based on the structural confirmation by the total organization (of which the group is a part), and the informal part, based on the acceptance of the leader's direction by the group. It's the familiar combination of legal power and personal influence as was noted by Blau:

> Compliance can be enforced with sufficient power, but approval cannot be forced regardless of how great the power. Yet the effectiveness and stability of leadership depend on the social approval of subordinates. . . . Effective authority, whether in formal organizations or outside, requires both power and legitimating approval,

but the one is more problematical for the informal leader, and the other, for the formal leader in an organization. (Blau 1967, pp. 201, 210)

The idea that the approval of the group itself affected leadership success was understood by the classical theorists, but only as a potential source of conflict which was to be minimized if possible. They maintained that formal approval by the organization was all that was necessary for adequate organizational leadership. The subordinates' effect upon the leader was to be minimized and, if possible, eliminated, because it conflicted with the needs of the organization as communicated from the senior management by the leader. Those organizational needs coincided with the needs of the leader, who represented the organization. The Hawthorne experiments (see prior chapters on structure) confirmed this through the suggestion that the social or informal evaluations by the workers of their management should be manipulated in order to obtain agreement between organizational goals and those of the workers.

However, there were those who recognized that this could not always be done unilaterally by management. Workers (and especially "knowledge workers") think for themselves, and when they perceive a difference between their own needs and those of management, their informal social organization often becomes strong enough to assist in mitigating those management needs. In that case, authority "rests upon the acceptance or consent of individuals. . . . The decision as to whether an order has authority or not lies with the persons to whom it is addressed, and does not reside in persons of authority or those who issue the orders" (Barnard 1938, p. 163.)

You can see why the armed services would not voluntarily adopt this suggestion in many of their organizations. In combat organizations, the authority must lie with the symbols of office or the uniform, not the wearer, because there are often rapid and unexpected situational and personnel changes, and subordinates' compliance must be an almost automatic response. However, in modern industrial organizations, the independence of the knowledge worker suggests that these ideas of automatic obedience are not quite valid, and the worker can accept or reject the leader in charge. As one engineer told me, "When they sent out the memo announcing my promotion to chief, it was like getting a hunting license for lions. Now all I have to do is get the lions to cooperate or else I could get badly hurt."

In this case, getting the "lions" to cooperate was defined as obtaining the approval of the group or, in more formal terms, achieving a positive interaction between group and leader behaviors. That happy situation can result when there is a very strong psychological bond between the leader and subordinates or, conversely, when there is none. A strong bond results from a coincidence of interests such as when the leader provides either psychological support (as when he has charisma) or very tangible and valued gains to the subordinates. With charisma or rewards, the leader can easily gain the group approval. But leadership can also be effective when there is little emotional bonding with subordinates. For example,

> . . . consensus leadership is the relative absence of strong emotional bonds between leader and follower. The leader is a first among equals; and calculated self- and group interests are the ties that bind men to the structure. However, men are willing to compromise in order to reach some satisfactory consensus in which interest groups neither win nor lose. . . . The reversal of the usual dependency pattern is especially marked in complex bureaucracies where the leader knows less about any particular issue than selected subordinates. (Zalesnick 1984, p. 121)

Therefore, if the leader can provide immediate positional rewards, subordinates will probably follow him or her. These rewards could be material or charismatic. In my opinion, however, the charismatic extreme is not likely to occur in technical operations because of the kind of training that we all receive in the university. A "charismatic" answer, for example, in a physics or chemistry class, would not be received well. More likely the situation supports either social consensus or expectations of material rewards.

2.3.1 Path Goal Theory

How this interaction of group and leader behaviors was to be gained was the subject of much research. One important question in that research was, "Why should the group accept the leader?" The potential answer was the path-goal theory (House 1971). This theory suggested that the interaction was based on behaviors exhibited by the leader that aroused subordinates to perform and to achieve satisfaction from the job to be done. The leader did this by clarifying the goals of subordinates and *defining the paths* to achieve them. Perhaps more importantly, the leader also controlled the rewards that subordinates valued and *cleared an organizational path* for them so that they could achieve those rewards.

This brief description of path-goal theory shows its roots in many theories of motivation and organizational structure design. The motivational roots include the appeal to the self-interest of the subordinates and the ability to organize the structure that clears a path for the achievement of these management-valued goals. The theory's effectiveness lies in the specific definition of the methods by which leaders can achieve the compliance of their subordinates. Those methods are clarifying goals and defining paths to achieve them.

2.3.2 Contingency Theory

However, the leadership theories that, in my opinion, seem to apply best to technical organizations are the contingency or situational models. These theories essentially say that there is a fit between the situation (i.e., the organizational environment) and the structure (i.e., the interactive behaviors of people). In other words,

Much of the research conducted with contingency theory as its base has attempted to develop theoretical principles. Contingency theory states that, for an organization (or its sub-units) to be effective, there has to be goodness of fit between its structure and environment, thus focusing on the objective perspective using a macro level of analysis. (Gattiker 1990, p. 28)

If you change "environment" into "situation" and "structure" into "leadership behaviors," it is possible to use these theories in building improved personal leadership theories.

For technical operations, it seems to me that one of the better specific theories within the general category of these theories is the contingency leadership theory developed by Fiedler (1967). It describes measurable aspects of the leader's personality and relates those aspects to unique characteristics of the situation. It predicts which leadership personalities will succeed in which types of organizational situations. The measurement of the leader is operationally determined by a test called the *Least Preferred Coworker* (LPC) test. The leader is requested to complete a paper and pencil test checking off the characteristics of someone she or he knows or did know in the work environment whom she or he would define as her or his least preferred coworker. See the example shown in Table 11-1.

Now think of the person with whom you can work least well (emphasis is mine, not the author's). That person may be someone you work with now or someone you knew in the past. The person should be the person with whom you had the most difficulty getting a job done. Describe this person as he appears to you (Fiedler 1967, pp. 268–269).

These four items in Table 11-1 are typical of the total of sixteen items on the LPC test. The leader checks the appropriate answer for each of those items. The favorable pole of each item is rated as 8 and the unfavorable as 1. These scores don't appear on the actual LPC test but are added by the test scorer after the appropriate check mark is made on the scale. The sum of the item scores constitutes a person's LPC score. A high LPC score denotes a person who is relationship oriented and a low LPC score, one who is task oriented.

. . . a person who describes the Least Preferred Co-worker (LPC) in a relatively favorable manner tends to be permissive, human relations oriented and considerate of the feelings of his men. But a person who describes his least preferred coworker

Table 11-1. LPC Test (a sample of the complete test)

Pleasant	Unpleasant
Friendly	Unfriendly
Accepting	Rejecting
Helpful	Frustrating

in an unfavorable manner, who has what we have come to call a low LPC rating, tends to be managing, task controlling, and less concerned with the human relations aspects of the job. (Fiedler 1976, p. 485)

Research on LPC scores suggests that they are relatively stable over time and that low LPC people receive satisfaction from successful completion tasks, whereas high LPC score people are concerned more with successful interpersonal relationships. You might consider this to be a possible extension of Trait theory, but it is much more than that. It does not measure parts of the person's personality, the individual traits, but is instead intended to measure the person's entire or global frame of thinking about others. Measuring a global characteristic such as LPC is more applicable to a particular person than measuring pieces of that person, such as traits, and then attempting to reassemble those pieces back into the whole person. The test reliability and validity for the whole is greater than that of the sum of the tests of the parts.

The definition of the situation was also operationally defined by questionnaires resulting in three measurements. These are:

1. Leader-member relations: The leader-member relations measurement determines how well the group and the leader interact—how willing the group is to follow the leader's guidance and direction.

2. Task structure: The task structure measurement determines how well the task is spelled out in a step-by-step methodology. It is well defined or is it nebulous and poorly defined?

3. Position power: The position power is a measurement of the power inherent in the position itself aside from any influence that the particular leader might bring. An analogy would be that a general in the army has a high position power, regardless of the personal influence of whoever happens to have that job.

The research shows a correlation between the LPC score of the group leader and the group performance on the vertical y axis and the three determinants of the situation on the x axis. (See Figure 11-2.) It was found that low LPC leaders (task oriented) performed better and managed more effective groups when the quality of leader member relations, the task structure, and the position power were either very favorable or unfavorable for the leader. The measurement used was the correlation between LPC score and group performance in specific situations. The high LPC leaders (relations oriented) were more effective when neither of these extremes appeared. In other words,

A positive correlation (falling above mid-line) shows that the permissive, non directive and human relations oriented leaders performed best; a negative correlation

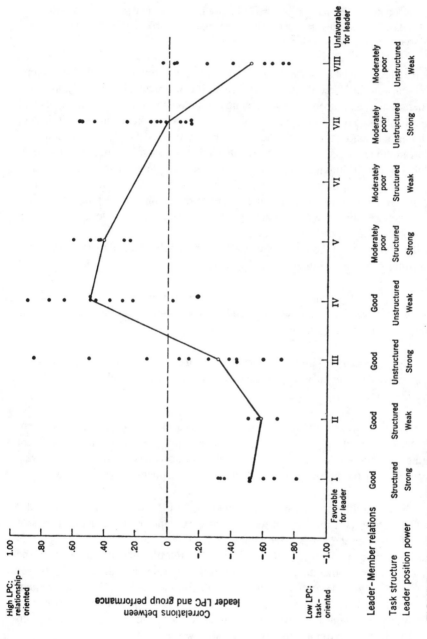

FIGURE 11-2: Leader LPC and Group Performance Correlations. (Courtesy of Fiedler, F. E., *A Theory of Leadership Effectiveness*, McGraw-Hill, New York, 1967, p. 146.)

(below the mid-line) shows that the task controlling, managing leader performed best. For instance, leaders of effective groups in situation categories I and II had LPC [versus] group performance correlations of −0.40 to −0.80, with the average between −0.50 and −0.60; whereas leaders of effective groups in situation categories IV and V had LPC [versus] group performance correlations of 0.20 to 0.80, with the average between 0.40 and 0.50. (Fiedler 1976, p. 488).

The implications for leadership prediction and control become clearer with this approach. There are suggestions for almost all kinds of leaders in Fiedler's model—those who are at the extreme of being totally task oriented (in Blake and Mouton's theory that might be similar to the extreme concern for production) and those who are totally relations oriented (concern for people). But there are also implications in Fiedler's models that there are potential leaders who can operate successfully between these extremes. Even two leaders with either a task or a relations orientation can exhibit temporary behavior that is supposedly atypical. The successful task-oriented production foreman who would be concerned about the family problems of one of the workers would be exhibiting atypical behavior. However, that behavior could still represent the particular manager adequately, because this measurement indicates an *overall tendency* and doesn't pretend to describe all possible behaviors. The LPC score indicates the *dominant mode* of behavior; it is not intended to be exclusive. Other recent research findings (Kennedy 1982) apply to middle LPC managers, who seem to be the best all around managers, according to these data.

> . . . the results . . . provide strong support for the hypothesis that middle LPC leaders perform well in all leadership situations . . . these results suggest that the middle LPC leader who . . . is least concerned with the task and the opinion of others, appears to be most capable of performing leadership tasks in an effective manner, regardless of the situations . . . more flexible, not overly constrained by any one goal orientation, and therefore better able to employ the behaviors that will maximize performance. (Kennedy 1982, p. 7–9)

This has some aspects of universal theory, as it suggests that there is one type of leader who can perform well in all situations. However, the measurement of the leader's performance was determined only in the kinds of situations noted in research about this contingency theory. Although these situations certainly describe a broad range, it's possible that there are other situations not covered. This research, therefore, is only indicative, because the researcher is describing only limited experimental situations. Although Fiedler's work has been replicated many times, we still don't have a method to manipulate the results to make them uniformly applicable to anything that's unique in our own organizational situations. In other words, there's a general fit but it may not be an exact fit for you.

Now, assuming that we try to use these data in building our own theories, we should start with some typical questions. For example,

1. How do the experimental situations compare with those that we perceive in our own organizations?

2. Do the extremes of our situations apply; i.e., does task orientation describe our technical organizations better than descriptions of relationship orientations or vice versa?

I'm sure that there are other questions you have considered. This contingency model, as noted before, has been supported through replication many times, so although the situation might not match perfectly, it's closer than many other theories to the typical situation in technical groups. It's a good beginning for your model.

Another model within the contingency-situational framework is that proposed by Hersey and Blanchard (1969). They suggest that the model of the leader group interaction should be responsive to, as they call it, the "maturity of the organization." (In my opinion, the "maturity" of the organization is represented by the organizational situation since it refers to the situation as defined by the subordinates, not by the leader. That's why this theory is presented here. It also seems to be applicable to the situational changes in many technical organizations.)

These authors also have a graphical representation of the theory (see Figure 11-3), but they are concerned with a "life cycle" approach.

> According to Life Cycle Theory, as the level of maturity of one's followers continues to increase, appropriate leader behavior not only requires less and less structure (tasks) but also less and less socio-emotional support (relationships). . . . Maturity is defined . . . by the relative independence, ability to take responsibility and achievement motivation of an individual or a group. (Hersey and Blanchard 1969, p. 29)

The components of maturity typically include level of education and amount of experience. As an interesting aside, maturity is not related to chronological age. The researchers consider chronological age almost irrelevant. They are only concerned with psychological age, and there may be no relationship between the two.

The life cycle theory, within the overall contingency theory framework, suggests that the leader's behavior should change as the organization matures with a consequent change in the situation. It should move from high task and low relationship behaviors (quadrant I in Figure 11-3), to high task and high relationship behaviors (quadrant II), through high relationships and low tasks (quadrant III), to low relationships and low tasks (quadrant IV) as the followers move from immaturity to maturity. Life cycle theory seems to be analogous to the parent-child

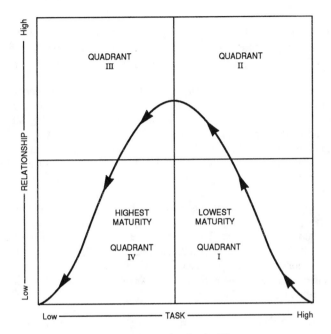

FIGURE 11-3: Life Cycle Theory

relationship and is similarly expected to exhibit the slow, gradual developmental process that is evolutionary, not revolutionary, in nature.

2.3.2.1 Advantages of contingency-situational theories. These models with the general contingency-situational category of theories have attempted to solve the measurement problems within the limits of accuracy of the social sciences, and they seem to have made a step in that direction by considering both the person and the situation. Although we won't go into further explanations in this book, it appears that the major research problems of construct and predictive validity and test result reliability seem to have been met; additionally, the data have been replicated many times. Fiedler's situational constructs do not agree exactly with Hersey and Blanchard's but, in my opinion, there seems to be sufficient coincidence to indicate that in the future there may be larger contingency theories yet to be formulated within which these models will fit.

Therefore, although these models within contingency-situational theory have not solved all the problems in defining leadership group interactions, they seem to be an important step forward in describing and explaining more of the intricate relationships between the leader and the group. These are complex models within complex theory, but sometimes complex explanations are required to describe complex phenomena. It assists in generally predicting how the behavioral (i.e., observed behavior) characteristics of a particular leader will interact with the particular characteristics of a group and/or situation and it points to better solutions than we have been able to use previously. It begins to provide the general framework into which many of the smaller theories can be fitted. For example,

> Leaders differ in their concern for the group's goals and the means to achieve them. Those with strong concern are seen as task oriented (Fiedler 1967), concerned with production (Blake and Mouton 1964), in need of achievement (McClelland 1961) and production oriented (Katz et al. 1950) . . . Such leaders are likely to keep their distance psychologically from their followers and to be more cold and aloof. . . . When coupled with an inability to trust subordinates, such concern for production is likely to manifest itself in close controlling supervision (McGregor 1960).
>
> Leaders also differ in their concern about the group members in the extent to which they pursue a human relations approach and try to maintain friendly, supportive relations with followers (Katz et. al. 1950), concern for people (Blake and Mouton 1964), and in need of affiliation (McClelland 1961). Usually associated with a relations orientation is a sense of trust in subordinates, less felt need to control them, and more general than close supervision (McGregor 1960). (As noted in Bass 1981, p. 331)

2.3.2.2 Disadvantages of contingency-situational theories.   These theoretical models are helpful within the technical organization, but as is the case with all theories, there are unique situations that no theory can handle exactly. These unique situations might be manifested as changes that affect the technical organization in different ways. For example, changes in technology (e.g., from a typewriter to a computer terminal) or in the level of economic uncertainty (e.g., crude oil prices have tripled within one year and we make petrochemical-based plastic parts) are not handled well so far and these can occur more quickly, causing sudden situational changes in the technical operations.

It seems to me that the Hersey-Blanchard model approach tends to handle this problem of rapid change better than most, but it proposes that the total organization itself changes and matures, with required consequent changes in the leader's behaviors as it does so. However, technical organizations are not monolithic; they can have different degrees of maturity at the same time within themselves. There is probably more certainty or maturity in the manufacturing engineering department and less in the advanced R&D group. Manufacturing (which might be defined as "maturity") requires high task orientation and lower human relationships orientation, whereas the unit production of the R&D group requires the opposite (the organic operations of "less maturity").

3.0 THE LEADER'S BEHAVIOR AND THE CHANGED SITUATION

And now we must consider the effect upon the leader's behavior of a changing situation. It takes a long time to change anyone's behaviors permanently. We are all products of both our environment and our heritage. Changing human beings as quickly as the organizational situation changes is a difficult task indeed because of these semihidden mental processes that we all carry with us. It seems to me that individual change can occur with

1. Trauma (either physical or mental),

2. Therapy, or

3. Changing of organizationally required behaviors by transforming the person's job in the organizational structure.

However, since physical trauma is not usually acceptable in most technical organizations ("The flogging of engineers will continue until morale improves around here.") and mental trauma can occur only with the permission of the person being traumatized (i.e., If you don't recognize that somebody called you a "bad" name, it doesn't affect you. You've got to participate in mental trauma. No participation—no trauma.), trauma is an uncertain tool, at best. Therapy (providing assistance to the person in order to support better mental health) is not considered by me to customarily be a part of management's prerogatives. (Although I've had some disagreements both with many technical supervisors and with human relations professionals about this, I've always felt that "therapy" should be administered only by those who have been appropriately licensed; that is, the "state" qualifies them, not the company, and it should be received only by people who agree to undergo those processes.) Therefore, changing required behaviors by transforming the person's job is the only practical path left. The job may be transformed in several ways.

1. The leader may be transferred to a different technical group that has a different organizational culture.

2. The leader may be promoted into another job with a higher level of uncertainty.

3. The leader's job description, authority, and/or responsibilities may be changed.

Assuming that the transfer of a leader to a different technical group is *not* intended by senior management to radically change that group, when the organizational culture of that new group supports those who wear casual clothes to work, the transferred leader who shows up *consistently* in a double-breasted suit with matching shirt and tie is sending a message to the situation (i.e., the new group). The message may be that the leader is not about to change his/her leadership behaviors to fit the new situation. A perceptive leader would probably begin to wear casual clothes within a short time period as that is what this situation requires.

But if the intent of the transfer *was* to change the internal culture of that group, the leader might maintain his/her dress code and thereby send another message. "I am not going to change but this group will." Of course, the transfer of any leader into a new situation causes changes in the perceptions of subordinates (i.e., the situation), but the perceptive leader should determine before the fact what the purpose of the transfer was and act appropriately. This means defining the part of your personal management theory that might apply such as Fiedler, Hersey and Blanchard, etc., or any other theory that you feel would fit. If no theory, in your opinion, fits closely enough, try the "scientific" method; that is,

1. Define the new management job.

2. Develop a general explanation/forecast of what has to be done.

3. Try it and then correct your behaviors as time goes by.

If the leader is promoted, obviously the situation has changed. In some cases, the promoted leader may not recognize the reason for promotion. Even when politics and social success are considered in addition to technical competence, etc., the promotion probably occurred because of someone's forecast that superior performance (however measured) in a prior position would be an indication of future superior performance in the promoted position. The "old" job is gone; there are new behaviors to be exhibited because the new job has *more uncertainty*. This is the time to change one's behaviors. The above suggestions still apply.

If the leader's job description is changed, the "new" job should be treated the same way as if a promotion had occurred. The responses therefore are similar.

Therefore, when there are changes in the internal environment or in the culture within the technical organization, there may be equivalent changes necessary in the leadership behaviors. But in order to change leadership behaviors, the leader must be able to define how the organizational changes affect required leadership behaviors. You have to first define what it is that you are required to do.

3.1 Change Behaviors: Change Thinking

Obviously, when using the ideas in contingency-situational theory, we recognize that subordinate behavior is a major part of the situation, and therefore the situation will change because of the changed perceptions of organizational participants. In other words, in all cases change your leadership behavior and you change both your and others' thinking. That is a bit different from the usual approach of change thinking to change behavior, but we can never be sure if thinking has changed. We can observe behavioral change. I suggest, therefore, that if you can change behavior, thinking will follow.

3.1.1 Prescriptions: Technical Operations Leadership

The following prescriptions are specifically intended to match typical situations in technical functions. They assume that if it is possible to define the general situation, it is possible to modify not only your own behavior/thinking but in some ways also the situation to fit your unique situation and improve all the consequent management behaviors. These prescriptions should be used as noted before in the scientific method. As one modification of that method:

1. Gather appropriate data.

2. Analyze the data to determine what the problem is (i.e., definition and diagnosis).

3. Modify those general prescriptions that are applicable because of your diagnosis.

4. Try your prescription. And finally, whether it succeeds or not, because there are always minor variances from your predictions,

5. Start over again at 1.

This process supports personal theory building (and prescriptions or hypotheses that flow from your theory) that is grounded in the empirical data from the organization. This personal theory-building process follows a type of "grounded" theory construction (Bass, 1981, p. 26; Glaser and Strauss 1967) that is often used by practical managers and leaders. It isn't the kind of process that theorists in the physical sciences might accept, because it has few experimental controls and is not really replicable. However, it has been very useful in the social sciences and can be very useful to you, because it applies to your situation. Situations are part of a social science system.

According to the concepts proposed earlier in the chapter on structure, technical operations can be generally divided into two related situations: functions and project operations. If we are to use leadership models that are applicable, we have to develop two generally prescriptive models applicable to these two different but typical structural situations. However, we understand that these two prescriptive models are only prototypes, and your real situation may require extensive modifications. These models are also based partly upon my background and experience in technical organizations. We start with functional operations.

3.1.1.1 Functional leadership. This involves defining the group and the situation, and then applying several of our researchers' theories.

Defining the group: I would assume that most people in technical jobs have been through some type of structured training process. This provides varying degrees of "maturity" (as defined in the Hersey and Blanchard model). They have probably learned to discipline their thinking and behavior (if you don't learn that, you just don't graduate from most technical curricula), and they are generally aware of the contributions that they are making to the firm. It is possible that they have the independence of the traditional journeyman technician. Generally, their group has some degree of group cohesiveness or identity. They think more like each other in their particular group than they do like members of other functional groups in the same organization (Lawrence and Lorsch, 1967).

Defining the situation: In technical jobs there is more uncertainty in the task-environments than in most of the rest of the organization. It may not be at the highest level for the entire organization (although that could happen in some marketing or advanced R&D functions), but it is higher than the organizational average. There are many tasks, however, that are repetitive, and these tend to lower overall uncertainty. Those tasks typically deal with production support, maintenance, standards, quality control, and evolutionary changes and improvements in products and processes.

By definition, most of us who have managed technical operations have been trained throughout our academic experiences to operate in a task-, rather than relations-oriented mode. The basic undergraduate courses in mathematics, physics, chemistry, etc., demand that the student provide, through logic, only one answer. The instructor isn't really interested in a creative approach to that answer to the calculus problem. Most of the time, there is a specific answer required on the test and it must be provided only one way. If this is the training path that you have followed in your academic period(s), we can assume that you, as the technical manager, approach the processes of leadership in a functional organization with an initial *disadvantage*. You have been trained in a task environment, but the functional situation does not have the higher levels of uncertainty of the project situation. Therefore the concentration on training and continuous improvement of technical subordinates generally appears to need a relations-oriented approach. Of course, emergency situations may be different, requiring prompt, logical, task-oriented behaviors. The following models seem to show this.

Defining the leader: We now apply our three theoretical models.

Using Fiedler: When the leader initially assumes a leadership position over an operating functional group, the leader-member relations could be moderately poor at first, the task well structured, and the leader's position power strong. This would be equivalent to category V (see the chart in Figure 11-2) and require a high-LPC or relations-oriented leader. However, when leader member relationships improve with time or design problems suddenly occur or there are changes in the organizational environment, with the tasks becoming less structured, a different leadership pattern is required. It's interesting that according to this theory, when the situation is very good or very bad for the leader, a task-oriented behavior is suggested. When the situation is moderate for the leader, a relations-oriented behavior is suggested.

According to Fiedler, the changed situation as noted in the previous paragraph would then require a task orientation in the leader. Usually, this task-oriented change is temporary in functional situations as new problems are solved and then filed into repeatable categories. If the relations-oriented leader recognizes the change as temporary, and is capable of temporarily assuming a task role and then resuming a relations-oriented role, that leader's relations behaviors are again matched with the situation in an optimum leadership pattern. This appears to be possible, just as it is possible for task-oriented leaders to become temporarily relations oriented, if required. The important word is *temporarily*. Fiedler's general model seems (to me) to be satisfactory when applied specifically to functional technical organizations. Here, leaders should be relations oriented, as a normal behavior.

Using Blake and Mouton: The initial approach for a newly appointed functional (i.e., "permanent") technical manager should be "concern for relationships" until a positive human interaction is established. That seems to me to be the first goal, because the manager will probably be with the group for quite awhile. The next step is a change in behavior to a concern for production until a satisfactory production output is reached, and then a concern for both, or, as the authors describe it, a 9,9 approach. The path of optimum leadership behavior is directed upward on the graph almost sinusoidally with the curve of behavior measured over time moving up and down with a dampening decrease amplitude until the leader gets to the 9,9 position. This model is also applicable to leading technical functions, but it doesn't provide for rapid changes in the situation or in the uncertainty as Fiedler's does. It is concerned with only two variables: relationships and production.

Using Hersey and Blanchard: Initially the functional group would require a high relationship and moderate task orientation if we were to follow our prior scenario. With time, "maturity" in the situation increases, thereby requiring less relationship, and leadership that provides subordinates with more self-direction. With increased uncertainty, there would be more emphasis on task behaviors. This model is the most difficult to apply, because it assumes a close and correctly interpreted diagnosis of the maturity of the followers and an equivalently correct set of behaviors exhibited by the leader. It provides fewer allowances for externally caused changes: just for those related to the people in the group being managed.

3.1.1.2 Summary: Functional prescriptions. As I interpret these models, the behaviors most appropriate for long-term success in leading functional technical operations would be leadership behaviors that are relations oriented, with an emphasis on supportive continual training and updating of subordinates' skills. There would be occasional limited leadership changes into a task-oriented behavior intended to solve specific short-term problems, but leadership behaviors would have to be consistent and predictable over time. This is primarily a mechanistic situation.

3.1.1.3 Project leadership. Project leadership also involves defining the group situation, and applying our theories.

Defining the group: Although we can assume that the same objective training has been imposed on the people who work in projects as in functions, we cannot always assume that they were very proficient in project operations because each project is different and includes different people, tasks and possible leaders. This always occurs.—Project operations are new for each project, and a new structural environment has to be created. Of course, if the same team is given sequential projects, they will be somewhat similar to functional groups as their set of people and structures are unchanged. Usually, however, new groups are formed for each project. Even if people are quite familiar with project operations in general, they may not have interacted in this specific project organization before, and each project is new by definition. Members of a project group are more independent, have lower group cohesiveness, and are probably more different from each other than functional employees (Lawrence and Lorsch 1967). In addition, there hasn't been enough time for all of them to develop an organizational culture or a relationship with the project manager.

Defining the situation: There is high uncertainty at the beginning of a project with respect to both the project tasks and the team-human interactions. If the declining backward S-shaped curve for uncertainty vs. time (see Figure 11-4) is reasonable, uncertainties probably tend to decrease more quickly at a point that is about 25–30% into the project. Then, after the middle of the project passes and project task unknowns continue to decrease as design and production problems are solved, uncertainty (as perceived by participants) may begin to rise for human interactions.

Therefore, since we are always concerned with human interactions, the uncertainty (as viewed by project participants) would probably initially be high in projects, drop to moderate at about midpoint, and, if successful, continue to drop *for the group* as the project ends. For individuals on the project, it might spike up as the project end approaches and people begin to wonder about the next project (or if there will even be one). The declining S-shaped curve of uncertainty vs. time (Figure 11-4) refers only to the total project situational uncertainty that is concerned with reaching specified project goals and not neces-

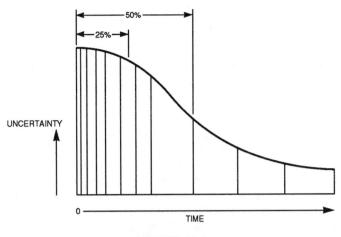

FIGURE 11-4

sarily that of the individual participants' perceptions of their own uncertainties. That, as noted, could be different as the project ends. The total project's uncertainty curve, which means meeting the project's goals, continues to decrease; some of the participants' uncertainties may rise depending on personal viewpoints. (By the way, Figure 11-4 also shows that there should be frequent feedback reports in the beginning when uncertainty is high and fewer reports as uncertainty drops.)

Defining the leader: Using the three theoretical models noted above, the suggested leadership behavior recommendations are quite different.

Using Fiedler: Because all the conditions at the start of a project usually seem to be negative for the leader—poor leader-member relations, task structure, and position power—a low-LPC leader would be appropriate. The expected *task orientation* of the technically trained person is necessary and an asset to organize, define, and, in general, get the project moving. This would probably fall into category VIII of the Fiedler graph (Figure 11-2). As project task problems are solved and uncertainty decreases in the middle of the project, the task structure increases, requiring a change to *relationship-oriented* behaviors. Then as the project begins to end, and if leader-member relations have improved, there is again a need for a *task orientation*.

Using Blake and Mouton: There is an initial need for concern for production. If that is satisfied and the project moves along, the general recommendations that Blake and Mouton make for a 9,9 (i.e., concern both for relationships and production) become useful in the middle; concern for relationships should occur toward the end. These predictions conflict with those of Fiedler's model.

Using Hersey and Blanchard: The behavior would be initially high task, then high relationship, and, as the project began to end, very little leadership is required. This also conflicts with Fiedler.

3.1.1.4 Summary: Project prescriptions. According to my interpretation of these models, and fitting them to my experience in project operations, I would suggest that the leadership patterns should be task oriented as an initial management behavior, change to those that emphasize relationships in the project middle, and then change back to task during the final phases. That's not easy to do if the project has an extended life. Deliberately changing one's behavior to fit a changing situation requires unusual personal discipline and patience, which are usually in short supply when uncertainty begins to rise. That would, however, be my prescription for the optimal leadership patterns in managing projects.

There is an interesting extrapolation to this prescription that would apply if you were managing several projects, all of which were in different stages of uncertainty. To handle all correctly within the same day would probably require a multiple personality in one person, and that surely is not the usual description of a technical manager! What it seems to mean is that whatever dominant type you are, task or relations oriented, there are occasions when you have to be able to use other behaviors, *at least for a short time*. So it might be useful to consider what is required when facing different stages of your projects. That is really the basis of this prescriptive hypothesis.

Before using these or other recommendations, you should thoroughly evaluate the applicability to your situation. There has never been a controlled, in-depth replicative research program or testing in all situations because this is part of social science with all its problems of changing time, people, and other variables. Thus, it is wise to consider that these general recommendations for changing effective leadership behaviors in functional and in project operations are based primarily upon my nonreplicable personal experience in the field and extensive anecdotal empirical confirming data gathered from others. If the data, the assumptions, and the descriptions of the people and the situation do not hold, the diagnoses and prescriptions will fail.

Several other considerations should be kept in mind.

1. Any relationships between leaders and subordinates are built over time and involve multiple exchanges between the two. Those exchanges include both tangibles, such as rewards and compensations, and nontangibles, such as commendations and approvals. They affect the relationships, which change

with time and affect both the leader and the subordinates. Behaviors, in most technical organizations, are difficult to permanently change, even though that's the suggested management route. Just as it is virtually impossible for a leader to change from task to relationship permanently, it is just as difficult for groups to build internally cohesive and predictable behaviors that approve a leader quickly.

2. There is no assumption in the above prescription of leader behaviors that an either-or behavior is required. Task-oriented leaders seem to be capable of short term relations-oriented behaviors to achieve their ends, just as relations-oriented leaders are capable of acting in task-oriented ways. The one research finding on middle LPC leaders (Kennedy 1982), indicates that a middle-of-the-road could also function well.

3. These prescriptions, in general, match two fundamentally different but related organizational models, one directed toward nonrepetitive major problem solving, innovation, or unique decision making (relationships and organic structures) and the other toward disciplined performance or production (tasks and mechanistic structures).

The situation leader-subordinate interaction is a "gestalt" or total process that cannot easily be dissociated for analytical purposes. The process may even be a creation of the observer. Those could probably be major factors in the inability of both universal and situational theories to explain all the observed behaviors. We have separated some of the parts here merely for instructional purposes, and have pointed out that contingency-situational theory is more applicable to technical operations than others. It can never be the entire answer—that answer depends upon you and you are somewhat of a variable.

3.2 After Diagnosis: The Training

It is very difficult to permanently change the behavior of leaders or potential leaders.

> A person's leader style . . . reflects the individual's basic motivational and need structure. At best, it takes one, two, or three years of intensive psychotherapy to effect lasting changes in personality structure. It is difficult to see how we can change in more than a few cases an equally important set of core values in a few hours of lectures and role playing or even in the course of a more intensive training program of one or two weeks. (Fiedler 1967, p. 248)

It seems to be much easier to change the organizational situation than the leader's style. The job can always be changed by organizational rules. As an example, we have the newly appointed project engineer who has had his or her job changed drastically from dependency in a functional group to relative independence in the project situation. The engineer has just been given the authority to add or remove people from the project, after serving in a staff position with none of this power. The position power has been increased, and according to most of the models, the leader requirements have become more relationship oriented.

Then, considering what was said before about projects, the engineer's behavior should be changed during the course of the project life. Initially it should be task oriented (project beginning), then relationship oriented (project middle), and finally revert back to being task oriented (project end).

Therefore, if the people on the project can feel confidence about new projects coming up when this one ends, they will probably continue to be relationship oriented when this one ends. According to these new contingencies, the project manager is supposed to behave like a functional manager in the middle of a project after the initial organizing or task phases of the project beginnings are completed. Changing behaviors in organizations is usually attempted on a formal basis through training. After the analysis of the difficulties of changing behaviors, the question becomes: Does training really work?

There are reports from all the contingency-situational researchers that we have discussed (Fiedler, Hersey and Blanchard, and Blake and Mouton) of successes in training managers to fit the requirements of the situation as they have defined it. Fiedler et al. (1976) concluded that there were supportive results for participants who had gone through the four- to six-hour self-paced programmed instruction workbook titled "Leader Match." Hersey and his associates provide industrial training for many organizations, and there are anecdotal reports of success as a result of it.

In a more controlled research study, there was an interesting report about using the Hersey and Blanchard system for leadership training (Hart 1975). The report stated that the trainees perceived themselves as better able to take control over their lives, become more active, and make better decisions. Blake et al. (1964) concluded that there was an increase in positive managerial attitudes and an improvement in departmental productivity as a result of grid training to achieve a 9,9 framework.

One major problem is reported with all this leadership training. It is nicely quoted as follows:

> There is evidence that leadership training programs result in some behavioral and attitudinal changes, although these changes are dependent upon a supportive organizational climate. (Schein and Bennis 1965).

A supportive climate includes the ability to practice what has been learned on the job and to receive valuable feedback or corrected reinforcement for these different practices. Perhaps the most important kind of reinforcement is that received from one's immediate supervisor. Researchers (Haire 1948) have often made the point that for management or leadership training to be effective, the entire management group should be trained. It is nonproductive and even possibly destructive for lower levels of management to be trained in a supervisory style that is incompatible with that of upper management.

This was further supported by some applicable research on the practicality of committing funds for individual management development that includes leadership training. A major conclusion was that the primary focus should be on promoting the development and institutionalization of specific organizational practices before committing resources for individual development purposes, because the environment of the organization itself constrains what leaders can do (Dreilinger et al. 1982, p. 70).

In one organization with which I consulted, the lower levels of leaders were trained to be supportive, creative, and relations oriented when dealing with their technical personnel. The positive effects of this training stopped short when a memo arrived from the engineering vice president stating that he wanted as much creativity and innovation as possible, but that any changes that did not agree with existing product lines or personnel procedures had to be approved beforehand by him. In other words, ". . . beginning immediately, you are directed to be more creative, but only within the limitations that I have established."

To put this in more appropriate research-oriented language, the results of training are unsatisfactory when they collide with a top management group that is insensitive to the suggestions for change that the training was intended to initiate (Sykes 1962). Therefore, it seems reasonable that training to achieve better leadership behaviors can be effective provided two general criteria are satisfied.

1. The best job situation is developed as perceived by the organizational participants.

2. Top management either goes through the same training or commits itself to supporting the results of that training.

In following chapters, we will deal with training again, but from a different viewpoint: that of achieving total organizational change rather than individual change. As with many of the other subjects that we have covered, training can be analyzed from several positions.

4.0 ETHICS AND LEADERSHIP:
DEFINITIONS AND SUGGESTIONS

Leadership involves decision making. Although there may be parts of the organization that are relatively less structured, the technical groups have certain restraints on their decision-making processes that these other groups may not have. We deal primarily with one opponent, Mother Nature, and she doesn't care about us nor will she change her rules to fit our needs. Therefore, in most cases involving technical considerations our decisions have to be clear and defensible with no modifications that are primarily due to requirements of the company. No matter what senior management wants, if those needs conflict with the laws of nature, it won't happen. But if they conflict with the law of the land, the technical professional may be held personally responsible. It is sometimes difficult to get this across.

That special responsibility occurs because technical personnel and especially technical managers can be held personally liable for decisions that they make on the job even though those decisions were made while the person and/or manager was on some corporate organizational payroll. It is the definition of the "professional" that is crucial in this situation. The "professional" is bound to respect the scientific laws, the legalities of the country and the ethical rules of the profession. Therefore the "professional" can be exposed to personal liabilities if any of these constraints are violated. While many of us are aware of this, we are not familiar with the fact that our decisions can be measured and are determined to be acceptable or not depending upon the competency of our peers. (whoever they are) For example,

> The professional can usually be held liable for damages only if he or she performed below the level of expertise in the immediate community. Increasingly "community" has come to mean the entire inhabited world. (Kolb and Ross 1980, p. 85)

With a "community" that large, it becomes obvious that technical concerns cannot be modified by corporate directives (even when we're not sure of all the answers ourselves). Occasionally, technical personnel may be pressured into making poor or even unethical decisions. This acceptance of the direction of seniors was explored in a famous study by Milgrim (1974).

Briefly, the experimental situation involved two people and the experimenter. The experimenter was dressed in a white lab coat and was expected to be the "authority" figure. One of the two people was a confederate (i.e., the "learner") and the other was the "subject." The subject was unaware that the other person, the learner, was an associate of the experimenter. The experimental results were unusual. When the subject was in an adjoining room separated from the learner only by a window, the subject was willing to follow the "orders" of the experimenter and turn a dial on an electrical device to give the learner electrical shocks when the learner misspelled a word. The learner was strapped in a chair and writhed in (obvious) agony whenever the electrical shock was administered. Of course, the dial did not administer any electrical current but the subject didn't know that. The subject was allowed to hear the loud complaints of the learner and see the learner's struggles.

Milgram explained this by saying that people have a strong tendency to willingly abandon their personal accountability when they receive orders from persons of apparent authority. Although there may be little physical similarity between this experiment and the organizational situation, the response to authority can often be a problem for technical personnel when the orders received conflict either with their technical competence or their knowledge of the law.

Another problem for technical personnel arises because we never completely know what all the "unknowns" in any new design are. Every new design, process or product has a potential for some future failure that we are unaware of during the design stage. If the automotive engineers were seers, they could forecast problems and design them away, thereby eliminating any of our automobile recalls to correct defects found in the field. But they didn't eliminate these hidden defects because they, as we, didn't know what they were at the time. We do know that every new (and even some old) product might have a hidden problem that could not be foreseen. We just do the best we know how and are then measured by the level of expertise of our peers. A prime example of this is the emphasis on product safety.

> Product safety is no longer a matter of conscience and good will. It is a professional specialty founded in knowledge of prediction techniques, fault tree analysis, failure modes, effects analysis, and operator-design interface. Some experts have estimated that the cost of poor quality can amount to 25 percent of sales at manufacturing companies and up to 40 percent of operating costs at service companies. (Florman 1987, p. 93)

Although product safety is obviously a major concern, there are other areas where we might have to make decisions that are ethically based. In some instances, the situation is not clear. The following questions could be guidelines and therefore might be helpful.

1. How did this situation occur and how would you define it now, accurately?

2. Could your decision injure someone?

3. Would there be any problem in discussing this with your boss, the CEO or an outsider?

4. Why did you make this decision?

5. To whom are you loyal—the company, the user, your family—society and how would you handle conflicting loyalties?

6. When would you agree with exceptions to your decisions?

The typical professional with whom many of us can relate would be our family physician. Even though that professional may know you for years, at every professional visit, the physician updates the notes on your individual diagnosis, progress and general condition. It is the way of the professional and is, to a great extent, replicated in well-managed technical groups. Documentation in these groups includes:

> "minutes of Design Reviews, basic calculations and preliminary designs, drawings, design histories of related products, testing data, records on the establishment of quality control protocols and production specifications, service manuals, installation instructions, correspondence with marketing and production and feedback from product users." (Kolb and Ross 1980, p. 91).

In rare instances, there can be a problem between the directions of the organization and the ethics of the technical professional as you understand them. Being caught between two forces is difficult, and occasionally the technical manager must make a decision. When that decision conflicts with the directions of senior management and involves going public with the problem, it is called "whistleblowing."

4.1 Handling the Impossible Situation: Whistleblowing

There are some general recommendations for processes to be followed when "whistleblowing" is an alternative, according to the following authors.

> Ideally codes of ethics and other policies that relate to wrongdoing should spell out more specifically—but without appropriate rigidity—what activities are considered wrong. (p. 67) If a supervisor is directly involved in the wrongdoing or feels compelled to take management's side, then the employee would be forced to think twice before proceeding higher in the hierarchy. In such cases, communication systems that circumvent the chain of command, such as use of an Ombudsman can help. (p. 68) . . . the most powerful incentive a company can offer potential whistleblowers is its willingness to correct wrongdoing. (p. 69) (Miceli and Near 1994)

Unfortunately, "whistleblowing" (although protected in most cases by law) can have negative consequences. According to one engineer, those consequences included losing his job and quite a bit of his financial resources. This is a summary of what happened.

One of the problems that this individual faced in a new position was the unwillingness, as a government employee, to continue to allocate work to preferred contractors without competitive bidding, as required by law. He was demoted for this. Fortunately, he was called before a government subcommittee and after relating his problem, he was returned to an equivalent position. In this new position, he was responsible for the safety and repair of buildings and certification that they were safe for human occupancy. When he refused to approve unsafe conditions in those buildings because the funds allocated for repairs were never released during a cost reduction program established by his superiors, he was fired. He has classified several retaliation methods that he feels others can use against "whistleblowers." These are:

1. Make the whistleblower the problem and fire him.

2. Isolate the whistleblower by blocking access to required information.

3. Give the whistleblower an "impossible" task and then fire him for nonperformance.

4. Make outrageous charges personal conduct of the whistleblower.

5. Eliminate the whistleblower's job even though the company is hiring other people.

6. Withdraw any data access, staff and research capacities. (Berube 1988)

And according to the following research, this general problem of ethical decision making continues.

Virtue may be its own reward, but it seems to stop there, or two Columbia University business school professors discovered in surveying 25 years of alumni experience, seeking to demonstrate the value of ethics training.

Of the 1,070 alumni responding from the classes of 1953 through 1987, 40% said they had been implicitly or explicitly rewarded for taking some action they considered to be ethically troubling—twice as many as were rewarded in some way for refusing to something that was ethically wrong.

What's more, 31% of those who refused to take some ethically troubling action said that they had been penalized for their choice. The penalties ranged from outright punishment to a vague sense that the person's status within the company had been diminished. . . .

"I was really surprised to see these results," says John Thomas Delaney, an associate business professor at University of Iowa and at Columbia, and one of the study's authors. Despite a flood of interest in ethics education and in corporate value statements, Dr. Delaney says that the survey demonstrates that "there are cases where people get rewarded for doing these things, and companies don't want to acknowledge it." (*Wall Street Journal* 25 Jan. 1990)

It can be a difficult situation. We are considered by our society to be "professionals" and therefore we must be responsible for our own actions even when employed within a corporate framework.

5.0 REVIEW AND PRACTICAL TIPS

5.1 Leadership

This chapter has reviewed some of the pertinent leadership theories that seem to apply to technical organizations. It has attempted to describe them considering both the situations and the leadership patterns that should fit the two principal operations of most technical departments: functions and projects. Guidelines for organizational and personal diagnosis have been suggested and general prescriptions developed. The pluses and minuses of training individual people in leadership behaviors have been covered. As usual, the application of this material is up to you.

But it is not possible to train people to be leaders in all situations, according to the following author.

> I do not see any form of leader behavior as optimal for all situations. The contributions of a leader's actions to the effectiveness of his organization cannot be determined without considering the nature of the situation in which that behavior is displayed. (p. 502). . . . Our findings strongly suggest that decisions made by typical managers are more likely to prove ineffective because of deficiencies in acceptance by subordinates rather than deficiencies in decision quality. (p. 508) (Vroom 1983)

But we knew this from our studies of the contingency-situational theories. All of those theories include the behaviors of subordinates as part of the situation. There are some other "obvious" findings that many of us have experienced, but I find it interesting when there is some research that is confirming, such as

> Effective leaders are very good at building alliances and creating commitments so that others will share their vision. They possess great team building skills . . . Leaders make the empowerment of followers seem deceptively simple. The trick is to express high performance expectations. (p. 75) . . .
> Clinical observation confirms that even the most successful organizational leaders are not exactly rational, sensible and dependable human beings, but, in fact, are prone to irrational behavior . . . (p. 79) . . .
> In their personal relationships, leaders who are wary of the dangers of hubris should bear in mind what I term the three "H's" of leadership: humility, humanity and a good sense of humor. (p. 88) (Kets de Vries 1994)

There are other equally interesting research findings about leadership behaviors in general. For example, when under stress, leaders often rely on past experience and don't make the best use of their intelligence. And because most assumptions of rationality demonstrated by calm, unhurried, clear and logical decision making are not supported in actual practice, decision makers' attitudes and commitment are occasionally created retrospectively by consequent justifications (House and Singh 1987).

5.2 Ethics

The need to define and to follow ethical decision-making processes is a constant in every organization but is especially important in technical groups. We are professionals and therefore are responsible not only to make the best decisions possible (as measured by the expertise of our peers) to our organizations but also to our customers and associates.

> . . . an engineer who starts out by being conscientious must end up by being honest, since competent engineering, excellent engineering, is in its very nature the pursuit of truth. A conscientious engineer, by definition, cannot falsify test reports or intentionally overlook questionable data, cannot in any way evade the facts. (Florman 1987, p. 104)

Developing precise prescriptions to effect improved leadership and ethical decision making for others is almost impossible. For example, although we think the laws on theft are very clear, every day juries and judges make decisions that implement these laws. If they were absolutely clear for every instance, those juries and judges would not be necessary, but circumstances change. Therefore, a straightforward procedure that I have often found helpful in developing my own theory and consequent behaviors is:

1. Try to define what presently exists.

2. Determine what should exist.

3. Determine how to go from what exists to what should exist.

For example, you might want to quantify the "what is" and the "what should be" subjectively, so that you'll have some guidelines on what to do first, then second, etc., to attain your goal. Using a subjective value scale from 1 to 10 (with 10 being the best) might give you the following guidelines (Table 11-2).

This could mean that problem 2 is most important. One possible way to improve would be to redesign the management information system to get faster

Table 11-2. Quantification of What Exists and What Should Exist

Typical Definitions	What is	What Should Be	Difference
1. Projects are well-defined initially	7	10	3
2. Reports are delivered promptly	5	9	4
3. Problems are uncovered immediately	4	7	3

data. You may have your own interpretation, but the idea is there. This method is repeated for you in Chapter 12.

This procedure is less costly to put on paper than one that is attempted without any kind of a blueprint or plan. Define your theory, set up your hypotheses on paper, then try them in the organization. Just the process of writing things down helps the behavior-cognition interactive process, which is more simply called "learning."

6.0 "ONE-SENTENCE" SUGGESTIONS

1. It is not what the leader thinks that is important, it is his/her behavior and how others interpret that behavior.

2. It is possible to behave "differently" for short time periods. For example, a predominantly relations-oriented leader might act in a task-oriented manner during a project definition meeting.

3. Leadership behaviors should fit the situation. The difficulty is in deciding what the situation requires.

4. Lead and manage so that others can generally predict how you will act and what you expect of them.

5. Tell subordinates how you intend to behave as a leader. If you don't follow through, they'll remind you.

and finally: *The important thing is to know how to take all things quietly.* (Michael Faraday) and remember, honesty is the best policy because there is less competition. In addition, if you tell lies, you have to have a good memory. I don't have one so I tell the truth.

SUGGESTED ANSWERS TO CASE QUESTIONS

1. The leadership requirements seem to have moved from a task orientation to a relationship one (Fiedler), from an immature or young group to a mature one (Hersey and Blanchard), and from a strict concern for production to concerns for both production and people (Blake and Mouton). The leadership requirements involve more coordination of others and coaching tasks and less of actually doing the work. The R&D group has moved from a functional to a project orientation, because the tasks are no longer concentrated in one area, but are spread across departmental boundaries.

2. a. Defining the problem: Problems should be documented in some type of standard format, perhaps classified by major customer or industry, because requirements change across different product lines.

   b. Project leader's authority: The project leader should have the ability to select and relinquish the personnel needed to do the projects assigned. This is one of the better times to install project controls, such as the ability to open and close financial aspects of projects.

   c. Intergroup cooperation: One mechanism would be to install dual reporting. Another would be to schedule total project design reviews at predetermined intervals during each project life. These reviews would be formal meetings during which project problems and progress could be discussed.

3. Implementation of change requires communication and expertise. With regard to communication: What will the proposed design mean to participants and how will it benefit them? Ask the people affected for answers in order to determine what the proposed change means to them. With regard to expertise: The "experts" in this case would be Melanie and Mike. They should be selected as an ad hoc design team to develop proposals for the new system. What would happen next and how would they get their answers? The success of a system implementation is determined the same way that any forecasted achievement is measured: according to predetermined standards. What would a possible measurement be that would mean the successful achievement of a new scheduling system? How could it be measured?

If this had happened in your company, what would have occurred in Bob's office? At the staff meeting?

Do you have alternative answers?

REFERENCES

Barnard, C. I. **The Functions of an Executive.** Harvard Univ. Press, Cambridge, MA. 1938.

Bass, B. M. **Stogdill's Handbook of Leadership.** Free Press, New York. 1981.

Berube, B. G. "A Whistle-Blower's Perspective of Ethics in Engineering." **Engineering Education** February 1988, pp. 294–295.

Blake, R. R. and Mouton, J. S. **The Managerial Grid.** Gulf Publishing, Houston, TX. 1964.

Blake, R. R.; Mouton, J. S.; Barnes, J. S.; and Greiner, L. E. "Breakthrough in Organizational Development." **Harvard Business Review.** 1964. 42:133–155.

Blau, P. **Exchange and Power in Social Life.** John Wiley & Sons, New York. 1967.

Burns, T. and Stalker, G. M. **The Management of Innovation.** Quadrangle Books, Chicago, IL. 1961.

Doyle, J. "Executive Summary." **Academy of Management Executive.** 1994. 8(2):89–91.

Dreilinger, C.; McElheny, R.; Robinson, B.; and Rice, D. "Beyond the Myth of Leadership Style Training: Planned Organizational Change." **Training and Development Journal.** October 1982. 36(10):70–74.

Fiedler, F. E. **A Theory of Leadership Effectiveness.** McGraw-Hill, New York. 1967.

———. "Styles of Leadership." **Current Perspectives in Social Psychology,** Hollander, E. P., and Hunt, R. G. (eds.). (4th ed.). Oxford Univ. Press, New York. 1976.

Fiedler, F. E.; Chemers, M. M.; and Mahar, L. **Improving Leadership Effectiveness: The Leader Match Concept.** John Wiley & Sons. New York. 1976.

Florman, S. C. **The Civilized Engineer.** St. Martin's Press, New York. 1987.

Gattiker, U. E. **Technology Management in Organizations,** Sage, Newbury Park, CA. 1990.

Glaser, B. G., and Strauss, A. L. **The Discovery of Grounded Theory.** Aldine Press, Chicago, IL. 1967.

Haire, M. "Some Problems in Industrial Training." **Journal of Social Issues.** 1948. 4:41–47.

Hart, L. B. **Training Women to Be Effective Leaders.** Dissertation Abstracts International, 1975, 35, 1977.

Hersey, P., and Blanchard, K. H. "Life Cycle Theory of Leadership." **Training and Development Journal.** 1969. 23:26–34.

Hollander, E. P., and Julian, J. W. "Contemporary Trends in the Analysis of Leadership Process." **Current Perspectives in Social Psychology,** Hollander, E. P. and Hunt, R. C. (eds.). (4th ed.). Oxford Univ. Press, New York. 1976.

House, R. J. "A Path-Goal Theory of Leader Effectiveness." **Administrative Science Quarterly.** 1971. 16:321–338.

House, R. J., and Singh, J. V. "Organizational Behavior." **Annual Review of Psychology.** 1987. 38:669–718.

Katz, D.; Maccoby, N.; and Morse, N. C. **Productivity, Supervision and Morale in an Office Situation,** Univ. of Michigan Institute for Social Research, Ann Arbor, MI. 1950.

Kennedy, J. J., Jr. "Middle Lpc Leaders and the Contingency Model of Leadership Effectiveness." **Organizational Behavior and Human Performance.** August 1982. 30(1):1–14.

Kernberg, O. F. "Regression in Organizational Leadership." **The Irrational Executive, Psychoanalytic Studies in Management,** Kets deVries, M. F.R. (ed.). International Universities Press, New York. 1984.

Kets de Vries, M.F.R. "The Leadership Mystique." **Academy of Management Executive.** 1994. 8(2):73–92.

Kolb, J., and Ross, S. S. **Product Safety and Liability: A Desk Reference.** McGraw-Hill, New York. 1980.

Kotter, J. P. "What Leaders Really Do." **Harvard Business Review.** (May–June 1990) 103–111.

Lawrence, P., and Lorsch, J. **Organization and Environment.** Harvard Univ. Div. of Research, Grad. School of Business, Boston, MA. 1967.

Likert, R. **New Patterns of Management.** McGraw-Hill, New York. 1961.

Litwin, G. H., and Stringer, R. A. **Motivation and Organizational Climate.** Harvard Univ. Press, Cambridge, MA. 1968.

Luthans, F. "Successful vs. Effective Leaders." **The Academy of Management Executive.** 1988. 2(2):127–132.

Machiavelli, N. **The Prince and the Discourses.** Modern Library, New York. 1940.

McCall, M. W., Jr. "Leaders and Leadership: Of Substance and Shadow." **Perspectives on Behavior in Organizations,** Hackman, J. R.; Lawler, E. III; and Porter, L. W. (eds.). McGraw-Hill, New York. (2d ed.) 1983. pp. 476–485.

McClelland, D. C. **The Achieving Society.** Van Nostrand, Princeton, NJ. 1961.

McGregor, D. **The Human Side of Enterprise.** McGraw-Hill, New York. 1960.

Miceli, M. P., and Near, J. P. "Whistleblowing: Reaping the Benefits." **Academy of Management Executive.** 1994. 8(3):65–72.

Milgram, S. **Obedience to Authority.** Harper & Row, New York. 1974.

Mischel, W. **Personality and Assessment.** John Wiley & Sons, New York. 1968.

Schein, E. H., and Bennis, W. G. **Personal and Organizational Change Through Group Methods: The Laboratory Approach.** John Wiley & Sons, New York. 1965.

Sykes, A.J.M. "The Effect of a Supervisory Training Course in Changing Supervisors Perceptions of the Role of Management." **Human Relations.** 1962. 15;227–243.

Thompson, J. D. **Organizations in Action.** McGraw-Hill, New York. 1967.

Vroom, V. H. "Can Leaders Learn to Lead?" in **Perspectives on Behavior in Organizations,** Hackman, J. R.; Lawler, E. III; and Porter, L. W. (eds.). McGraw-Hill, New York. 2d ed. 1983. pp. 501–509.

Woodward, J. **Industrial Organization Theory and Practice,** Oxford University Press, London, 1965.

Zalesnick, A. "Charismatic and Consensus Leaders; A Psychological Comparison." **The Irrational Executive, Psychoanalytic Studies in Management,** Kets deVries, M. F. R. (ed.). International Universities Press, New York. 1984.

———. "Doing the Right Thing Has Its Repercussions." **The Wall Street Journal.** (25 January 1990).

CHAPTER 12

Repetitive Decision Systems Information: Tell Me What's Happening in the Organization

Case Study:

The Confused Manager

Cast
Tony Ogard: General manager
Sam Greenshades: Controller

The company was the hydraulic controls division of Monolith Industries. Tony Ogard, the new general manager, was sitting in his office. He was trying to make sense of the reports that he had received from the division controller through the interoffice mail. He couldn't understand what they were all about, and the instructions that accompanied them didn't seem to be consistent. He picked up his phone and asked Sam Greenshades, the controller, to meet with him that day. Later in the afternoon, at the exact time scheduled for the meeting, Sam came in and sat down.

Sam: Before you start on the analysis of these sheets, let me tell you that I had nothing to do with designing them. That's the way they want them up at corporate headquarters, and you know that I report to the corporate controller as well as to you. It's not easy, having two bosses, but I'm trying to satisfy you both. Now that we have that out of the way, what can I do for you?

Tony: Very interesting speech. I find it encouraging because you seem to believe that something is wrong, too. But let's go on and maybe we can sort this thing out. My reports show that our inventory in some of the smaller valves has increased over 100% since last month, and some have dropped by about 10%. Why should I care about percentages? An increase in a valve that sells for $20 to $40 indicates a 100% improvement, and one that sells for $300 whose price increases by 15% equals a change of $45, but that seems to mean less because of the low percentage figures.

Another thing that I don't understand is the detail in these reports. They cover everything that has happened in the division for the last month. Not only that, but when I phoned our chief engineer, Maria Wheatley, up in engineering and asked her about some variances in her overheads, she said that she'd get back to me as soon as she got the reports from you. Apparently she gets them after I do.

Sam: Well, that's how the previous general manager wanted things. He said that he wanted to keep on top of everything, and that he couldn't do that if the people who reported to him had more details than he had. He also wanted percentages because that's the way the president of the company reports to the stockholders and he felt that being consistent in his reports to the president was very important.

Tony: I also have this package of instructions from corporate finance that lists in detail how to do my business plan and then the budget to put the plan into operation. Apparently this is a one time per year operation. As part of the instructions, I found this memo from the executive vice president telling all the general managers that they spent too much money last year and that they've got to cut down by at least 20% this year. What's going on?

Sam: Well, we do follow an annual budgeting plan because we are publicly owned and we have to report to the stockholders on how their company is doing. The budget is often used in the president's address that goes along with the annual report. You know the sort of thing: "We'll be doing big things in this division or that division next year." Of course, for the last few years we've been hit by large increases in raw materials costs, so we haven't been able to hit those optimistic forecasts. When that happens everybody gets all excited. The general manager before you (who, by the way, is now a group vice president) used to fire people when they didn't meet the budgets he set out for them. He said that he wanted to run a tight ship and he did. He got things going.

Tony: I understand what you're saying, but I noticed that this division is being charged for the corporate expenses that have been allocated to it. Do they ever get cut? Can you find that out for me?

Sam: Look, Tony, you're new around here. Questions like that are not answered very well at corporate. If I called up my boss at headquarters and asked about it, I'd never hear the end of it. They say that they provide us with coordinating services and other valuable things and we should be grateful that they're around, especially when we can't meet our budgets and we have to go to them for money.

QUESTIONS

1. What should Tony do about designing an information system that can help?
2. Any suggestions for the frequency of the budgeting process? Should it be done more often than once a year? How? What would the additional administrative costs be?
3. What defense do managers have against budget cuts and/or directed budget expenditures that fit into the decision-making process as it is laid out?
4. What should Sam do? How can he do it? What would the results be?

1.0 OVERVIEW AND INTRODUCTION

So far, we have defined and described the three central components of our technical organizational model, developed some ideas about how to measure them, outlined several important interactions that they have with each other, and finally suggested how you might adapt them (if you choose to do so) in your own personal management theories. We have also discussed the overall concept of leadership and/or management as the behavioral (and inferred mental) processes that guide and support those three components and integrate them into a total organizational framework. That framework was then (for illustrative purposes and perhaps artificially) separated into two different managerial areas: that of functions and that of projects.

But the differences and similarities between these two managerial areas, functions and projects, have become apparent only within the past several decades. In the early part of this century (as noted before in this book),

1. People were considered as interchangeable, therefore organizations used this resource without attempting to upgrade or train.

2. Technology was based on a machine-like model, therefore management always concentrated on manufacturing by improving the "machine-controlled" time and decreasing the "human-controlled" time to decrease costs and improve profits.

3. Structure followed a *"one best way"* in a power-based, hierarchical model.

4. Leadership was dependent primarily upon the organizational position, and that position directed orders downward and received information on achieving those orders that flowed upward.

The organizers and the managers of these arrangements believed that there were universal designs that were the best possible to use for the formal structure. These designs were modeled on the social mores of the then existing master-servant culture. However, we now know that there is no longer only "one best way" but many alternative ways. Many semi-independent variables such as market growth and diversity, the passage of time, and changes in the economic, social and political environment have resulted in enormous changes in the successful technical organization. These changes began within the recent past, according to the following author.

When World War II came, however, we had no choice; we had to ask the workers. In the plants we had neither engineers nor psychologists nor foremen, they were all in uniform. And when we asked the workers, we found to our immense surprise as I still recollect that the workers were neither dumb oxen nor immature and maladjusted. They knew a great deal about the work they were doing, its logic and rhythm, the tools, the quality and so on. . . . In making and moving things, partnership with the responsible worker is, however, only the best way after all. Taylor's telling them worked, too, and quite well. In knowledge and service work, partnership with the responsible worker is the only way; nothing else will work at all. (p. 227)

. . . Consequently, all the managers in a plant will have to know and understand the entire process, just as the destroyer commander has to know and understand the tactical plan of the entire flotilla. . . . managers will have to think and act as a team member mindful of the performance of the whole. Above all, they will have to ask: What do the people running the other modules need to know about the characteristics, the capacity, the plans, and the performance of my unit? And what, in turn, do we in my module need to know about theirs? (p. 242)

Finally, a good deal of work will be done differently in the information based organization. Traditional departments will serve as guardians of standards, as centers for training and the assignment of specialists; they won't be where the work gets done. That will happen largely in task focused teams. (p. 348) (Drucker 1993)

Organizations that succeed now must have a supportive mix of the repetitively planned (i.e., the formal organization), and the unplanned, but acceptable (i.e., the informal organization), behaviors of people in an organizational situation. For example, the organization of a traditional production-oriented group could have a structure that was very formal, mechanistic, or machine-like, logical, and effective in delivering products and services at the lowest cost with the highest quality for the market. However, that organization could then have a nontraditional, engineering or technical management organized in an organic, creative, project-oriented structure in order to handle unpredictable problems and rapidly changing environments. In general, that management would include the novel, problem-oriented parts of the organization that are more susceptible to change, and able to respond effectively and quickly to rapid variations in the environment.

But responding to change, solving novel problems, maintaining existing organizational health, and developing new products for future organizational health is only possible with the advent of prompt and clear communication channels within the organization. Those channels are no longer bidirectional. Carrying order downward and responses upward is inadequate. Now, information flows are

1. *multidirectional:* You don't have to go through your boss's office so that he/she can get permission from someone else's boss to communicate with that other person in another division. You just pick up a telephone, send a fax, or even walk over there.

2. *socially based:* Friendships and social groups within a company can get information very quickly to other interested people. When you want to know if

that supplier has shipped those important components, you call your friend in the purchasing department.

3. carrier independent: If the usual way of getting data is through monthly printed reports, that carrier is now too slow for most technical groups. This means that written reports, oral reports, meetings and informal conversations are almost equivalent in their ability to process data and provide feedback.

These different information flows are not hierarchically or formally based. An effective information system provides every manager-leader-decision maker with data to help reduce uncertainty. And because uncertainty is the unknown in every decision, it is imperative that the information system be designed to be flexible because each manager may need data unlike that of anyone else. In one sense, information might even be considered to be part of the organization's "technology" component; however, in the interests of clarity, it is handled separately in this chapter. It is a vital tool of the company.

2.0 DEFINITION: WHY AND WHAT ARE INFORMATION SYSTEMS?

Information systems are the patterns used to communicate both among various knowledge workers in the internal organizational components and between the organization itself and its environment. These systems include both those patterns sanctioned by the formal organization and those used by the informal organization. There are two major, but intertwined, reasons for information systems: they support and communicate decisions, and their interpretation modifies human behavior.

Typically, information systems include formal documentation such as budgets, drawings, material requisitions, time sheets, standard operating procedures, internal memos and minutes of meetings, in addition to all the nonmemorable trivia that often avalanches into the "in" basket (and then, if you are well organized, goes expeditiously into the "out" or wastebasket). They also include the external documentation through which the company communicates with its environment, such as purchase orders, financial statements, employment applications and the press releases that forecast the company's "ever-increasing sales and profits."

In addition, there are all those oral messages including the "never-ending" phone calls, informal hallway discussions and formal meetings. These oral messages, as opposed to those that are documented, can be vital as they are more flexible and amenable to changes very quickly. This is illustrated by the astounding speed with which rumors about an impending management change, restructuring, or anything else of interest to "people" can spread throughout an organization. It is truly astounding.

2.1 Oral Information and Meetings

A fellow consultant told me about an information flow that occurred while he was working with a major aerospace firm several years ago. Upon entering one of the company's manufacturing plants, he was overheard discussing a new product design that another division of the company was contemplating. He intended to announce the new design at that meeting because the decision had been made only minutes before he reached the door of this factory. However, by the time the meeting started, almost everyone in the room had already been informed through the company "grapevine": the informal rumor processing network of "people."

Although the grapevine or the rumor mill is useful because it can tell everyone what the "knowledge workers" are talking about, it is more advantageous for management to meet with the knowledge workers to correct misinformation, gather data to decrease uncertainty and to maintain support for decisions that are reached. Meetings are also necessary to provide interaction among individuals in ongoing problem solving and decision making. But meetings are not intended to be haphazard (even though they occasionally seem that way to the attendees). They should be planned well. For example,

> A key question to consider when planning a meeting is simply this: Is this meeting really necessary? . . . The most important question you should ask is: "What is this meeting intended to achieve?" You can ask it in different ways—"What are the likely consequences of not holding it? When it is over, how shall I judge whether it was a success or a failure?" (Cleland and Kocaoglu 1981, p. 137)

When meetings are organized well, they are marvelous tools. When they are not, consider the following which was found recently on a company bulletin board:

> Are you lonely? Hold a meeting! You can see people, draw up organization charts, discuss vital topics, feel important, eat doughnuts, drink coffee and all on company time! Meetings—the practical alternative to actually doing work. (Anon.)

In contrast, however, the mechanics of a successful meeting are fairly straightforward:

1. *Distribute the agenda* beforehand and follow it during the meeting.
2. *Start and finish on time.* But provide enough time. A major problem may not take the ten minutes that you allocated to it. Be practical.
3. *Assign action items* to specific individuals with a follow-up date. Then follow up.
4. *Sum up orally at the end.* "Have we all agreed to the following wording for the minutes?"

And then read them.

An important point to consider is that a meeting is only one tool to transmit information. Its primary purpose is to secure interaction among people. If only one or two people have the needed data, don't call a big meeting. Talk to those few who can contribute.

When data passes through the many information systems, the variety of descriptions and the consequent descriptions could be almost infinite. Each system is a function of many variables, typically the operations of the particular organization, the people in it, the environment in which it operates, and the company's past history. The satisfactory information system accounts for many variables. Some of those that are generally applicable are (according to the following author):

—Commitment. Control of engineering activities depends a great deal on self imposed control through the personal commitment of the individual contributors. Without this commitment, controls will focus on dealing with excuses for slipping performance. The same commitment to the plan, its resources, and priorities must also come from senior management.

—Competence: The people implementing the plan must be professionally competent. The best plans and strongest commitments are useless unless the people have the capacity to perform.

—Measurability. If you can't measure, you can't control. Activities and milestones must have measurable results, which should be compared against the planned resources and schedules on an ongoing basis.

—Focus on key objectives. . . . Trivia should not be measured or controlled. It only masks the principal objectives and leads us into the "activity trap."

—Simplicity and adaptability. Management controls must be simple. The purpose is to identify problems early on and take corrective actions.

—Early problem detection. Controls should focus on problem prevention and early problem detection.

—Controlling authority. Management controls should be in the hands of the people who are accountable for the results. (Thamhain 1992, p. 183)

With all these suggestions, it would be difficult to design an information system from the beginning. That is not necessary in most cases. No matter what size the company is, there is one formal group of systems that always exists in one form or another and it predates all others in every organization. That group is the accounting system. Obviously, we cannot do more than cover generalities here but additional information may be found in most *management accounting* (*not* financial accounting) texts such as Horngren (1972). We want to understand how these systems work now, how they should work to satisfy management needs and then be able to develop applicable systems that support our unique needs in decision making. Many of the concepts we shall cover are common to most effective accounting systems. Therefore we start, as usual, with definition.

2.2 Accounting Systems and Technical Management

The following is one of the better definitions.

> The accounting system is the major quantitative information system in almost every organization. It should provide information for three broad purposes:
> 1. Internal reporting to managers, for use in planning and controlling routine operations,
> 2. Internal reporting to managers for use in making non routine decisions and in formulating major plans and policies,
> . . . The data raise and help answer three basic questions:
>   a. Scorecard questions: Am I doing well or badly?
>   b. Attention directing questions: What problems shall I look into?
>   c. Problem solving questions: Of the several ways of doing this job, which is the best?
> 3. External reporting to stockholders, governments and other outside parties.
> (Horngren 1972, pp. 3,8)

2.2.1 What Is Now

This definition covers most of the intended uses of accounting systems. Unfortunately, too often the emphasis is on the third purpose, reporting to stockholders, governments, etc., and it operates to the detriment of the first two purposes. It is possible that this occurs because the developers of the information systems usually are not the same people as the users of the information carried by these systems. In too many cases, the development of the accounting system is, by default, left to the people in the organization who have historical, rather than decision-assisting, purposes in mind. The "scorecard" questions are the most important to them, and systems designed to answer those types of questions are typically not very useful to operating managers when they are not designed by the managers themselves.

2.2.2 What Should Be

The wrong scorecard questions include, "What were the mistakes that I made last month?", rather than the right decision-assisting questions, such as, "Why is there a variance between my plan and the actual results and does this variance indicate that a change in my forecast is necessary?" These latter types of questions, in frameworks, are those that guide the future. In general they are intended to help in answering the question, "What can I do better today to influence results tomorrow?"

The future is where we are all headed, and our expectations about it, as much as our behaviors in the past, can change what we will do when we get there. Of course, our expectations are a combination of our past conditioning and our present information. Accounting information systems that provide guidelines to assist in making that dimly seen and changing path to the future easier to trod are the ones that should be developed. Although tomorrow's potentials are not actually unlimited (after all, we are always limited by physical, mental, and material conditions), more of these potentials are available than we can ever conceive and use. Therefore, we need all the help we can get to choose the best one when we are making decisions, especially under conditions of uncertainty. As William James said,

> . . . the mind is at every stage a theatre of simultaneous possibilities. . . . mind, in short, works on the data it receives very much as the sculptor works on his block of stone. In a sense the statue stood there from eternity. But there were a thousand different ones beside it, and the sculptor alone is to thank for having extricated this one from the rest. (James 1890/1950, p. 288)

There are accounting systems that have been designed to assist in looking forward, but they always should be developed from inputs supplied by the users. They then can produce outputs to help in extricating the particular statue that the manager thinks is best from all the statues that are possible in his or her organizational stones. Accounting systems that serve this purpose will invariably include a most valuable device. It is called a budget. The budget is part of a plan. Here, we are not concerned with the grand, strategic plan of senior management. Too often, in my opinion, that plan has little relationship to how the technical organization is to operate. No, here we are concerned primarily with direct, relatively foreseeable activities that must be forecasted and with feedback accomplished against that forecast in order to determine what to do next. In reality, as the feedback will *never* match the forecast exactly because of variances that occur with time, there should be a limited strategy of the overall plan that indicates what to do next. For example, if the variance or the difference between forecast and actual cost is a predefined minimal amount, we will correct our actions to try to match the forecast. If the variance exceeds that minimal amount, we will change the forecast. The strategy part of the plan defines the amount and the even the possible causes of the future variances.

2.3 Budgeting and Responsibility Accounting

To briefly review, a budget is not a conceptually complex communication system. It is described easily; the difficulty comes in the prescription and implementation. It includes two parts of a three-part plan: a forecast and a measurement of the actual situation. As noted above, the third part is the strategy that defines any corrective actions.

The forecast is intended to be initially a qualitative, and then a quantitative, plan of action that reflects some previous qualitative inputs and decisions. It coordinates and assists in decision implementation. For example, "We have decided to design, develop and manufacture a new product called a "blodget" for the maritime industry. The engineering specifications, delivery schedule and manufacturing quantities have been defined. Now we need a budget."

If one selects the real definition of a forecast as part of a budget, it is possible to see that the resulting forecasting process is never wrong. That definition is the selection of a particular alternative today from those that are perceived with that alternative intended to achieve some future goal. It may be obvious to point out that no one can completely predict the future. Therefore, a forecast is not the same thing as looking into a crystal ball. It is only a selection of one of a number of alternatives perceived at the present time by the person who is doing the selecting. (I am dealing only with single estimates here, not probabilistic ones that are more complex and would add more heat but not much light to our present discussions.)

When the forecaster and the user of the forecast are different people, there is going to be less than optimal forecasting. This happens even under the best conditions because both people rarely perceive the same range of alternatives. There are also differences in interpreting which alternative is the best out of this unlikely condition of agreed-upon range of alternatives. If you have ever been involved in a forecasting process, I'm sure you'll agree that this happens a lot. Another difficulty is that there is less motivation from the user to attempt to meet this imposed forecast, which was used in the resulting budget. Few of us are objective enough to prove that we, as the users, were wrong when we insisted that the forecaster didn't select the best alternative. In other words: "The preparers (forecasters) focused on their technical responsibility and the users focused on the performance of functional tasks. This disparity has led to the neglect of the important task of fitting the forecast to the function required of it." (Makridakis and Wheelwright 1978, p. 647).

When the forecaster or the person selecting the alternative and the user are the same person, the inferred motivation would probably be to select the best behavior (or sculpt the best statue from those perceived in the block of stone) and thereby make the optimum forecast to fit the function or job required as *interpreted* by that user. But this occurs only when that user is reasonably sure that his or her plans will be accepted by the approving levels of management. (One learns very quickly if there is a repeated pattern of budget cuts and, of course, adjusts the original budget accordingly beforehand.) Another way to support the selection of optimum budgeting behavior is to provide for all accounting and other information systems to be developed in accordance with *responsibility accounting* concepts.

With responsibility accounting, "revenues and costs are recorded and automatically traced to one individual who shoulders primary decision responsibility for the item" (Horngren 1972, p. 158). This fits into the ideas on decision making in prior chapters—it explicitly states that if you can't control it or change it, it shouldn't belong to you, so why bother with it? For example, some budgeting systems assign a nondirect overhead or burden rate to a direct cost activity. A nondirect overhead or burden rate would include period costs such as senior management salaries, real estate taxes, and machine depreciation over which the technical manager has no control.

I recall the response of one manager when his budget was charged with these costs. His memo to the accounting department suggested "cost reductions" such as:

a. selling the corporate airplane to allow senior executives to fly commercially, tourist fare,

b. dropping machine depreciation costs by sub-contracting all production to outsiders,

c. selling the expensive corporate headquarters and allowing engineers to work at home using their personal computers.

There were other suggestions such as those referring to senior management salaries, but you can figure those out for yourself. Unfortunately, the accounting department couldn't agree with the logic of these suggestions, so they just kept adding these costs "downward" with every monthly budget report. This solved nothing because these costs were ignored by those who received them in their reports.

If the manager of project A or engineering design, section B, has no control over the costs in that time-period-based overhead cost, the forecasting and the measuring of actual overhead costs is a waste of time. The managers of A or B can't change them anyhow.

The overheads, however, do belong to someone. The organizational position that can change that cost is somewhere in the organization, and that is where it belongs. Using our example of overhead, let's take one of the least variable elements such as real estate taxes on the plant or offices; this can be assigned to the Chief Executive Officer (CEO). That is the person who can decide to change those taxes by suing in court for tax relief or deciding to move the company to another location with lower taxes. Because the decision maker, in this case, is the CEO, it follows that those taxes should be forecasted and become part of that officer's budget in this example. That follows the concepts of responsibility accounting. Other aspects such as costs are covered later.

Before moving on into a model of the larger information systems that include departmental budgets, we define the second phase of a plan, the one that follows the first phase of forecasting. That phase is "measurement." "Measurements are the yardsticks that tell us how we've done and motivate us to perform." (Zaire 1994, p. 5). This author emphasizes a point made earlier in this book that applies to "measurement."

Measurement for the sake of measurement will not lead to a change in behavior if it appears to be threatening, if intended to apportion blame and if it is to lead to reprimands (i.e. Who did it?). For measurement to be effective it has to be non-threatening, focusing on the process, geared toward improvement and positive action. Measurement has to be hard on the problem but soft on the owner. (Zaire 1994, p. 6)

Then the third phase of a plan, "strategy," defines the amounts of variances that initiate either of two alternatives: change the actions or change the plan.

The incorporation or prescription of these ideas cannot be more specific now because each situation is different, but the general prescription that follows should be useful for you in beginning to design your own system. The model of the information system prescribed here is presented as a sort of general hypothesis of forecasting, measuring, and managing limited strategy that can be modified to fit your situation. It follows the ideas of responsibility for one's plans and is intended to be a cooperative rather than a competitive system, because "a competitive system can almost always be beaten by cooperation" (Haberstroh 1965, p. 1184). It is a system that supports the future orientation of managers, providing the data today to help make the decisions intended to solve uncertainty-based problems. It provides managers with the ability to respond to environmental and internal change by allowing repetitive problems to be classified and resolved through a prior decision matrix, as we have noted before. Managers can then do what they should be doing: solving nonrepetitive problems by absorbing uncertainty.

3.0 BACKGROUND: DESIGNING THE INFORMATION SYSTEM
(adapted from Silverman 1970)

We will provide some background on general communication problems and company environments.

3.1 More Description

The scene is probably all too familiar to the experienced manager. A communication problem arises, and in attempting to solve it, the manager gets the feeling that it has all happened before, that the same problem has been solved dozens of times in the past. It may have appeared in different disguises, but beneath the changing surface there lurks the same tired beast that should have been put to rest long ago. Too many of our management tasks fit this description. They have to be done over and over again simply because their real causes are not recognized, and therefore there is no effective means of preventing their recurrence. The only reliable way to avoid having to solve the same problem repeatedly is to establish a system that permits the manager to compare the present problem with past solutions, determine if they fit, and allow only really new problems to occupy his or her time.

A company grows and prospers in direct relation to its ability to accept change. Changing markets, sources of supply, consumer tastes, and organizational needs all require flexibility and fast response. The marginal company, or even the one that grows only as fast as its markets, is one that responds to change only when the mismatch between organizational actions and those needed by the changes in the environment is very obvious. The response is usually a major "crash" program of some kind. As examples, lost markets or dropping profits can trigger responses in the form of "establishing a field office to improve communications about quality problems" or restructuring the sales and marketing reporting systems. These are often nothing more than creations of the moment or, at best, merely stopgaps until the next crisis occurs. These triggered reactions treat the symptoms without curing the disease, and the underlying disease in each case is slow response to change.

A more effective and prompt response cannot occur with information systems that are historically oriented because many of them assume that the future is an extension of the past. When information systems can provide timely, accurate, and relevant data to support management response to constant change, providing guidance through indications of variances that help in separating novel and repetitive problems, the first step in solving problems under conditions of uncertainty is successfully taken. That first step always involves definition. If the problem can't be defined as new or old, important or trivial, it is treated as new. This wastes the most valuable assets the technical organization has: its managers' time and efforts. Definition of new and important change, therefore, is a major criterion in designing information systems. This type of change for most organizations usually begins in the organization's environment.

3.2 Environment

Companies don't exist in vacuums. Events that can have the greatest effect on company growth often occur outside of the particular company's organizational model. The promptness with which the managers within the company detect and respond to events or changes in its environment can determine organizational growth patterns. Conversely, if the company's internal environment is isolated from or nonresponsive to these outside influences for too long, the chances of growth are minimal and eventually its chances of survival are affected.

As an example, when there are great differences between perceptions of managers and customers about the value of the company's products measured against those of the competition, those differences are going to affect the company. If the effect is decreasing sales, it may be due to a lack of organizational response. This could be quite destructive when customers change their evaluation of the company's products as greater than those of the competition. This scenario might be considered to be quite satisfactory, but it too requires an organizational response. If there is no equivalent increase in production when present levels of inventories are gone, customers might be willing to buy from others and potential income could be lost. Although lack of increased production is not quite as destructive as turning out less desirable products, the company still suffers. Response to change depends on appropriate management action as a result of information exchange and understanding between the organization and its environment and within the appropriate parts of the organization.

The foundation of control is based on an act of reciprocal communication. Not only must the managers communicate easily with the external environment, they must also be able to communicate easily internally within the company. This includes communications up and down throughout the company hierarchy and horizontally across it. It requires cooperative systems connecting cooperating departments. Responding to an increased or decreased sales demand for a product often cannot be done by only one department of the company. Engineering, sales, manufacturing, quality, procurement, and so forth must be able to coordinate their responses. Therefore, perhaps the most important single ingredient in the company's ability to plan and direct its growth (or response, as the case may be) is adequate systems that provide effective, relevant, and timely information to minimize uncertainty and support decision making.

4.0 DESIGNING THE COMMUNICATION SYSTEM

The design of the information system seems to be as important as defining effective organizational components. It cannot be a haphazard response. Even though internal departments (and the components) are "different" from each other, they must be able to feed information to each other and receive it in such a way that changes can be accepted quickly. This usually means some type of information or communication system in which most interactions can be forecast and can be measured when they occur, and the variances fed back to each responsible manager for appropriate actions. There are design parameters for these systems and, as we have noted before, responsibility accounting is one of the most important ones. It provides that the user, if not the same person as the forecaster, at least has the right to accept or reject inputs that he or she has not approved.

The following list cannot be complete because of the varying types of organizations that exist, the rate of environmental change, and the specific goals the company or department has set. It does, however, provide some important considerations for the design of effective information systems that you can modify as required.

The system must be a closed loop. Acceptance of change is a continuous process. Therefore, an information system must be operational at all times and oriented to respond to change. There must be provisions for management to alter the course of the company at any point in the loop without disturbing the continuity of the systems. In effect, when there is a change in course, at that time the manager must be able to compare the new direction with the old one in order to determine if she or he is on course. This requires some type of feedback of the difference, or error between the new course and the old one. The greater the difference, the greater the amount of management action that is required.

The system must permit maximum use of all appropriate management systems, including those with different philosophies coming from both inside and outside. Certain external institutional restrictions are placed on some company operations; e.g., antitrust, interstate commerce regulations, and community relations. The information system should permit the company to operate well within these restrictions. It should also provide for free operation of different management methods and philosophies in different parts of the company.

We know from our analysis of the way change is perceived in different parts of the company that some managers have perceptions that are consistent within their own groups but very different from those in other groups. As a minor example, the application of predetermined output standards as used in mechanically organized production machine shops is probably not appropriate and would not be tolerated in the advanced engineering design department. There should be standards—just different ones for different groups. Similarly, although there might be some commonalities due to legal, social, etc., requirements, all systems do not have to be exactly alike in every detail.

The system must have a predetermined set of measures of efficiency. The most familiar measures are revenues and costs. The arithmetical differences between the two is usually called profit and is generally supposed to be maximized, but sometimes profits are put aside in favor of making a long-term investment in new products, improving production machinery, satisfying changed political environmental demands for pollution controls, or any other organizational goal of the moment. The system should be able to include the measures of efficiency in achieving any of these or other goals.

The system must be applicable to all forecasting time periods. It should permit the use of data generated by the system for making short-term, medium-term, and long-term plans. This assumes that the detail with which the system deals shall change with the time period involved. I have seen budgeting systems that require weekly forecasts of activities when those activities are not scheduled to take place for several months into the future. That is a waste of time. In my opinion, the detail in forecasting generally should be inversely related to the time period being considered. The longer the time in the future, the fewer the details that are forecast. As an example, weekly forecasts may be appropriate for the first six months, monthly from seven months to a year, quarterly from a year to two or three, and semiannual for several years beyond that.

Within these general parameters, an information system can be constructed that provides the kind of information managers need to make effective decisions. In other words, although the specific system may vary from company to company, the parameters that limit the theoretical design should remain fairly constant. The proposed general design shown in Figure 12-1 is built within the parameters mentioned above and can be adapted to many kinds of organizational structures. It is the process of adapting this general design to your specific situation that becomes complex. This general system consists of one main closed loop that represents the total information system, intersected by three subsidiary closed loops.

The forecasting loop: The forecasting loop is the primary connection between the internal and the external environments of the company. Its function is to translate environmental changes and requirements, consumer needs, or vendor qualifications into the standards that apply to forecasts. Within this loop, management's perceptions of the environment are converted to company objectives, the objectives become standards, and the standards become the basis for an operationalized plan against which progress will be measured. I want to emphasize that the forecasting loop is not a beginning in the strictest sense of the term. All the loops are continuous and closed. There is a continuous feedback through all of them, and managers should be able to institute change in any of them at any time.

The translation of environmental change into changes in the repetitive decisions behind the company policy and the possible reverse translation of changed company policy into environmental changes should be equally likely. Usually, however, the heavier flow of data is the movement of external change into the company. Therefore, being able to accommodate to this influx is one of the first tasks of the well-designed forecasting loop. It helps the managers involved in forecasting to perceive which data being received are consistent with prior standards and which are not. That is the purpose behind company statements of the premises on which objectives and plans will be formulated. These company premises are probably some of the better data screening tools that the company can provide. They provide answers to typical kinds of very important questions, such as What business are we in? and What business should we be in? Answers to these premises are inputs to this loop and help to determine where the company will be going (i.e., which statue will be sculpted out of the block of marble).

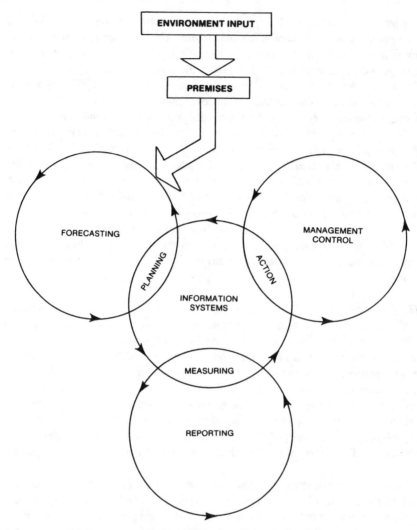

FIGURE 12-1: Proposed General Information System

For example, a railroad company that says it's in the freight business will have a different set of premises and competitors than one that says it's in the tourism business, even though both businesses require the tracks and locomotives of the railroad. The major freight business competitors would be trucking firms. Therefore, the railroad company setting up data in its forecasting loop would be very concerned with minimizing operating costs. One outcome could be the abandoning of tracking that does not immediately produce sufficient freight revenue.

On the other hand, tourism has different organizational premises for a railroad company. The major tourism business competitors would probably be all the buses and airlines. A possible outcome could be an expansion and upgrading of trackage that hadn't been used very much, provided that trackage went into very scenic areas of the country. Buses have to stay on major highways, which don't go through some of the most beautiful country scenery, and there isn't much to see from an airplane at 30,000 feet. In this obvious example, the selection of the company premises defined the nature of the organization and its direction. They were different, even though in both cases the physical assets were much the same.

The definitions of the company's premises, which help managers to perceive relevant data, are usually followed by definitions of company objectives. Objectives are a more specific delineation of the goals that the company must reach to keep pace with change. They are the goals resulting from testing external and internal data against company premises; e.g., premises mean "freight" and objectives mean "abandoning trackage."

The next step beyond company objectives are the specific standards, which must be determined and against which interim progress toward goal achievement will be measured. They may be direct departmental subgoals of the overall company goals or they may be departmental subgoals that have only a general relationship to overall company goals. This last statement may hold with a high internal differentiation among departments. However, the standards are generally the sequence of steps that must be taken to achieve the objectives within the restrictions or premises imposed by the environment and/or the company management. They apply primarily to the internal departments of the company. Usually they are formulated first and are followed by the detailed plans for the operations of various departments, which support standards.

These steps constitute the complete cycle of the forecasting loop: environmental input through limiting premises, selection of objectives or goals, detailed departmental planning of subgoals, and setting of specific standards. With this cycle completed, managers can segregate nonrepetitive problems, anticipate the repetitive ones, and eliminate the reappearance of many of those repeaters through development of a decision sometimes known as company policy. New problems that have not been caught in this forecast loop, that have no standard or plans for their solution, should be the only ones to appear in the future. The final product of the forecasting loop, the plan, includes a forecast of the company's position at some time in the future and a yardstick of expected performance against which the company's actual performance can be measured. The next step is to put the plan into effect.

The reporting loop: Putting the plan into effect means moving the output from the forecasting loop into the total information system. That total system then carries the forecast into the reporting loop as an input. In other words, the total system connects the forecasting loop (i.e. future performance) and the reporting loop (tabulates data on actual performance and compares that with the forecast). It then carries the comparisons of the forecast and the actual data, usually called the variances, as an output from the reporting loop into the next loop, called management control, as an input. That control includes the predefined strategy for taking actions, or not taking them, depending upon the size and the cause of any variances.

Management control loop: The comparisons or variances produced by comparisons of the first two loops give the manager the starting point for implementing any required change. The management control loop is partially in the head of the manager and in addition to the predetermined strategy consists of the mental processes used to guide decision making. The second or reporting loop usually includes the standard accounting system that interacts with stockholders, governments, and others in the outside world. As part of the "scorecard" system that was described earlier, it may be the end result for historical accounting reporting but it is only an interim step for any effective management information system. It provides the actual in the comparison of actual vs. plan. As noted before, that comparison may also be known as a variance.

Variances are important when they are used as a problem classification tool. Of all the assets that every manager has, the most limited is time. There always seem to be more problems to solve than time to solve them. Variances can help the manager select which problems to work on first. A general rule, which may be part of the strategy, is to select the largest variances as the first (but not the only) indicator of the size of the problems. As an example, it is probable that an unfavorable variance of $50,000 in standard manufacturing costs is going to be tackled before a favorable variance of $5,000 in materials that have been received.

This is true even though both variances may prove to be equally unsatisfactory if a complete investigation of both is made. The excessively high costs of manufacturing may be due to unavoidable labor spent on a power failure in a loading device. The favorable variance may be due to a vendor price reduction that has not been included in the system yet. Favorable dollar variances don't necessarily indicate favorable management consequences. That example of a favorable variance of $5,000 in materials received may indicate a major problem in the documentation for the receiving department. Maybe the department really got it but doesn't know where it is. On the other hand, both variances might be predicting some terrible problems in the organization. Usually, the size of the variance is only the first criterion for problem selection. The direction, either favorable or unfavorable, is the next. However, these criteria are not the only way to categorize problems.

A small, continuing, favorable variance in the maintenance budget may be a major problem if it is an indication that the machines and the plant are not being repaired adequately. Conversely, large variances may not be as much an indication of potential problems occurring as they are of poor planning or setting of standards. The plan, the standards for the maintenance department, or the budgetary strategy itself might be wrong and need correcting. A variance is only a difference between some predetermined standard and an actual cost; it is possible that both the standard and the cost may be incorrect. There could also be as many problems lurking with no variances as there are with major variances. If compensating errors that are both plus and minus occur, they could cancel out any variance, keeping the manager in the dark about some problem building up until a disaster strikes. Going back to the original point about the limited amount of problem-solving time, I suggest that variances are useful first tools but not the only ones to use in establishing decision-making priorities. They tell us where to look first.

Data that are generated in the reporting loop fall into two general categories: achievement and cost. Data on achievements are often self-explanatory. "We shipped 389 widgets yesterday against a planned shipment of 390." Conversely, data on cost usually represent only the shadow of accomplishments or lack thereof. These accomplishments are the central concern of the manager. The costs themselves, with no explanation, are not very useful.

4.1 Accuracy vs. Speed

Although information is the principal product of the reporting loop, the methods (i.e., technology) by which it is reported can be equally important. It takes time to gather data and to report it, even when using the latest computerization techniques. Data can also include errors or even become mislaid. Typically, invoices for shipments are lying on somebody's desk longer than they should be, and as for the results of the latest market survey, that always takes "forever" (and "we didn't like the results anyhow").

You can make a choice between data accuracy and promptness. Following up all those labor sheets, invoices, surveys, or what have you will give you accurate data for management decision making, but in many cases the length of time after the fact that it takes for you to get it tends to make it stale. There are those who say that if you wait long enough, you'll find out everything about something that happened too long ago to be of any use to you. When there is a choice to be made between accuracy and speed, I always choose speed. It is useless to know January's loss down to the last penny if that information is received the following May. It's too late to do anything about it then. I feel that inaccurate reports can be valuable if they are provided on a timely basis. However, they are only useful when these few rules are followed. You are informed beforehand that the report contains errors because it contains estimates and there is only *one* consistent source for these prompt reports. With only one source, you can make repeated evaluations of the report's credibility and accuracy over time. When the same person(s) makes the estimate week after week, both the report writer and the receiver can account for repeated errors that begin to crop up. Most inaccuracies are repetitive. After making the same labor cost estimate for three or four months, your cost accountant can include the errors that she or he knows will occur every week and correct for them. The reports then become very useful and timely tools. However, timely means different things to different people.

4.2 Time Delay

The manager with primary responsibility for an activity should receive the reports about that activity first, and for a limited period of time should be the only one who has the report. This manager is in the best position to correct any shortcomings that the reports show, because they are supposedly within his or her direct area of responsibility. There is no need to give them to the manager's supervisor at the same time. In fact, doing so could be detrimental. The responsible manager should have enough time to correct a situation.

Assuming that the problem falls into the manager's decision area (i.e., it's nonrepetitive), is within the manager's authority to solve, and seems to be important, why not allow him or her to solve it by providing enough decision time? If it's beyond the manager's decision area, the manager then has the time to push it up the line to senior management or down the line to a subordinate for action. On the other hand, if the manager does not or cannot solve the problem within a predetermined time period, condensed reports including this problem should be passed up to the next higher level of management.

A similar situation should occur at this next higher level. Let's assume that there are managers who are either unable or unwilling to correct a situation (rather unlikely, but we are designing the closed loop system that includes feedback as positive action). The unsolved problem will eventually appear in a condensed form in the report of the CEO some time *well after* the problem occurred. That CEO is now aware of a string of management failures, beginning with the manager below him in the organization chart and continuing down to the manager directly responsible for the original problem. It's a person-oriented concept, designed in accordance with the rules of responsibility accounting. The amount of information varies almost inversely with the organizational level to which it is reported. Next we see why and how that happens.

4.3 Level of Information

The inverse relationship between the detail contained in a report and the level of management for whom the report is intended means that the higher the organizational structural level, the more condensed, less detailed and later the report should be. When this rule is not followed, all managers can get voluminous data that help very few except those at the bottom, who have to know the details in order to manage. You, as the technical department manager, are not really concerned with the status of one particular project. The project managers are responsible for that. Your concern is with the total of all the projects in your department.

This doesn't mean that all the data doesn't exist, just that they are "owned" by different individuals in different levels of detail. The same applies to groups, committees, and any ad hoc teams that are established. Groups or teams do not own their assigned tasks and do not require detailed data per diem. Only a specific person can be responsible. Either that person manages it or it shows up on the next higher level as a failure. But either individuals take action or no one does! The next design question could be how often these data should be transmitted.

4.4 Frequency of Reporting

The frequency with which reports of "actuals" are received is different for functional and for project operations. Functional organizations are supposed to be immortal; they go on forever. Therefore, the frequency of the reporting should be a constant for them. That frequency should be established by the shortest time span in which corrective action can occur; in other words, when a relatively constant, acceptable level of uncertainty applies.

For example, the engineering manager might be concerned with weekly reports on the status of all designs in the drafting department; she or he may want a monthly report on all purchase requisitions in the administrative department and a quarterly report on all performance evaluations for her or his people from the personnel department. In each case, the frequency of the reports is set to a minimum time span that provides sufficient time for that engineering manager to take corrective action.

The frequency of reports for projects is quite another matter. Projects are not "immortal"; and the amount of uncertainty generally is inversely related to the life span of that project in a nonlinear way. Therefore, because uncertainty decreases with time and we want the reports to cover time periods of relatively equal uncertainty, it follows that the frequency of reports is very high initially but must diminish in frequency as time goes on. Initially, for example, those reports typically can be produced on a weekly basis. After a reasonable number of reports are issued, they can then be issued on a biweekly, then monthly, quarterly, and finally semiannual basis. The amount of uncertainty covered by each of these reports is assumed, in this example, to be fairly constant for the project; because uncertainty is dropping, the time period between reports can increase. We are not concerned with cost, but with uncertainty. If we could measure and plot the curve of uncertainty vs. time (we can't, of course; uncertainty by definition is not measurable in the usual sense of the word), it would probably be a decreasing curve, as shown in Figure 12-2.

For example, a project manager might even request daily reports for the first few weeks of the project, weekly reports for the next few months, and, finally, a monthly summary report. But if these reports are expected to point out variances and are to guide the project manager in corrective action, why decrease the frequency of reporting? Is that not an implicit assumption that variances occurring at the middle or the end of a project are less important than those at the beginning? The answer should be yes. Variances at the middle and the end are not as important as (even though they may be the same size as or even bigger than) those at the beginning. The decisions made in the beginning of most projects will determine the middle and end results anyhow.

In the beginning, there is greater uncertainty. Management input is much more effective during this phase than later on. When you are controlling projects, if you have to make a choice, be more concerned with beginnings than with endings. If the projects don't get started and controlled right when they are born, they will never reach the goals set out for them as they mature, and may never mature at all. On the other hand, all projects that are started correctly may not end correctly but when they are started right, at least they have a reasonable chance of ending the way they are supposed to. The timing of variances is more important than their size when you're dealing with problems, and early reporting is therefore one of the most important management criteria of a project reporting system.

This change in the frequency of reporting can be accommodated fairly easily with computer reporting. Just set the frequency of reporting at the smallest

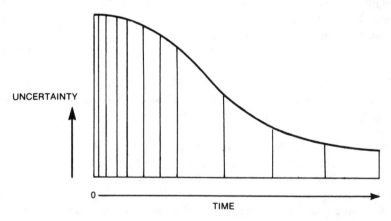

UNCERTAINTY

0

TIME

FIGURE 12-2: Non-linear reporting for projects—Frequency of reporting vs. time.

interval (say, one day or semiweekly at the beginning of the project) and then at some predetermined point such as the first design review have the computer programmed to suppress and summarize those daily or semiweekly reports and produce these summarized reports on a weekly basis. Do the same after the next design review, but make the weekly reports monthly reports. It's easy for computers to do; it's more difficult when people have to do it. People have trouble following the rigidly disciplined (and completely uncreative) path required for condensing data. The reports are now available. Therefore, the next step is to use them correctly.

4.5 Correct Usage

Reports should be used for only one purpose, as guides for actions due today that are intended to influence the future. The past cannot be changed, but the variances in reports can be used to correct those present activities that are affected and to preclude the same kinds of errors in the future. It's analogous to a "sunk cost" concept in economics. You can't "unspend" yesterday's money. When you hear someone say, "Why, we've spent so much time and money on this problem, we really should spend another two months or $2 million because of what we've got invested," you're listening to someone with little understanding of economics. The choice of spending the additional two months or $2 million *today* or of not spending them is the one that has to be made, and not the choice of whether or not to follow on what was spent before. Today is the setting for making choices, not yesterday. Yesterday may be useful to learn from but it doesn't help in absorbing uncertainty today.

Management has to be existentially oriented. The reporting loop encourages action to be taken at the point at which it is most effective. It also assists in fixing responsibility for such action firmly. The first-line manager of a function or an activity is the first person to receive a report on that function's performance. When a problem is brought to light in this report, it is his or her responsibility to solve it. If the reporting loop were not used properly, however, the first person to be informed of a problem would probably be an upper level manager. Before that upper level manager could correct it, he or she would have to seek out the source then deal with the responsible manager, losing valuable, irreplaceable time in the process.

4.6 The Management Control Loop

The forecasting loop provides the plan against which the performance data generated in the reporting loop was compared. This comparison process is then carried by the total information system into the next loop—the management control loop. That is where decisions are made and actions taken. If actual performance has not matched planned performance, there is a reported variance. In a simplistic sense, managers can only choose one of two available alternatives: *change the action or change the plan.* (I have not dealt with changing nothing, because that action results from a decision that there really is no variance and nothing has to be done.) In the first alternative, changing action, the manager has decided that the plan is more desirable than the actual performance and something has to be done to bring the performance into line with the plan. In the second alternative, changing plan, the actual performance is more desirable and the plan has to be changed. The intent in either case is to match actual and planned performance in some *future* time period, based upon what we know *today.* Which alternative is chosen, of course, depends upon your diagnosis of the causes of the variances.

Most of us have been involved with management control loops when it is always the actions that are supposed to be changed. ("We've got to work harder to get rid of that production slippage showing up on the charts.") But the other alternative is really just as viable. There are many reasons for changing the plan, some of which are:

1. The plan might be unachievable. ("That production slippage should never have been on the charts because there's no way that we can get those mainframe castings into the shop on time to meet that schedule. Who set that thing up anyway?")

2. It might be arbitrary. ("We're cutting the expense budget by about 25% this year, but I'm sure that you'll make it because you're such a great manager.")

3. It could have been inadequate because of unforeseen conditions. ("I know that we're in the petrochemicals business, but how were we to predict the tremendous increase in the price of crude oil during the early 1970s?")

4. Probably the most important reason of all is based on our definition of a forecast. That definition referred to is only a selection today of an alternative from all the alternatives presently perceived to solve some future problems. That selection may change as time goes on.

Because we cannot actually predict the future, when experience later indicates that the objectives of the plan cannot be reached, the plan itself must be changed. Changing numbers on paper is easy and plans that cannot be used must be changed into those that can.

5.0 THE APPLICATION (PRESCRIPTIONS) OF THE DESIGN

To show how an information system works and to clarify the terms used in the foregoing discussion, it might be helpful to examine the system in operation.

This limited case study is intended to illustrate such an information system and show how it operates.

A major consumer electronics firm, closely directed by its founder for over thirty years, had grown from a small radio shop into a multimillion dollar operation. Its products were well known and reliable, and the company had always produced a respectable profit. In spite of its reputation in the marketplace, its profits had declined steadily in recent years. The founder died suddenly, and his replacement as president was recruited from a competitor. The new president brought along a new vice president for engineering as part of a general management reorganization.

The first task of the new team was to determine the reasons for the decline in profits. They found that the company had failed to keep pace with changing consumer tastes. The founder had always been able to do market research informally by communicating directly with his major distributors, but the company had grown too large and unwieldy for one man to serve as an adequate pipeline for marketing input as customers' needs changed.

An additional factor was that these favorite distributors no longer seemed to be able to supply accurate data. When the company's competitors changed their distribution systems successfully by using their own warehouses as retail stores to sell products at a discount, the company's independent distributors did not follow suit but moved away from the eventual customer by expanding their retailer-serving (not customer-serving) warehouses and moving out of the retailing end themselves. They became middlemen rather than the partial direct retailers that they had been before, as some of those prior sales were direct to consumers. They lost all direct contact with the eventual customer. Therefore, the marketing data that they fed back were not as accurate as that data had been in the past.

This deterioration of the quantity and quality of marketing feedback data was not the only problem, but the new president knew that problems are often symptoms of underlying company diseases. He decided to treat the disease, not the symptom. He diagnosed it as the lack of an adequate overall communication system including marketing and internal operations. He therefore started the cure by requesting the vice president of marketing to evaluate the company's present distribution systems, set up market information-gathering processes, and, in general, determine the environmental needs upon which design engineering would develop the next year's line of products, as the product designs would be based on those needs.

In more typical situations, the establishment of a management communication system (it's the same here as information system) requires lengthy management training if the system is to be effective. We know that it is difficult to change organizational culture, and the information system is a part of that culture. Modification requires either drastic action or long-term training. Drastic action could mean wholesale replacement of personnel or other draconian measures to make new systems work. Training is a more complex change tool. It usually includes setting different behavioral standards and rewarding personnel who exhibit those new behaviors.

In this case, however, the new president felt that immediate action was needed for a product redesign and cost reduction, with management training of all the management staff to increase their sensitivity to external change. In the long run, most of the burden would fall on the new vice president of engineering, who was familiar with the forecast reporting control system concept and could implement it in production. A request to the vice president for marketing initiated these changes as far as reviving the flow of current, accurate market information.

Using the forecasting loop: A market research program directed at past and potential users of the company's products was completed to determine what changes these users would like to see in the company's products. The company also set up wholly owned retail stores in strategic locations as sources of these marketing data. As we are concerned here primarily with information systems within the organization, we'll deal only with the findings that applied to product design. The findings were that customers wanted:

1. Simplified controls with more pushbutton operation

2. Decreased external radiation in anticipation of proposed restrictive legislation

3. Longer product life with fewer defects over a five-year expected use, even at the expense of slightly higher purchase price.

With this information, the president knew that new product development was one change that had to be emphasized. However, the funds available were limited, because the past president had always insisted on high dividends being paid out (he was a major stockholder). Therefore, the development program had very definite restrictions on spending. The president consequently placed these restrictions into the planning process by stipulating that although there was going to be funding for new products, those products had to generate enough profit to pay for their development costs in two years. To ensure successful use of the applicable market research data, he also required that no new markets for the company's products would be entered until the newly developed product line was first made available to presently existing markets. One suggestion that he made to gain the simplified control that customers wanted was that the new designs would incorporate the latest developments in component miniaturization.

Within these limitations, the president and the vice president of engineering established the company's objectives under the new program. First, they directed that two models be developed: a red and a blue. The red model would have miniaturized components, pushbutton controls, a shielded chassis for radiation protection, and modular construction with heat sinks for longer life. The blue model would essentially be an improved version of the red model with a slightly higher selling price; it had wood instead of plastic cabinetry and was intended for a different market. Estimated sales were 30,000 red units per year for three years and 20,000 blue units per year for two years. Management set a profit objective of 30% of sales after charging off development costs. Finally, it directed that all new models should be ready for pilot production runs by a fixed date in the next year.

Many simultaneous changes occurred all over the company to meet these new goals. For ease of understanding, we will follow just one change that affected the technical group involved in manufacturing operations. The vice president of engineering called a meeting of the various engineers in charge of mechanical, electrical, and manufacturing services. He told them of the company's new objectives and requested that they develop a plan to reach those objectives.

The chief manufacturing engineer, using the data generated by the chief of electrical and mechanical engineering services, established standards of manufacturing upon which detailed manufacturing plans would be based. These standards included a breakdown of all product assemblies into manufacturing units requiring approximately the same manufacturing elapsed time. This was difficult, because the new miniaturized components took more time to handle in the assembly process. Additionally, they included the purchasing department's estimates of material deliveries from various vendors and the establishment of testing procedures and projected rejection rates from quality control. The manufacturing managers used this information as the basis for establishing detailed manpower loads for each project line, material requirements by time period and designation of delivery points for raw materials.

With the publication of detailed plans, the forecasting loop in the information system was complete (at least at this point in the cycle). Up to this point, the information system had given management valuable information about potential market demands, provided specific targets for production and sales over an extended period of time, and produced a set of forecasted standards against which the company's operations could be compared.

The reporting loop: The plan was then put into effect. The two new models were designed, tested, and prepared for manufacture. Materials were ordered, production lines prepared, and the work begun. Following the ideas of time delay, level of information, frequency of reporting, and correct usage, the reporting worked as follows.

Time delay: The accounting department prepared daily production reports, which were submitted to department foremen. A weekly summary of these reports was sent to the plant manager and a monthly recap sent to the chief manufacturing engineer.

Level of Information: Daily production reports showed actual achievement and costs vs. planned achievement and cost. The data in these daily summaries included production units, labor hours and materials used. Weekly summaries were exception reports, as they included only summarized actual performance over or under planned performance and beyond acceptable limits of tolerance. The monthly recaps were similar to the weekly summaries; they, too, were exception reports.

Frequency of Reporting: Because this was a functional operation involved in mass production, the frequency of reports was established at a constant rate but modified to fit the needs of the particular functional managers. Daily reports were needed for line foremen, but they were useless to the plant manager as he couldn't change the actions of the assembly personnel directly. Because the plant manager had to work through the line foremen, his report actually measured those foremen's performance as managers, not the performance of the assembly personnel.

Correct usage: The major purpose of an information system is to give managers the data used in decision making. Reports save management search time and, if designed correctly, pinpoint the areas to be attended to. The timing of the reports is such that the manager immediately responsible for the department in which a problem exists knows about it first and has a reasonable time either to solve it or to pass it on to the appropriate upper management level.

In this example, a foreman noticed a slow daily increase in the number of labor hours expended in chassis production. He found that the chassis metal being used was harder than that specified in the print, a condition that caused unforeseen shortening of drill life and more time spent in sharpening tools. He took what he thought was the only course of action available to keep the line moving by increasing the drill-sharpening schedule, but he also informed the plant manager and requested assistance.

At the end of the week, the plant manager's summary report showed a slight decrease in chassis output. Seeing that the variance was becoming unacceptable to his overall plant performance, the plant manager requested that a team made up of purchasing and engineering personnel investigate the problem and report suggested corrective action. He also forecast to himself when the variance would exceed the allowable limit of his own level of uncertainty, at which time he would have to bring the chief manufacturing engineer into the problem. For now, the prompt action by both the foreman and the plant manager indicated that there should be sufficient time to resolve it without going further up the organization. During the investigation, the metal hardness problem did not show up on the chief's report.

Had the reporting loop not been properly designed in accordance with the principles of time delay, etc., this might not have been the case. In this example, the foreman directly concerned with the problem was informed first, and was the first to begin work on its solution. Without this proper design, the plant manager and the chief manufacturing engineer would probably have become involved and valuable time would have been lost while both of them tried to find out where the problem was coming from.

The management control loop: The problem of the metal hardness, having been solved temporarily by shortening the drill-sharpening schedule, was now in the hands of the purchasing and engineering team. Members of the team found that a design change had increased the strength requirements for the chassis at the last minute and the tooling, which was on order during that change, had not been altered to meet the new requirements. They recommended that production switch to carbide rather than high-speed steel drills as the metal hardness couldn't be decreased.

This change in the manufacturing standards raised the cost of tooling slightly but decreased the production time to meet the prior standard, because carbide drills could handle the harder chassis metal as easily as high-speed drills handled the softer chassis metal. Changing the drill material changed performance, because it brought man-hour expenditures back down to planned levels. Changing the processing print to show the new drill material changed the plan; i.e., raised the planned cost of tooling to match actual cost. This completed the management control loop and solved the problem that the information system had brought to light.

(Note: With the permission of the publishers, some of the materials noted above were borrowed from an article that I wrote, published by the Society of Manufacturing Engineers in *Modern Aspects of Manufacturing Management*. See references for details.)

6.0 REVIEW AND PRACTICAL TIPS

Information systems tie the organization together and they take many forms. Oral communications, such as meetings, can be useful tools. When they have formalized guidelines as suggested in the initial part of this chapter, they provide interaction among people that results in lowered uncertainty that, in turn, supports better decision making. There are informal meetings that sometimes work just as well. For example:

We just didn't seem to be getting anywhere. Finally I had an idea. I started one Friday afternoon. I brought all the top people involved in the program to a cramped, sweaty conference room near my place. These people were from all over. It didn't matter where. After that, every week, on Friday afternoon, we'd sit there and thrash around, from early afternoon until the wee small hours if necessary. We'd go over every problem. We'd figure out what the heck it was each of us was going to do in the next seventy-two to ninety six hours. We'd review where we'd been in the last ninety-six. It was as simple as that. . . . It's real hard to sign up for an objective and then come back to that smoke-filled room like clockwork next Friday and report that you couldn't make it to the guy sitting four feet away across the table to whom you made the promise. Truth and honesty win out in such a setting with remarkable regularity. And that was our routine. Every Friday afternoon. And things began to occur! In a period of about ten weeks we made more progress than we had in the prior eighteen months. Now, that's problem solving! (Peters and Austen 1985, p. 27)

The documented information system that we covered was a prescription for a classically simplified management accounting system. The three loops in that information system, forecasting, reporting, and managing, operated continuously in every functional and project group served by the system. The foregoing example was applied to manufacturing, but any technical department or group of departments could have been used in the illustration. Properly applied and maintained, the information system supports the continual adjusting to changing environmental and internal conditions to which all organizations must respond or die. It provides timely and accurate communication of all the managers in forms that they can use; and it also allows errors to be picked up equally on all organizational levels. It supports the ideas of exception reporting, responsibility accounting, and a continuously self-correcting system. The company is able to direct its growth along those paths that it considers optimal, having full opportunity for the incorporation of environmental and internal change. Change is no longer a problem but an opportunity, as is management in general.

7.0 "ONE-SENTENCE" SUGGESTIONS

1. Budgeting for functions is a repetitive task. Therefore, history can be a basis for forecasting the future. Budgeting for projects is a onetime job; therefore, history about other (perhaps similar) projects should be viewed with suspicion.

2. In "responsibility" accounting," if you can't affect it, it shouldn't be in your budget. Who does "it" really belong to? Find out and the problem is solved. The accountants can move it.

3. The individual concerned with operating from a forecast must be the same person (or accept responsibility for) forecasting it.

4. Management accounting is not the same as financial accounting. There is no "depreciation" for example. In management, once you buy it, the total cost is immediately shown. Depreciation is an accounting problem, not one for operating management.

5. The cost for operating a multimillion machine (except for electricity, etc.) is exactly the same as operating a small $300 drill press. It's the cost of the operator, that's all.

6. "Cost" is not the same as "price." Cost is what you pay in cash. Price includes everything plus something called a "profit."

And finally: *One is only properly motivated to achieve a goal when the person is able to define it.* (Anon.) and obviously, everything should be as simple as possible but no simpler than that.

SUGGESTED ANSWERS TO CASE QUESTIONS

1. In this case, it seems clear that the overall direction and general goals are top-down generated. Managers can be given direction as to general goals, but they have an equal responsibility to tell how much they will cost and what kinds of resources are needed to achieve those goals. The units used (in this case actual dollars, not percentages) should be part of the forecasting and measuring loops. If the corporate staff likes to see percentages, perhaps one answer would be to give them a special report. Let Sam's group program a computer to convert actual results to percentages before sending the data upward. There's no reason not to give them what they want, as long as the operating managers can also have their own information needs satisfied.

2. Budgeting processes are primarily documentation of forecasts and measurements against those forecasts. Forecasting can be learned, and it should be used as such an opportunity. I would change the budgeting system from a once a year exercise to a rollover exercise. In effect, budget for a year with the first three months planned in weeks, the next three months in months, and the third and fourth quarters laid out as one number each. After three months, I would repeat (i.e., roll over) the budget and plan for another year. That is, I would add three months on the end, but this time the three months that were planned in months would now be planned in the greater detail of weeks, the following three in months, and the final two quarters as one number each. This way, a manager is required to inspect his or her own planning process at least four times while moving from a single number plan for a quarter to a monthly plan, and back into the weekly plan. The numbers may change from rollover to rollover, but that can mean that conditions have changed and/or the manager has learned how to budget better. Administrative costs might increase slightly, but not very much. As an example, do you remember how long it took and how painful it was to complete the very first budget that you ever did and how relatively simple it is now? Cognitive change or learning progresses very quickly, and although there is some increase in time that is spent planning, the returns are much greater than the cost because you, as the manager, are able to

evaluate your learning four times as often with this method as with an annual method.

3. They have the best defense possible, which is a complete lack of responsibility for the numbers. ("We really tried to increase production with your directed cutback of machine maintenance, but some of that equipment is pretty old and those continual breakdowns just held us up.") This statement can be translated into: "We don't have any responsibility for the numbers anyhow, because you told us what to do." When the details of a budget are top directed, the manager receiving those numbers faces a no-win situation. If the manager meets the numbers, the director often takes the credit, and if she or he doesn't, the manager is blamed for the failure to meet them.

4. Sam should assist Tony in designing a new information system and then present it to corporate headquarters with the suggestion that if necessary he can convert the data to meet corporate's needs. Because there is no change as far as the corporate offices are concerned, he should be able to do it. The result would probably be better and less expensive output from Tony's division, because each manager would have a greater capability for managing.

Have you ever been in any of these situations? What did you do? What happened?

REFERENCES

Cleland, D. I., and Kocaoglu, D. F. **Engineering Management.** McGraw-Hill, New York. 1981.

Drucker, P. F. **The Ecological Vision.** Transaction Publ., New Brunswick, NJ. 1993.

Haberstroh, C. J. "Organization Design and Systems Analysis." **Handbook Of Organizations,** March, J. C. (ed.). Rand McNally, Chicago, IL. 1965.

Horngren, C. T. **Accounting: A Managerial Emphasis** (3d ed.). Prentice-Hall, Englewood Cliffs, NJ. 1972.

James, W. **The Principles of Psychology.** (vol. 1). Dover, New York. 1950. (Originally published 1890.)

Makridakis, S., and Whcelright, S. C. **Forecasting: Methods and Application.** John Wiley & Sons, New York. 1978.

Peters, T., and Austen, N. "Managing by Walking Around." **California Management Review.** 1985. 28(1):9–34.

Silverman, M. "Directed Growth Through Management Communications Systems." **Modern Aspects of Manufacturing Management,** Vernon, Ivan R. (ed.). Society of Manufacturing Engineers, Dearborn, MI. 1970. pp. 305–316.

Thamhain, H. J. **Engineering Management.** John Wiley & Sons, New York. 1992.

Zaire, M. **Measuring Performance for Business Results.** Chapman & Hall, London. 1994.

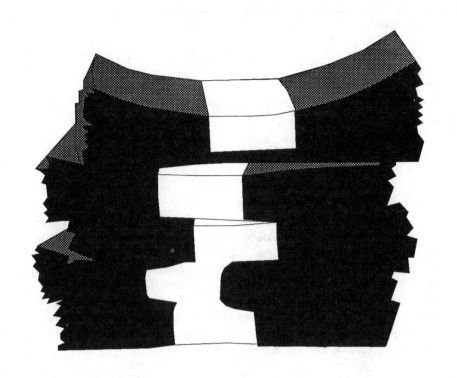

CHAPTER 13

Costs and Measurement: What Do I Need as a Manager?

Case Study:

Cast:
Kyle Rainer: Project manager, Green missile
Liz Miller: Designer, Guidance
Al Birney: Reliability engineer
Stephen Ledder: Vice president, Systems Company

The Systems Company had been awarded a contract to develop a commercial missile that would lift various payloads into space for scientific studies. The missile itself was named "Green" after the first president of the company. Kyle Rainier was the project manager and he felt that he was prepared for Monday morning's status meeting. As he walked to the meeting room, he whistled a little tune. Everything had been reported "on schedule" according to his management diagram. He was confident that after this meeting, he would be in line for the next promotion. The meeting began on time with Kyle, Liz and Al. Kyle noticed that Stephen was there. What a pleasant surprise, he thought.

Kyle: Before I start the meeting, I'd like to welcome our vice president. I'm sure he'll appreciate all the work that we've done so far. Anyhow, let's review the reports.

Al, when do you think that you'll finish the reliability studies on the guidance system? Liz notified me that she was finished with it.

Al: Kyle, I'm not sure when we can even start on it now. We were supposed to get a sample system from Liz three weeks ago but we haven't gotten a thing. I've had four engineers just waiting for it since then. I'm going to have to charge your project for all that wasted time because your project didn't deliver when they were supposed to.

Kyle: I don't understand. According to the reports I've received, Liz says that she delivered a sample on time. Liz, what's going on?

Liz carefully avoided Kyle's question and regarded some documents she had in her hand. Then, she spoke:

Liz: Kyle, do you remember when I asked you for the latest weight specifications on the guidance system? It was two months ago and I haven't gotten them yet.

Kyle: That's because the missile body design is 8 pounds overweight and I'm working with the materials group to find a way to get rid of that extra weight.

Al: Eight pounds—that's terrible. I won't be able to determine reliability because the thruster motor may not be able to handle that extra load.

Kyle: It's not a problem yet. The materials group thinks that they can handle it and I should have the answer in time for you to check the whole missile design. Besides, that's not supposed to be done for four months yet.

By the way, Liz, getting back to the guidance system; didn't you report that it had been delivered?

Liz: Well, yes, I did, but Al insisted that he wanted not only the sample system but all the paperwork that goes with it. I've been working on that for the past three weeks. I don't see why he won't just take the system as it is now and get the paperwork later.

Stephen: Kyle, your meeting is disintegrating. I'd like you to meet with me in my office in half an hour to see what can be done about it.

After a few closing remarks, Kyle ended the meeting. As he walked back to his office, he wondered if he'd ever get a promotion.

QUESTIONS

1. What is the major underlying problem?
2. What system can you think of that would prevent someone from saying they delivered an achievement without them actually having done so?
3. Is Kyle covering up when he talks about the overweight body? What about Liz?
4. Is Al correct in charging the project for the "downtime" that his engineers spent in waiting for Liz's delivery?
5. What should Stephen do now?
6. Has this ever happened to you? What did you do?

1.0 OVERVIEW AND INTRODUCTION

In Chapter 12, there was a brief discussion about budgeting as part of the departmental or tactical planning process. That planning process included the forecasting loop (i.e., the budget), the measuring loop (i.e., the feedback procedure), and the management loop (i.e., the strategy: the variance needed to either revise the actions or the budget). In addition, we covered forecasting/budgeting by the managers who would be responsible for the results and responsibility accounting. We will now go on to provide some detail to support that broad overview of accounting and cost systems. A promotion into management is always accompanied by a closer connection to these systems because the company will live or die depending upon its flow of funds.

It is this flow of funds that is of major concern whether the company is organized for profit or not. Even nonprofit technical organizations, such as those supported by associations or governmental agencies, are concerned when it comes to the amount of money available to them for operations. Many industrial companies confuse the term "profit" with organizational success. According to Drucker (1993, p. 101), there are really no profits, only costs. His point is that although companies report something called "profit" in their financial statements, these so-called profits are really made up of three different "costs": the costs of capital, (i.e. the insurance premiums paid against quantifiable risks or uncertainty of doing business) the cost of jobs and the cost of future pensions. The only true surplus, he states, is a monopoly profit that is extorted when the organization, such as OPEC, controls.

Of course, it might be difficult to convince your senior management that this is a logical viewpoint. However, it is an interesting way to regard those costs. In this chapter, we'll briefly cover some basic accounting concepts and then review cost systems for the more flexible division of technical operations: projects. When you are familiar with the complexities of project costing, the application to functional costing should be quite straightforward. Functions generally don't have the multiple changes that projects do and they also have the advantage of yearly budget reviews and updates. Projects usually have intended a single plan, but many consequent budget reviews can change everything.

2.0 ACCOUNTING: SUPPORT FOR DECISION MAKING

Three areas of accounting will be discussed: basic finance, management accounting, and costs.

2.1 Basic Finance

I have always believed that accounting methods can be divided into two different but related areas: financial and management accounting. Financial accounting is intended to provide a historical picture of the organization. There are two important documents upon which this history is written: the balance sheet and the profit and loss statement. The balance sheet is a record of the organization at one point in time. The simple equation for the balance sheet is: assets (what is owned) minus liabilities (what is owed) equals net worth (what is left for the owners). The profit and loss statement is a record of what happened financially to the organization in a given time period, usually a year. It can be visualized as the changes that occurred to the balance sheet from the beginning of the year until the end. Its formula is equally straightforward. The equation is: revenue (what did we receive) minus costs (what did we spend) equals profit (what we have left over). A sample balance sheet and profit and loss statement are shown in Figures 13-1 and 13-2, respectively.

2.2 Management Accounting

Management accounting, on the other hand, is intended to highlight existing operating trends that managers can use to assist in reducing uncertainty in their decision making. It includes the expected expenditures or budgets for the various operating groups and the measurements against those budgets. In many cases, there are supplementary inputs that assist in explaining variances from forecasts.

The purposes of the two divisions, financial and management accounting, are not the same. In my opinion, attempting to use the historical financial reports for management purposes such as guiding future actions is like trying to steer a boat by watching the wake. Accounting, however, can be useful in assisting decision making when it primarily reports current costs against forecasts.

2.3 Costs

Cost reports can support decision making, they are also a major measurement of achievement; they serve many other purposes as well. For example:

1. Valuing inventories: How much labor and materials are stored in our inventories? Inventories include both "finished goods" and "work in process."

2. Setting selling prices: What shall we sell it for? Be careful of using this application too broadly. Usually, selling prices are determined by what the customer will pay, not by what it costs you to make the product.

3. Matching performance against standards: This is one feedback against the budget.

4. Make or buy decisions: Shall we make it ourselves or should we buy it outside? The cost comparison should **not** include your (time) period overheads.

ASSETS—LIABILITIES = NET WORTH

BALANCE SHEET

ASSETS	$200		
CURRENT ASSETS			
Cash			
Receivables	400		
Inventory	550		
TOTAL CURRENT ASSETS		$1,150	
FIXED ASSETS			
Machinery & Equipment	$100		
Land	150		
Factory	200		
TOTAL FIXED ASSETS		450	
Other Assets		50	
TOTAL ASSETS			$1,650
LIABILITIES			
CURRENT LIABIITIES			
Accounts Payable	$125		
Employee Taxes Payable	25		
TOTAL CURRENT LIABILITIES		$150	
LONG-TERM LIABILITIES			
Long-Term Debt	$100		
Mortgages	75		
TOTAL LONG-TERM LIABILITIES		$175	
TOTAL LIABILITIES			$325
CAPITAL STOCK			
Common Stock		$500	
Retained Earnings		?	
TOTAL CAPITAL STOCK			?

FIGURE 13-1: Sample Balance Sheet Can you calculate "Retained Earnings" and "Capital Stock"?

REVENUES—COSTS = PROFITS

PROFIT & LOSS STATEMENT

SALES
Sales	$1,000	
Returns & Adjustments	100	
NET SALES		$900

COSTS
Direct Labor	$300	
Direct Material	50	
TOTAL PRIME COSTS		$350
GROSS MARGIN			$550

OVERHEAD & PERIOD COSTS
Indirect Labor	$ 75	
Expenses	55	
TOTAL OVERHEAD & PERIOD COSTS		$130

SELLING & SHIPPING COSTS
Labor	$ 10	
Material	5	
TOTAL SELLING & SHIPPING COSTS		$15
TOTAL INDIRECT COSTS			$145
CONTRIBUTION				$405

GENERAL & ADMINISTRATIVE COSTS
Miscellaneous Costs	$ 5	
General Management Labor	20	
TOTAL GENERAL & ADMINISTRATIVE COSTS			$ 25

PROFITS BEFORE TAXES   ?

FIGURE 13-2: Sample Profit and Loss Statement Can you Calculate "Profits before Taxes"?

Your period costs will go on whether you make or buy. It should be a comparison of the out-of-pocket costs only (i.e., how much will you pay for materials and the direct labor to make it inside vs. the vendor's quotation?).

5. Pointing out trends: Why is the material cost increasing? Why is the labor cost decreasing? These are typical questions that cost systems can bring to your attention.

6. Assisting product comparisons: Where shall we concentrate our efforts, assuming, of course, that the sales and revenue potentials are equivalent?

7. Determining cost of goods sold: How much did we pay to procure and process that product that we sold to a customer?.

And to briefly review, when creating the budget within the total plan, do not overlook the following.

(a) Responsibility accounting
- If you can't affect it, it shouldn't be in your budget.
- Level of information: if you can't use it, it's just data overload.
- Promptness before accuracy: you can always become accurate later.

(b) Budget plan links
- Closed loop system: always measure against specific forecasts.
- Iterate if the budget and the actual separate too much.

When these concepts are not included in cost systems, many problems arise, such as

- Unrealistic, low original estimates, budgets and bids
- Management decisions to reduce bid prices and budgets to meet or offset assumed padded estimates
- Uncontrolled, unnoticed increase in the scope of the work
- Extra work for changes/extensions responding to customers and/or management
- Unforeseen technical difficulties
- Schedule delays
- Inadequate budgeting, reporting and control practices (Archibald 1976, pp. 196–197).

Even with these problems in costing, you should understand how useful costs may be produced within the management accounting system for both functions and projects. There are differences between the two groups.

3.0 CONCEPTS FOR COSTING SYSTEMS: PROJECTS AND FUNCTIONS[1]

Technical tasks usually are set up to achieve these by-now classical goals:

1. Meet the specifications,

2. Within a specific time limit,

3. Without overspending some predetermined budget.

Functions, or the ongoing operations, achieve these goals within their organizational boundaries. When the task is too large, too novel, too important and (most important) crosses normal organizational boundaries requiring a manager for the entire process, a project is established. It would seem to be obvious that accomplishing these goals within the limited life of a project requires different organizational structures and cost systems than those used in ongoing organizational functions. This difference between projects and functions is important when cost systems are designed.

Because functions support some type of continuing organizational effort, their cost systems (budgets) are usually based on some forecasted time-based level of effort. For example, "We need to budget the labor and associated overheads for four more engineers to support vendor evaluation efforts for the next year." Functions rely upon continuing and relatively predictable human interactions.

Projects vs. functions: Whereas functions involve long-term interactions, projects are relatively short term, nonrepetitive, and less predictable. Because of this decreased predictability, which often results in increased levels of conflict, the project manager must be able to respond faster, using more subjective (and less restricted) management techniques than those used by his functional contemporaries. These techniques might include personal influence, intraorganizational bargaining, or other nonformalized management processes to get project tasks accomplished. These are all based on receiving fast, accurate information about the project's status at any time.

Effective project cost systems must therefore be more flexible than functional cost systems, and be able to provide faster, clearer data in support of the subjective management "style" of the project manager.

The project cost system primarily aids in defining and solving novel or onetime tasks, rather than supporting a continuing level of effort as its larger cousin, the functional yearly forecast, does. Project cost systems must also typically be more flexible because both management-directed changes and those required when

---

[1] Adapted from "Common Sense in Project Cost Systems," by M. Silverman, American Society of Mechanical Engineers, Winter Annual Meeting, 84-Mgt.-9, NY, 1984.

serious problems occur are expected facts of life. In comparison, it usually requires a really large change to effect a modification in a functional budget. These "differences" between functions and projects are not handled well in many organizational control systems and, in fact, they are often ignored in "order to get a standard reporting format." This happens even though it is well known that there are many organizational contingencies affecting functions and projects differently (Lawrence and Lorsch 1967; Nadler and Lawler 1983; Perrow 1973). These research findings implicitly suggest that the most effective project structures and cost systems are those that provide the greatest personal freedom to the specific manager to manage and control the project in his or her own way. When financial controls and cost systems that are developed for relatively standardized functions are imposed upon projects, usually there are less than optimum goal achievements in these projects.

There seem to be three cost concepts in the budgeting process for functions that become major problems when applied to projects. These are historical cost systems, overhead and allocation of costs, and fixed periodic reporting of results.

3.1 Cost System Descriptions

3.1.1 Historical Cost Systems

Cost systems based on historical data do not help evaluations of the future unless one assumes that the past will be repeated. Historical data may be useful for the accounting department's profit and loss statements, but they are not useful for control of future project operations.

In other words, historical cost systems are relatively useless for projects because they do not provide what is really needed—assistance in projecting the unique tasks to be completed *in the future*. The effective project manager is almost always concerned with answering the question: What do we do next?, rather than: Did we do alright in the past? It is an existential attitude concerned with solving the problems of today to minimize those of the future.

3.1.2 Overhead and Allocations of Costs

When overheads, corporate cost "downloads," or other nondirect costs that cannot be affected by the project manager are assigned to a project, that assignment is irresponsible and even destructive. If motivation in technical people is primarily dependent upon their expectations for the personal achievement of future rewards (Perrow 1973), this would definitely be a demotivator. This also applies to functional managers, but they don't have the project manager's disadvantage of "one time through to complete." With enough time and repetitive periodic budgets, functional managers may find that it is often possible to "adjust the follow-on forecast" upward to absorb these kinds of expected nondirect costs.

To emphasize the point, the assignment of overheads or other nondirect costs to project work breakdown is a misapplication of some basic behavioral guidelines. Those overhead assignment guidelines violate a basic management idea: that of "responsibility accounting," which is defined as assigning and providing information on costs only to the managers who can be held responsible for them. Responsibility accounting typically occurs when a "direct costing" system is used in which only those costs that are directly applicable (or variable) are charged to a particular project. Many larger organizations operate under "absorption costing" in which a charge is made for "overhead" or "burden" for every variable project cost. Under an "absorption cost" system (Horngren 1972, p. 943), there is a fixed manufacturing charge assigned to some variable such as labor, materials, or units produced, and this charge is called burden or overhead. In some cases, there is even an "allocation" or downloading of central corporate costs to the project. Under this latter system, the theory is that these costs are assigned according to the supposed benefits received.

3.1.3 Direct Costing

A "direct cost" system using responsibility accounting minimizes these problems and supports improved motivation and flexibility in managing projects. Then, for example, manufacturing or engineering overheads (and budgets) become the responsibility of the responsible manufacturing or engineering managers, not the project managers. Similarly, corporate "downloaded costs," such as real estate taxes become the responsibility of the corporate officers who can control them. The general rule should be: if you can't change it, it doesn't belong to you. In these systems, instead of loading costs downward, "contributions" (which are the difference between the revenue that the project brings in and the direct costs that it incurs) can be sent upward. If there are not enough "contributions" to pay for the period overheads, those overheads should be reduced by the managers who are responsible for them.

3.1.4 Costs Are Not Prices

Before going further, let us make a small diversion and clearly differentiate between "costs" and "prices." Determining a price at which to *sell* a particular product or service is primarily dependent upon what the customer will pay. It might be higher than all costs, or management may decide to "buy into a market" and absorb the losses for a while. In any event, it may or not be related to all the elements of *cost* that the organization expends. This is so whether the sale is a true one, as to an outsider, or even a developed price, such as a transfer price between departments in a company might be to an insider.

That price then becomes revenue, which is first charged for direct costs incurred such as labor, materials, etc. Then the *difference* between revenue and direct cost is used for allocated costs such as overheads and corporate expenses. If there's anything left over after that, it's profit.

When cost systems do not provide this clear view of the flow of revenue and costs, the standard answer seems to be to apply conventional (and functional or "absorption"-oriented) systems by assigning overheads (i.e., burden rates) and allocating the costs of the corporate offices (i.e., general and administrative expenses) to project managers. That confuses rather than clarifies project management actions. Responsibility accounting and direct costing, however, provide a clearer view of management actions. When project direct costs exceed the forecast in the work breakdown, the responsibility is the project manager's. When there are insufficient contributions to cover overheads and allocated costs, it is the responsibility of others.

These systems also apply when projects provide no contribution, such as those that are funded internally. With no revenue, there can be no contribution. In this case, projects are measured on a standalone, direct cost basis.

3.1.5 Periodic Reporting

The regular financial reporting of the progress or costs against a plan provides some minimal, but usually insufficient, assistance to the project manager over the life of a project. Many, if not all, of the project's major decisions are made in the earlier phases of the project. Therefore, more frequent reporting is needed during those initial phases to support better decision making. Project cost systems must provide reports on a nonlinear schedule with more reports in the beginning and fewer at the end in order to assist this kind of decision making. This is very different from the repetitive administrative requirements of functional managers—their needs for decision-supporting data are relatively constant over time.

3.1.6 Decision Making

Managers are expected to make decisions. Those decisions usually involve other human beings; otherwise, a manager would really not be required—just a computer with a very logical, optimizing set of programs. Managerial decision making is a never-ending process that is expected to operate within an organizationally predetermined range. Each functional department therefore requires a financial reporting system that reports periodically to support its never-ending tasks even though each has its own specialized needs for the kinds of data to be reported.

Project managers also have to make decisions involving other human beings, but the boundaries of the project structures cross various functions. Therefore, the range of problems is not as clearly defined as in most functions. The project managers, therefore, are usually exposed to more diverse situations with consequent requirements for diverse kinds of reporting. The nature and the size of the problems faced also changes more in projects than in functions as both activities proceed. The project manager may initially be concerned with engineering design, then with procurement, then with manufacturing, assembly, field service, etc. There are few regular project activities that extend over the life of the project that can be really compared with the long-term activities of functional groups.

Other complications involve the importance of decisions made during the early part of a project's life vs. those made later on. Earlier decisions are probably more important because they affect the whole design. It is probably a truism that many more dollars have been made or lost on the drafting tables than have been made or lost in the manufacturing shops. Because costs for a project do not become *cumulatively* large (generally) until the designs are completed and major purchase orders are placed, the project manager would be very concerned with the effects of these smaller costs in the beginning and less with those larger ones incurred as the project moves through the manufacturing and delivery part of its life cycle. In effect, the few dollars spent up front in engineering and procurement are more important than the many spent later on. Therefore, regularly delivered financial reports are not as useful as reports that are delivered more frequently in the project beginnings and less frequently as the project matures.

3.1.7 Proposed Solutions to These Problems: Historical Cost Systems

When the emphasis requires providing reasons for failure to meet *past* budgets, the rational response is to increase forecasted contingencies on *future* budgets. This prevents uncomfortable meetings in the future. This type of budget, of course, is no longer a control device because it becomes so bloated that few errors or changes can cause an appreciable overexpenditure. As an interesting aside, it is probably just as unsatisfactory to underexpend ("If we don't spend it all this year, we'll just have to give it back"); there is still little attention paid to minimizing costs. The intent is then to just "hit it right." Historical cost systems become even less effective (if that is possible) when reports are late. The older the report, the staler the news. The staler the news, the less useful it is.

3.1.8 Implementing Solutions: Opening/Closing Work Packages, Milestones, and Estimates to Complete

Project cost systems should be oriented in an existential framework; i.e., what do we do *now* to solve these novel problems? Therefore, planning phases of most projects attempt to separate the "known" from the "unknown" by forecasting how tasks will be done and progress measured. They are typically based on Gantt charts, PERT charts, and various kinds of work breakdowns (Silverman 1976, pp. 59–101), but there are limitations in using these types of tools (Vazsonyi 1970). And although they help to get everyone's thinking understood at the beginning, during project operations they suffer limitations similar to those of historical reporting because they refer back to the total original plan showing variances from it. They are also costly. Updating a PERT chart is expensive, the value of the update received is often not worth the price paid to get it, and, as noted before, the need is for fast and frequent information at a project beginning, less towards the middle and end. As few of us manage multithousand task projects with hundreds of staff members, it would seem reasonable that continual sampling of inputs from the project staff could supply the prompt, clear data needed.

It is rare when the direct staff on most projects exceeds ten people. (When was the last time that you knew of a project where more than this number, or even as many, reported *directly* to the project manager?) Each staff member therefore can provide the forecast for the area under his or her control. This agrees with the concept of responsibility accounting. These forecasts are then summarized into the project plan by the project manager (whose job, by the way, is primarily managing the interfaces among staff members).

Although a periodic update of the PERT or Gantt charts and/or the work breakdowns can provide revised critical paths or other areas to concentrate on, these are usually summaries rather than analyses. Analysis requires clearly detailed existentially oriented, free (or creative) problem-solving thinking of the particular staff member concerned with the particular tasks for which he or she is responsible. The following tools help to provide timely analysis data for project-staff members and the project manager:

1. Opening and closing work packages,

2. Correct definition of milestones

3. Reporting "estimate to complete" for work packages.

3.1.8.1 Opening/closing work packages.
Work packages are specific bundles or groups of costs that accompany specific parts or predicted accomplishments of the project (see Hajek 1977, p. 257 for a similar definition). They are usually related to each other in some sort of hierarchical fashion and are intended to control various parts of the project. When the entire project is opened for charges and the historical cost systems show expenditures against plan, it provides little guidance for the project manager unless there is some way to determine when and if a particular work package is being worked on and if incorrect charges have been prevented. Opening and closing work packages within the project breaks down the work completed and to be completed into manageable-sized portions. It fits the concepts of responsibility accounting, limits potential overruns only to those work packages open for charges, and helps to point out internal inconsistencies in scheduling within the project itself.

For example, if the functional manager of the hydraulics design group is expected to run some life tests on a particular component next March for a given project and the particular component is not delivered until the following April, should the hydraulics design group charge the work package for the time lost? Conversely, if the hydraulics design group's tasks are completed and the work package is slightly underexpended, should the hydraulics functional manager to whom it was assigned use that residual time for other purposes? Finally, while these problems are being solved by the project manager, would those solutions be relevant if the information about them arrived several weeks after the fact?

The project cost system must be able to open work packages when they should be worked on and close them when they are completed. This is the responsibility of the project manager. In addition, only those work packages that are open can be charged. This prevents projects from being considered as an open checkbook. Closed work packages will not accept any more charges. Those rejected charges will then, quite properly, become the concern of the functional manager in whose department they originated and his overhead charges. This eliminates a lot of search and destroy missions for bad project charges for the project manager. The next tool (milestones) provides a control to support the correct use of this open/ close ability.

3.1.8.2 Milestones. Milestones are usually defined as some significant event that is intended to occur sometime within the project life cycle. That is a useless definition. However, when milestones are defined as *tangible deliverables,* they decrease some of the problems with historical reporting systems because at least two people on the project, the milestone deliverer and the receiver, know if a delivery has been made to schedule. If milestones are mutually defined during the project planning phases, it is in the best interest of both deliverer and receiver to agree what that milestone will be and how they both will know it has been delivered. This decreases conflict when work packages cross functional boundaries. Milestone receivers want to receive them in excellent condition and as soon as possible because then their work packages are open for charges. In effect, they informally monitor the progress of the milestone deliverer. Milestone deliverers cannot hold onto them too long because their work packages will be overexpended. An acceptable milestone is approved by both parties before the next work package is opened. The accepting of an opened work package implies the ability to work; thus, few milestone receivers will approve opening their work package unless there are adequate inputs. A tangible milestone used as an interpackage deliverable is a very useful project management tool. But projects are always concerned with "today" and how it will affect where they will end rather than with where they have been. Therefore, we now turn to the next tool, the "estimate to complete."

3.1.8.3 Estimate to complete. Attempting to use actual costs vs. planned costs as a planning process is not helpful. Yesterday's problems may not apply to today. The regular financial function can still provide the actuals, but each submanager on the project who is responsible for a work package should report the estimate to complete (ETC) for the particular work package at the same time that the actuals are reported. When the ETC is added to the actual, the project manager is monitoring the "estimate at completion" (i.e., EAC = ETC + actuals). This is not the same as requiring a report of the "work completed or expended vs. plan." Even if we can assume that the estimators of the work completed are correct (and that's not always a reliable assumption), when those reports are concerned with "percent completed" or "actual amount completed," they are still historically oriented. Management decisions based on this data implicitly assume that the future will be an extension of the past. That is not always the case. The "percent completed" report suffers from an additional limitation—percent figures are rarely helpful. For example, if you can assume that two work packages are equally important and both are reported to be 75% completed, a report on them would still be insufficient because package A has only 50 engineering hours in it and will be finished in two days, whereas package B has 5,000 hours in it and will take the rest of the year.

The ETC requires the responsible submanager to evaluate what the situation is today and reforecast future requirements. It is another, updated estimate *now* of the future. It involves revising and re-estimating future tasks in every reporting period. Because people improve in this ability as they do it repeatedly in the open forum of a project meeting (Hagafors and Brehmer 1983, p. 223), they become better managers. With changes in the EAC over time, there is always a current ability to modify the future. That future can be changed either by modifying plans or by modifying present actions. We can still do that *now*.

There is an additional advantage, or problem, depending upon how it is evaluated. What is the project manager to do with the EAC for a specific work package that varies from report to report? A project manager's report that plots the sequential EACs over time for each staff member provides evaluation data on how a particular staff member thinks about progress on her/his particular work package. Finally, corrective action can be taken "now" when the EAC exceeds the amount of funds available.

3.1.9 Summary: Improvements

Opening/closing work packages, defining milestones as tangible deliverables, and using ETCs mitigate many of the problems of historical reporting. Only opening/closing work packages affect the overall organizational accounting system. The other two tasks do not, therefore they can be selected as the initial cost system redesign improvements.

"Responsibility accounting" or the idea that no cost should be assigned to a manager (functional or project) unless that manager can change that cost can be a major change in an organizational accounting system that uses "absorption costing." Projects are primarily concerned with direct costs; therefore, loaded overheads or allocations of charges from corporate headquarters for various administrative expenses are inappropriate. This concept of direct costing vs. absorption costing (Horngren 1972, p. 94) is quite familiar to the accounting group. It means that projects providing products or services that are sold to outside firms either have to provide sufficient differences between revenue received and direct costs to pay for these corporate costs or else *those corporate costs must be reduced*. Of course, the individual project differences are totaled into one overall contribution from which these overheads and allocations must be paid. The functional management responsible for overheads and corporate management responsible for those costs would then have to live within the contribution generated by all the projects.

3.1.10 Conclusion

Responsibility accounting is a very difficult concept to install in some organizations. Although the basic premises are clear and easily understood, there is often a top-down functional management reluctance to implement this financial system. This concept, with its accompanying emphasis on direct rather than absorption costing, clarifies the qualitative differences in the decision-making processes of various managers and it tends to differentiate those managers who learn from those who don't. This might be difficult to accept. Top-down direction is easier than responsibility accounting, but this type of direction in technical projects usually provides for less than optimal decision making. It minimizes the creative, problem-solving thinking process that is a central need of projects.

3.1.11 Suggested Solutions: Periodic Reporting

Fast information is always better than slow information, and reports at the beginning phases of a project are more valuable than those generated later on. In engineering, fast data comes from "educated guesses." In accounting, they call it "accruals." If the data always comes from the same sources, educated guessing or providing accruals may provide less accuracy for the first very few reports than waiting for all the data to be included. However, these inaccurate first reports can then be compared with the more accurate "actuals" that appear later in order to correct the "guessing" or "accruing" process. Eventually the relatively repetitive errors in this speed reporting are accounted for, and fast reports approach "actuals" in accuracy at least enough to support timely decisions.

This is easily implemented. If the frequency of reporting is originally set at some minimum level, say once a week, it would be possible to program the computer or instruct the report preparer to hold and summarize the reports after the first, say, ten reports are delivered. Then do it again for the next ten and so on. In this way, the reporting frequency would be weekly for the first ten weeks, then biweekly for the next ten weeks, then monthly, then bimonthly, etc.

The development of fast, nonperiodic reporting structures is fairly straightforward. It does, however, require somewhat more training for managers who are unfamiliar (and probably uncomfortable) with this idea. It can be implemented informally, on an intraproject basis, before requesting data processing or accounting to reprogram its reporting periods. When there is a choice between speed and accuracy in reporting, go for speed every time—only don't make any major decisions until you have had a chance to debug the first few reports and find the repeating errors in estimating. Set up reporting periods to a minimum schedule and then have them condensed as the project progresses.

The relatively short life cycle of a project requires many management solutions that are different from those that functional managers can use. Solutions involving cost systems can be more easily solved by following several well-established behavioral and accounting concepts. These include open/closing of work packages, defining milestones as tangible deliverables, requiring ETCs, and varying the reporting periods. Responsibility accounting and direct costing ideas are well established and if properly employed can be a tremendous asset to the project manager.

3.2 Comparison: Absorption vs. Direct Costing

Although it should be obvious at this point that "absorption" costing is not helpful to either functional or project managers, it might be instructive to compare how a profit and loss statement would appear in both instances. See Figure 13-3.

The end result of profit before taxes is the same, but the treatment of overhead (i.e. period costs) is different. In absorption cost systems, the overhead travels with the product and is inventoried. In direct cost systems, it is charged off in

COMPARISON: ABSORPTION VS. DIRECT COST SYSTEMS

	Absorption	Direct
Sales	$100	$100
Cost of Sales		
Materials	$ 10	$ 10
Labor	10	10
Variable overhead @100% of labor	10	10
Period overhead @50% of labor	5	0
Total cost of goods sold	(35)	(30)
Gross profit	$ 65	
Total contribution		$ 70
Period costs including fixed overhead		(5)
Contribution to G&A and profit		$ 65
General and administrative	(20)	(20)
Profit before taxes	$ 45	$ 45

FIGURE 13-3: Comparison of Absorption and Direct Cost Systems

the time period incurred. Direct costing *clarifies* management decisions. It is a rational tool for evaluating performance.

By the way, direct cost data is easily converted into absorption data by allocating a portion of period costs into inventory depending upon actual sales during the time period. (Don't worry about this detail. Either you can puzzle it out by reading any good management accounting text or else let the accountants do it. They know how.)

3.3 Responding to Cost Inputs and Other Cost Systems

Strategy is the third part of any three-part plan; i.e., forecast, measurement and strategy. These are the predefined guidelines that recommend appropriate action when variances occur. Variances may be positive or negative. Neither is indicative of a problem until the variance itself is thoroughly investigated. Because we are always concerned with "variances" from forecasted costs (i.e., exceptions) rather than total costs, "standard" or budgeted direct costs can be helpful tools in decision making.

"Activity-based" costing involves attributing direct costs to a product at the operation level. Each cost center is treated almost as if it were an independent organization with its own overheads, salaries and direct costs. Some technical organizations have found that their estimating processes are easier to follow using this cost system. This system proposes that "it is activities which cause cost, not products and it is products which consume activities." (Jeans and Morrow 1989). In other words, the cost centers are monitored to ensure that each operation performed does not exceed predetermined standards. Then as a product moves through these various activities, the costs are transferred to it. As yet, the "activity-based" system has not attained the popularity of "absorption" or even "direct" costing. For further data, see the reference noted.

4.0 REVIEW AND PRACTICAL TIPS

In this continuation of our description-prescription process we have covered some crucial concerns about measurement of costs that concern both functional and project managers. It may be difficult to convince the accountants to provide direct or activity-based costing data, but it is quite possible to establish your own system within your area of responsibility. By measuring only those costs that are directly attached to products, the manager has a clearer idea of how tasks are progressing.

The forecasting or budgeting process itself is another area that must include the responsible manager. Without this inclusion, there is very little commitment.

It is just as pointless to attempt to meet a budget into which you had no input as it is to be measured on overheads over which you have no control.

Throughout most of this book, our discussions have focused on change and how it affects the technical organization. In the next chapter, we evaluate some of the theories (and the applications) about the design and implementation of positive change in the technical organization.

5.0 "ONE-SENTENCE" SUGGESTIONS

1. A correctly designed direct-costing or activity-based costing system can be one of the most valuable management tools for a technical leader. It supplies information on specific activities against a predetermined "standard" or budget and if followed in sequence can provide an early warning of trouble.

2. Always use tangible "milestones" to open and close work packages. They measure "achievement" as well as cost.

3. When planning a project, always insist on nonlinear feedback. The beginning of a project is the most crucial part and there should be constant reports on cost and achievement at this time. Later, when the project is designed and underway, the report frequency can decrease.

And finally:

Using the better method is like swimming with the current; you progress faster with less effort. (Anon.) and remember it takes less time to do something right because of the data you get, than to explain why you did it wrong because of a lack of good data.

SUGGESTED ANSWERS TO CASE QUESTIONS

1. The underlying problem is that there is no system in place that uses a "work package" and a regular requirement for an estimate to complete.
2. If the deliverer of an achievement reports that a satisfactory delivery of a "milestone" was made, that work package should be immediately closed for any more charges. Then the expected receiver of that "milestone" should report an adequate delivery before the receiver's work package is opened. With no report from the receiver, the receiver cannot charge the project. A milestone is an achievement that must be "bought" by the receiver, as the customer. Therefore, each deliverer in the system has a "customer" to satisfy.

3. Probably not if Kyle can get the job done without overexpending his budget and if he makes a delivery on time. Every problem isn't a disaster. We get paid to solve problems and if that can be done without missing a future delivery, it's satisfactory.

   On the other hand, Liz is definitely covering up. She should have not notified that she was complete when the "customer" said that she wasn't.
4. If the "work package" could only have been opened by the project manager, the "downtime" could not have occurred. Al would have been required to report the problem to his functional boss then because he had no other charge numbers to place those "downtime" costs. The problem could have been dealt with immediately because those "downtime" costs would appear on the functional boss's budget as a large negative variance.
5. Stephen should require Kyle to develop a management "get well" program immediately and have a daily 15-minute standup meeting with Kyle until he is satisfied that Kyle has the new management program on track. For the long run, it looks like Kyle needs some training in developing project cost and control systems.
6. Has this ever happened to you? What did you do?

REFERENCES

Archibald, R. D. **Managing High Technology Programs and Projects.** John Wiley & Sons, New York. 1976.

Drucker, P. F. **The Ecological Vision.** Transaction Publishers, New Brunswick, NJ. 1993.

Hagafors, R., and Brehmer, Berndt. "Does Having to Justify One's Judgements Change the Nature of the Judgement Process?" **Organizational Behavior and Human Performance.** 1983. 31:223–232.

Hajek, V. G. **Management of Engineering Projects.** McGraw-Hill, New York. 1977.

Horngren, C. T. **Cost Accounting A Managerial Emphasis.** Prentice-Hall, Englewood Cliffs, NJ. 1972.

Jeans, M., and Morrow, M. "The Practicalities of Using Activity-Based Costing." **Management Accounting** (U.K.). November 1989. 67(10):42.

Lawrence, P. R., and Lorsch, Jay W. **Organization and Environment: Managing Differentiation and Integration.** Div. of Research, Grad. School of Business Administration, Harvard Univ., Boston, MA. 1967.

Nadler, D. A., and Lawler III. EE. "Motivation: A Diagnostic Approach." **Perspectives on Behavior in Organizations,** Hackman, J. R.; Lawler, E. III; and Porter, L. W. (eds.). McGraw-Hill, New York. 1983. pp. 67–78.

Perrow, C. "The Short and Glorious History of Organizational Theory." **Organizational Dynamics.** AMACOM, a Division of American Management Associations. Summer 1973.

Silverman, M. **Project Management.** John Wiley & Sons, New York. 1976.

————. "Common Sense in Project Cost Systems." **American Society of Mechanical Engineers, Winter Annual Meeting.** 84-Mgt.-9. New York. 1984.

Vazsonyi, A. "L'Histoire de Grandeur et la Decadence de la Methode PERT." **Management Science.** April 1970. 16(8).

CHAPTER 14

Getting Things Changed

Case Study:
Heisenberg The Uncertain

Cast

Hank Heisenberg: Chief project engineer
Mike Johnson: Project accounting
Lief Gilder: Vice president, engineering
Marvin Swerley: Chief draftsperson
Dr. Peter Hazbee: A consultant retained by the engineering division of the company.

Caring Steel Company was a major producer of advanced state-of-the-art aircraft radars. The company had always been able to design and build the best equipment that the industry could offer, but it had always operated under defense cost reimbursement contracts. When it started a commercial division to supply civilian airports with similar equipment, it consistently lost money. Lief Gilder had called Hank Heisenberg, Mike Johnson, and Marvin Swerley to a meeting in his office to try to find out why the performance of the commercial products was so poor. The time was 8:30 A.M. in Lief's office.

Lief: I've asked Mike to review our cost figures for the last three contracts that we've had in our commercial division. The numbers are terrible. Mike, you put them together. What do you see?

Mike: Well, in each case the estimates were reviewed by the project team and then top management and were finally accepted by the customer. I recall that the price had to be shaved a bit during final negotiations, and the project team never met its schedules. The preliminary design seemed to go well, and the initial design review occurred on schedule, but the development costs went right through the roof and the rework on the field installation was about three times what we estimated it would be.

Marvin: I can explain some of the overrun costs in development, because my group expended a lot of those funds in redrawing the designs several times. The engineers just couldn't make up their minds. They kept changing the design, and you know what that does to costs.

Hank: Well, I really can't explain it. We manage those commercial projects exactly the same way we do our defense projects. We've built that general kind of equipment several times before, so it's not really unusual for us. I just don't know why we haven't been making our budgets. Of course, when we worked with the defense department, any changes they wanted, they paid for. These commercial airports are different. They think that they can get anything added on that they want without paying for it.

Lief: Well, we've got to get to the bottom of this. I've asked a consultant to come in to help us. He'll be talking to you all fairly soon. Maybe he'll come up with some suggestions. This meeting is over.

Several weeks went by. During that time, Dr. Peter Hazbee had a chance to visit all of them and discuss the problem. He then inspected engineering operations, drafting, support functions, project management, and the manufacturing departments. There was another meeting in Lief's office several weeks later, at which Dr. Hazbee offered some suggestions.

Peter: I'd rather make my report orally so that we all can offer contributions and suggestions and in any other way comment upon what I have to say. First of all, I'll give you the obvious conclusion, and that is that the nature of the business changes when you work with commercial clients. The government will pay for a change in scope if it wants another thing added to the design. Commercial customers expect you to deliver the latest state of the art as a regular business practice. If you won't, they'll get someone else who will.

The project engineering design group seems to be reluctant and uncertain about charging the customer by writing an impact statement for several reasons, as I see it: when the changes are transmitted from this office, it feels that it's a top-down direction and the group might appear disloyal if it charged the customer because then it didn't meet its original budget. If it comes from the customer, the group feels that if it charges for it the customer will complain to this office and the group will get blamed. There are a lot of other situational factors, but the people in this company not only know what they are, they even have some ideas about how to solve them.

Lief: Well, fellows, do you agree with these findings? Why doesn't anybody say anything?

Hank: Lief, I run my projects according to company policies and I don't really know if there are any solutions in the company, as Dr. Hazbee says. Consultants are supposed to come up with answers. Where are they? I haven't seen a written report or anything.

Lief: Well, Peter, what do you have to say?

Peter: A written report would cover exactly the same things that I'm telling you, and whether it's in writing or not, it's the ideas that are important. I suggest that an ad hoc committee be appointed to evaluate the problem thoroughly, develop solutions and even implement them.

Lief: Well, I wish that I could be on it, but I'm too busy. Why don't you fellows work with Peter here and become the committee? Set it up just the way that Peter described it. This meeting is over.

Several months later, the committee had accomplished the following list of items.

1. It had sent out questionnaires to the technical group asking its members to list the five most important problems in order of importance.
2. It then sent out another questionnaire with problems listed according to the data it received and asked for suggestions on how to solve them.
3. All the suggestions were scored and a design plan to correct the worst three problems was set up. This was then circulated to the people who had responded to the questionnaires. There was further feedback and the committee started to plan to implement their suggestions.
4. It kept the entire company advised through periodic progress reports and requests for feedback.

One month before the design corrections were to be implemented, Lief called Dr. Hazbee into a meeting in his office.

Lief: Peter, I don't mind telling you that all of this committee stuff has the organization in an uproar. That's all everybody is talking about, and I'm not sure that I want to agree to the committee's suggestions. It wants to have project management issue an impact statement almost every day, it seems. And this idea of the project team having to approve a price after I've negotiated it is just not workable.

Peter: Lief, everybody is talking about it because they're interested in it and can see how they'll all benefit from the proposed changes. If you don't agree with the committee's suggestions, why don't you do something about it, like meet with them and assist them in their work? And why shouldn't the project people approve the final price? Although it is true that project people are responsible only for the cost, if they don't feel that they can keep costs down and give the customer what is wanted within the final sales price, why shouldn't they say so at the beginning? And Heisenberg, your project manager, always seems to be uncertain about his team being able to get the job done because of changes that are imposed by top management or the customer, without any equivalent increase in time or funds.

Lief: Well, I'm going to sleep on it. I'll let you know what I'd like to do next week.

QUESTIONS

1. What do you think Lief's alternatives are?
   (a) What do you think that he will do?
   (b) What is the best solution for the technical group?
2. Do you believe that Hank really is uncertain?
   (a) Why do you think that he wanted the consultant's report in writing?
   (b) What tools does he have or can he create to help him get the job done?
3. Why did Peter say that he believed that the people in the organization knew what the problems were and could offer suggestions for their solution?
4. The original meeting in Lief's office didn't provide any answers. Has this ever happened in your organization?

1.0 OVERVIEW AND INTRODUCTION

The last few chapters dealt with information systems and their use in the measurement of achievement. Those systems included provisions for reflecting modifications both to the organization and to the information systems themselves when changes such as innovation, competition, legal requirements, social needs, or any other of a vast number of possible alterations in either the organization or the environment took place. But those modifications to the information systems always followed some equivalent response to either internal or external organizational changes. The central issue for many managers is how to define and implement necessary change as almost an ongoing process. Change is the usual state of affairs in technical organizations and is therefore vital if the company is to survive and grow.

By accepting the definition of managing as "absorbing uncertainty," change becomes an integral part of leadership behaviors and of the knowledge workers themselves. Although functional groups have relatively limited problems to solve, the fact that there are problems also indicates an ongoing change is necessary. For example, "We have a new valve to add to our product lines. I guess we'll have to change all our customer manuals and training sessions to account for that." And it is clear that by their very nature, projects are always directly concerned with change.

In addition to the usual reasons for starting a project, which typically include something that is too large for the existing organization, something that crosses normal group boundaries, etc., there are the implicit reasons that something has happened in the environment, the organizational structure, the technology, or the people, and a "new" situation exists that has more than the usual amount of uncertainty. In the most difficult example, that "new situation" may affect the whole technical organization, not just a part of it as before. Then the standard sequence of trying to fit the new situation into repetitive decision matrixes no longer applies. There are new contingencies to consider, and the leadership concepts that worked well in more stable situations (even those of project operations) may not work well anymore. In a larger sense when the entire organization is concerned, "strategic change requires a basic rethinking of the beliefs by which the company defines and carries on its businesses." (Lorsch 1986, p. 97)

1.1 Descriptions: Major vs. Minor Change: Re-engineering and Total Quality Management (TQM)

Less than total organizational change, such as the minor change affecting individuals and small groups, is almost an everyday occurrence. This type of change can be generally defined as that which causes no major alterations among the relative positions or sizes of our model components (people, technology, and structure, plus the information systems surrounding them). However, when a change does affect the relationship or the size of these components, the magnitude and the frequency of internal interactions change quickly and the complexities of the situation seem to become geometrically larger. The leadership tasks are correspondingly increased.

Consider starting a project on the development of a special series of high-pressure, corrosion-resistant "widgets" in a company that produces many other kinds of "widgets." There might be some minor changes in the organizational components when the project gets going including the assignment of several designers, engineers, technicians, etc., to the project manager on a temporary basis (either directly or in a matrix form). This is defined as a minor change because the development of the new "widget" line is not usually expected to affect the total organization permanently. (It might, anything could, but it's not *intended* to.)

Conversely, if the project to be started is to redesign the functionally organized engineering department (design, production support, maintenance engineering) to a product oriented structure ("widget" products and "blodget" products), most of the engineering organizational model would have to be restructured. The shapes and relative sizes of the organizational components would be permanently changed. This would definitely be a major change and should occur less often. Of course, this change classification is not that rigid. Some organizations might even have both major and minor changes all the time. (I was once consulting with an aerospace company that changed its organization chart so many times that the inside joke was that the company was going to put it on the computer on a real-time basis. Another saying was, "If my boss calls and I'm not at my phone, be sure to get the phone number. I want to be able to call the right person back!")

On the other hand, current management concepts such as Total Quality Management (TQM) and re-engineering have "change" at their core. There is a major difference between them: see Fox (1994) for TQM and Hammer and Champy (1992) for re-engineering. TQM is built on continuous small improvements or change. Re-engineering is a radical restructuring of the entire way of doing business and that is major change. Perhaps a better definition of minor vs. major is the degree of repetition in the change that is perceived by organizational people and in the leader's ability to control the situation.

For further discussions here, let us assume that the degree of change will be major. If we can initiate, control, and implement major change, we should be able to do the same for the less problematical minor changes. Managing change is just another facet of improved leadership behaviors.

In prior chapters, we defined leadership as an interactive process. That means that there must be some type of relationship between the leader and subordinates or contemporaries. That relationship is not fixed. Multidimensional change and equivalent response in leadership behaviors with varying contingencies are thus an integral part of the change process. Many of us have seen the example of the take-charge, task-oriented leader operating successfully in a mechanistic, production-oriented organization who found that the heretofore successful leadership pattern failed when he was reassigned to the research and development (R&D) group. Creativity seems to be a fragile flower that responds best when supervised least. It seems to do well when exposed to a supportive, relationship-oriented leadership set of behaviors.

On the other hand, the personally supportive, relationship-oriented leader who was superb at managing subordinates in the research laboratories that created the new and improved "widget" can fail when transferred into the production department. Production usually depends upon consistent, repetitive, and disciplined actions, and creativity may result in lowered quality and output. Typically, this type of transfer of a successful leader from one environment to another as part of a major change process can often exceed the leader's control ability, unless the leader understands the changed situation and exhibits an appropriate changed set of leadership behaviors.

When managers are successful in one arena, they may perceive their new environment through a more or less fixed prism of past achievements. Because past behaviors brought success, a natural tendency is to continue to exhibit them. That is a problem. There usually has to be a casting off of old behaviors and the adoption of new ones for success in the new situation.

As noted before (see Chapter 11 on leadership), the way in which the leader is perceived by the subordinates is a central factor supporting success in managing technical operations. Perception affects all of us, especially in the way we work. When there is a perception that change per se is the responsibility of top management to *design and direct,* there is often resistance. The communication seems to be that workers do not have to think, merely do. There is another factor that deals specifically with the general way in which knowledge workers are trained. Change isn't easy for them sometimes.

> An illustration of resistance to change—to the acceptance of new ideas in this case—comes from the scientific community. . . . By inclination and training, scientists are committed to the search for truth, to discovering new facts and developing better theories to encompass important findings. Yet despite the commitment, their record of receptivity to new ideas is not impressive. Findings and ideas that challenged conventional understanding were generally vehemently resisted when first introduced. Copernicus' ideas were not accepted until almost a century after his death and Newton's work did not meet with general acceptance for more than half a century. The problem of converting scientists to new views has often been noted by scientists themselves. In his *Origin of Species,* Darwin wrote that he did not expect "to convince naturalists whose minds are stocked with a multitude of facts all viewed . . . from a point of view directly opposite to mine." Max Planck, whose quantum theory revolutionized physics, observed that "a new scientific truth does not triumph by convincing its opponents and making them see the light, but rather because its opponents eventually die, and a new generation grows up that is familiar with it." (Zilbergeld 1983, p. 229)

1.2 Prescriptions for Leader-Initiated Change

However, there are some suggestions that help the leader in changing the situation for technical personnel. Under conditions with ambiguous tasks, unclear policies, and less need for personal autonomy, such as in a critical or confused situation, the directive leader is usually received well, and there is a positive correlation between this behavior that he or she exhibits and the personal satisfaction of subordinates (House and Mitchell 1974). However, when the task becomes clear and policies become quite specific and are interpreted as being personally restrictive, this same leadership directness usually becomes a hindrance rather than a help to subordinates. This happens in leading smaller projects that don't have an impact on the total organization.

When we deal with major change, there is a corresponding increase in situational ambiguity, and we would expect the leader to be much more direct because of it—even more task-oriented—There always seems to be less clarity both in tasks to be accomplished and in the organizational structure during any change process that involves more than very small groups of people. When we deal with familiar people around us, most of us can fairly well define the problems that we have and can relate to them because of past experience. When dealing with entirely new kinds of things and different people whom we know less well, we're not as sure about the problems that occur.

Therefore, when you must oversee the design and implementation of a major change, you are in the same position (although greatly amplified) as the project manager at the beginning of an entirely new project: the behavior should be very task oriented and directive. That seems, however, to conflict with the idea that the psychological needs of technical subordinates or knowledge workers doing creative or novel tasks generally favor the supportive leader and maximum creative space in order for them to exhibit maximum creativity at work. In effect, major change seems to require directive leadership initially and major creativity seems to require supportive leadership initially.

The resolution of these two conflicting recommendations lies in an implicit assumption that we have made about the knowledge level of the leader. If we assume that the leader knows what to do with respect to the change-tasks, the leader is directive (i.e., "OK, although this group hasn't worked together before, we'll start the design of the new "widget organizational operations" similarly to the way that I developed the "blodget operations" and we'll see how this group adjusts those plans to fit this new situation. Therefore, these are the goals . . .").

On the other hand, if the leader doesn't even have the beginning answer for whatever reasons, the leadership behaviors should be supportive (i.e., "OK, although this group hasn't worked together before, the task of designing the new widget organization has only been generally defined. How shall we organize ourselves and what should the sequence of design be?").

The negative responses typically might include a few of these:

"We don't have the people to make changes."
"There's no money budgeted to test new products."
"We've always done it this way."
"I don't want to have to retrain my people."
"We've used this product for years."
. . . There is a general reluctance to change. Companies become familiar with doing things a certain way and believe that new technology will require a substantial investment on their part for education, maintenance, etc. (Mosier 1988)

When developing the change process, these objections can be solved by asking the affected people to participate, thereby dealing with any objections during the change design and implementation: but only the affected people and that may be only a small group at the beginning.

. . . in any organization, a little participation in decision-making was better than none, and that a lot of participation was better than a little. The second thing

. . . you can't start off with a lot of participation—that is impractical in a large organization. . . . Five percent of the people in an organization, if they work in a concerted way, can affect the culture of the entire system. (O'Toole 1985, p. 55)

Being prepared for these responses by requiring participation is one method. Another is to understand (because you've been in the company for awhile) that these objections could occur and be able to handle them. Examples of objections could be, "will there be funds and training available?" "Why do we have to change?" "What would be the probable consequences of *not* changing?" etc.

After preparing as best as you can, it is wise to exert caution in your first choice in your leadership behaviors. The support of subordinates is always a major situational variable when major change is to be accomplished. We know that people interpret change differently. Therefore, the experienced leader will always begin the change-design-implementation process carefully, in a supportive leadership posture if the situation is not closely directed. Conversely, if you know exactly what you want to accomplish and how to do it, a task-oriented behavior might be more effective.

1.3 More on Attitudes and Acceptance

The research on the way people interpret change is not entirely an unknown area, even if it is not completely predictable. There are some data on it, and it is perhaps belaboring the obvious to point out that everyone is always in some type of change process. Similarly, the organization and the economic environment in which the change exists are not static; there is always some motion among the parts of the organizational model and between the company and the environment. Therefore, the technical leader is responding simultaneously to many different contingencies, as are the subordinates. But there is research that suggests that there is a very limited range in the change that subordinates will allow the leader to impose or manage before major problems begin to occur (Barnard 1938). According to that author (and I paraphrase here),

. . . the subordinate will allow the leader to control certain of the subordinate's behaviors as long as the proposed change doesn't require the subordinate to exceed that range of behaviors that the subordinate has established. The subordinate will allow the leader a limited freedom to control the situation. According to this concept, the subordinate evaluates the proposed change and is probably saying (unheard), "I am in control of what I allow you to do to affect my behavior and, for a price (money, position, status, social conditions, etc.), I will allow you to direct some of the things that I do. This permission can be revoked at any time."

When the proposed organizational change is within the limits set (or understood by) the subordinate that affect his or her own behaviors, the change is probably evaluated as a "minor" one; otherwise, it is probably "major."

This is particularly applicable to knowledge workers in the technical organization. Therefore, understanding this unheard message and proposing "minor" changes such as in new product development in a task-oriented way, but "major" changes such as organizational reorganizations in a relations-oriented way seems to be appropriate. The relations-oriented behaviors support an easier extension of the "mental" boundaries that the subordinate might have.

In other words, leadership direction is applied either in a limited way, a way that does not violate the limits set by the subordinate extensively, or in a way that is intended to revise those limits by raising the price the leader and/or the organization pays to the subordinate to achieve his or her cooperation. When a major change is to be designed and implemented, it usually violates predetermined limits, and supportive leadership is necessary as part of the "price" to be paid. This is not a fixed process, because leadership behaviors are not institutionalized, but personalized. This does not, of course, minimize using other rewards such as money, increased job titles, etc., as another part of the "price" to be paid, but we are concerned with technical leadership at this moment.

The design and implementation of a successful change depend upon an (unstated?) agreement between the leader and the subordinate. When the leader has to exceed some pre-existing limit, as defined by *the person affected,* that leader is dealing with an automatic definition of "new or major" change. Accordingly, in this example, the leader has to consider how to modify the behavioral limits set by the subordinate or the person experiencing the change as part of the leader's design of that change process.

Pre-change design preparation by the leader to be able to answer expected negative comments has already been partially covered in terms of general leadership behavior suggestions. Some other guidelines could include the answers to: How does the subordinate evaluate the change? If it's evaluated as positive, the personal limits may automatically be lifted. If it's evaluated as negative, something should be done to obtain the subordinate's personal commitment, because the end result of the change depends upon it. As noted before, commitment is usually tied to some price (however defined) that the organization must be prepared to pay. In turn, that payment seems to be related to some reward that the person perceives will result for supporting the proposed change. The key to major change is therefore finally tied into personal motivation (see Chapter 6). The basic questions are: What's in it for me? and I know what I have, why should I do something else?

As noted before, the subordinates' behaviors is affected by how they interpret the proposed change. These interpretations can sometimes be predicted during the implementation of changes that are viewed by subordinates as being personally profitable.

What influences participants' interpretation is not an idea in the mind of the official but the behavior of the official in announcing it. . . .

Participants . . . initially find the innovation devoid of larger meaning. They estimate its utility as relatively low, changing their behavior as difficult, and situational problems as being the focus. (p. 47)

First and most important, do not assume you know what people are thinking: Systematically test your assumptions about other's beliefs by, for example, collecting . . . data. Second, do not assume that liking or positive evaluations will necessarily lead to behavioral change. Third, beware of vivid analogies, unless they come out of shared experience. And even if they do, they may be interpreted differently by different people. Fourth, if you really want people to spend time doing something new, give them the time to do it. Take away old activities, provide released time, ask them to work overtime—somehow demonstrate that time allocation patterns are to change. Fifth, beware of dispositional attributions. That is, do not focus only on trying to change people. Focus also on how their situation should be changed. (p. 58) (Sproull and Hofmeister 1986)

2.0 MAJOR AND TACTICAL CHANGE

It is unfortunate that, in my experience, there are too many cases where major changes that affect the entire company are inadequately communicated totally to the technical groups. However, these groups invariably must participate because of their central role in any change of product or process. This participation might be called a *major* change when it involves several technical groups. When the change involves only one group, it can be called a *tactical* change as opposed to *strategic* or major change involving many groups. Because major changes involve many groups, there are more variables to consider, more uncertainty in decision making, and probably more problems that must be solved. Change design and implementation in most operating organizations that I have seen are rarely managed in the standardized way that many textbooks seem to suggest. Just as in structural design as we discovered in prior chapters, there is no single theory or paradigm that works for all. By this time, we probably are all familiar with the following established and open management pattern:

1. Establish goals.

2. Forecast resources and needs to reach those goals.

3. Set up measurements of progress or achievement.

4. Develop detailed plans to coordinate the activities of the various organizational groups, and implement a small change first.

5. Re-evaluate achievements on a timely basis and go back to Step 1 to start the process all over again.

This is the scientific method dressed up in planning terminology. It is all very clear and direct, but it never really happens this way. This method is a top-down, unilaterally accepted direction that would follow the classical theory of management. Developing any major change for a technical organization involves more than this. The organization is not an army that happens to be involved in technical operations; it is a group of highly individual knowledge workers (including managers as well as subordinates) involved in a shifting, complex process, possibly including first supportive, and then directive, leadership strategies. Change involves a team. Without the commitment of the team, change quickly goes wrong, because no plan can provide for all contingencies. One way to bring an organization to a grinding halt is for subordinates to do exactly what they are told to do and no more. Therefore, beginning change cannot be a fast process.

> Change should begin with the slow laborious task of observing and diagnosing social behavior as opposed to technical behavior. This must be done rather carefully because skillful managers realize that they cannot express their opinions without understanding the potential effect of those opinions on others. Subordinates may have difficulty differentiating between the boss's feelings and his statements of fact. . . .
>
> Therefore, the introduction of any change requires a thorough analysis of the potential effect on the social interactions of others. This is as necessary as the more familiar, thorough analysis of the technical advantages that the change is intended to achieve. The first step is to provide all those affected by the proposed with as much information as available concerning the benefits to them, as individuals. Self interest is a prime motivator. (Miller 1988, p. 804)

Accordingly, major change planning may require a more adaptable leadership process within the general rubric of relations-oriented models. Following the suggestions of Quinn (1980), leaders can begin the process using nondirective, relations-type approaches. The stages are similar to a muddling through or "cut and try" system. They are:

1. Creating awareness. By alerting the informal networks to intended changes and obtaining feedback from them, leaders can circumvent the tendency to shield themselves from potentially negative responses too soon. The ideas can be circulated as trial balloons in order to elicit informal suggestions that would either support or mitigate future implementation.

2. Generating alternatives. The trial balloons may not fly well, but maybe someone has an idea that modifies those balloons into something that will. Just the process of evaluation and suggestion fed back by some subordinates to the leader increases the potential for co-optation of those subordinates who are against the idea or else provides modifications intended to improve the original ideas proposed.

3. Broadening the support for change. This may be done by allowing a reasonable gestation period for the ideas, setting up study or advisory committees that include both those who are for and those who are against the ideas (but mostly those who are for them), assigning "champions" to develop detailed plans, and making sure that discussions about the ideas are conducted under "no lose" ground rules. There is not supposed to be any personal stigma attached to having suggested an idea that is not accepted. At this stage, the relations oriented process tends to minimize the "hardening" of subordinates' behavioral limitations.

This is proposed as an existential, step-by-step approach that is expected to modify the personal psychological limits of persons affected by the proposed change. It allows the leaders and subordinates to define the proposed change (for and to themselves), contribute to it, and tentatively predict how their own behaviors will be modified by the change. This procedure is intended to be conceived broadly and to be under constant refinement. It is never supposed to be completely finished, and it usually means a great deal of organizational, in addition to personal, ambiguity affecting large numbers of organizational participants (another form of the scientific method.)

Although this existential procedure can be useful as one alternative for changing the limits of personal or psychological acceptance of change, it doesn't help us to determine the total direction of that proposed change, nor to implement it. That is defined by the difference between the existing organization-environment fit (as it is presently understood) and some expected optimum fit toward which the organization should be moved. In other words, when we can determine both the end point and where we are at present, we then can understand both the differences between them (i.e., the amount and the direction of the required change) and can begin to direct the existential process that not only changes psychological acceptance but also works to minimize this total difference. Because the process of defining where we are can be similar to defining where we should be, we can start the definition process with either point. But we can usually be more objective about defining where we want to be. Thus, I suggest that the first step in designing and implementing this kind of change should be defining our end point, optimum, or best situation. Goal forecasting is easier, and less threatening personally, than defining a present unsatisfactory situation. In summary, if you're not sure, gain allies before starting change.

2.1 The Optimum Fit?

Many factors can be used to define and measure the optimum. We have covered some of them in other chapters such as when we analyzed several improved designs for the technical organizational structure. Because structure is defined as repetitive behavior patterns in both the formal and informal systems, there are two different structural designs available to you, as the organizational designer. These are "functions" and "projects." They may be used in developing your change programs depending upon the amount and rate of change by which you want the structure to be modified. With slower change, the functional organizational structure has been suggested to best handle "immortal" or continuing operations such as product support, drafting, and design standards. These operations have a fairly low, but continuing, level of problem solving, with very little overall change. With more rapid change, project organizations would be best as they are most useful for "mortal" or relatively short-lived operations.

Technology is another variable component defining type of "fit." In prior chapters, it was defined as the methodologies used by the organization to transform inputs into outputs. Typically, those methodologies could be manufacturing processes, decision-making processes, or others, such as information processing. The transformation process in manufacturing generally involved "organic" structures for unit and process production and "mechanistic" structures for mass production.

Other chapters dealt with the environment, information systems representing repetitive decisions, and some of the tools of leadership to manage it all. The one conclusion was that all the optimums were contingency (or situationally) based and that there was no "one best way." There are many optimums, not only one, and there are many end points or goals that can be defined. As the amount of change needed to reach those defined goals increases or, in other words, as the difference between "today" and "what should be" increases, the change methods change. In the structural example, it goes from functions to projects. Similar changes are needed for technology, as noted before: organic or mechanistic. The "optimum" depends on where you want to go and if you can get there from here. But how should one select the best optimum at the moment? What are the contingencies that tie all these variables together? What is an end point or optimum design that we can at least start with, understanding that when we reach it, we may have another goal or end point at our new "optimum"?

2.2 Starting with People

One optimum based on the person-organizational interaction is described in Lorsch (1977). That model involves the relationship between the organizational tasks and the individual sense of competence motivation. It is very heavily influenced by human needs and expectancies, and because those types of variables are key ones in technical organizations, the model could be a very appropriate beginning to our temporary optimum design. The variables are defined by Lorsch:

> An individual's sense of competence is a self reinforcing reward. As an individual performs a job successfully, feelings of competence encourage continued efforts

Table 14-1. Dimensions that Can Be Assigned Ordinal Values (i.e., 1 to 10) for "Change" Measurements in Two Typical Groups in Technical Organizations

Design Dimension	R&D Lab.	Manufacturing Dept.
Span of control	Wide	Narrow
Job definitions	General and broad	Specific and detailed
Personnel measurement	Less frequent	Very frequent
Planning	General goals	Specific, detailed
Rewards	Professional recognition, career	Management, money
Selection	Technical qualifications	Process and cost analysis
Leadership	Relations oriented	Task oriented
Training	Professional conferences	Human skills, technology

to do the job well. Different tasks seem to be attractive to persons with different psychological make ups. For example, research scientists who work on uncertain complex and long-range tasks prefer to work alone, with freedom from supervision, and enjoy highly ambiguous and complex tasks. In contrast, factory managers whose jobs are more certain, predictable and short range, prefer more directive leadership, closer relationships with colleagues and less ambiguity. (Lorsch 1977)[1]

The organizational structure that is designed acknowledges the differences among people. Table 14-1 can be used to develop ordinal measurement of your optimum. Subjectively using ordinal scores, say from 1 to 10, give a score to each of the "what is now" situations and another score to the "what should be" situation. This process has been slightly modified and used successfully in other change programs (Lorsch 1977). Using these data can help in defining the organization's present and end points. Obviously, any discrepancies between the proposed end points and the present situation would be a major reason for a strategic or major plan of organizational change. Although these data are illustrative of a specific research point of view, they are only qualitative and involve the general fit between the needs of the economic environment and the requirements of the individual and the organizational culture.

The assumption that the organizational culture does fit most of the economic demands of the environment is not always entirely valid. Misfits do occur and may take a long time to cause failure. But there are occasions when the misfit is obvious enough for management to perceive it and take corrective action. That corrective action might even require changing the total organizational culture.

---

[1] Reprinted by permission of the publisher, from "Organization Design: A Situational Perspective," in *Organizational Dynamics*, Autumn 1977, © 1977 by AMACOM, a division of American Management Associations, NY. All rights reserved.

That is a major, lengthy operation, well outside the central scope of this book, even though we dealt with a small portion earlier in this chapter concerned with methods for changing subordinates' psychological limits. We therefore deal with it as a supplementary, rather than a central, concern.

2.3 Organizational Culture

The culture of a firm is defined as the unique and readily observable repetitive behaviors that they recognize, bind its members together and separate them from other organizations. That culture may be good or bad, functional or dysfunctional, or anything else, depending upon how we choose to evaluate it. Generally acceptable economic evaluations are not the only ones used. Companies may be losing money because of extensive R&D or market development activities and still be viable for long periods of time. If we choose to evaluate the culture using an anthropological framework (O'Toole 1979), we can use the term cultural relativism, which means that there is no single absolute standard against which to measure an organizational culture, but rather relative standards that can change over time.

For example, there may be an accepted code of behavior within a specific firm that determines all interpersonal contact and is a part of the culture. People may be encouraged to use everyone's first names, even when addressing the president of the company, but they are also aware that this freedom doesn't apply when meeting socially after work. That culture has then a "relative" standard of behavior. While the company will hire the maverick, he or she is quickly worn down into the accepted modes or is ejected (voluntarily or not) from the company. Although all types of people may be selected (and even this might be questioned in some organizations), only certain behaviors are rewarded. Culture may not wholly determine personality, but it greatly influences which personality patterns will succeed.

For instance, one of the major Fortune 100 manufacturing companies was very centralized, with most of the important decisions being made at the top levels. All the top executives were members of the same golf club, and it was their practice to discuss business and make decisions while dining or playing. That club excluded women, blacks, Catholics, and Jews from membership. In recent times the company found it harder to recruit and retain top young managers and the Board of Directors could not understand why the quality of those younger executives it did retain was not up to the standards of its competition (based upon O'Toole 1979, p. 20). That organizational culture gradually contributed to an organizational misfit of the company with the needs of the environment. The change required for a new optimum fit was certainly beyond the capabilities of any internal technical manager. There were gradual but important changes in its economic and social environment. Those environments had moved on past the familiar landmarks that existed when the company was started, and old answers no longer applied.

You can see how changing the organization's total culture would be beyond our scope here, because it involves extensive management actions and an extended time period (sometimes years) to accomplish. However, there are some general ideas that might apply in the very limited circumstances surrounding the technical group. (I'm not suggesting giving up. I'm suggesting handling the part of the problem that we, as technical managers, might be able to solve.) Ideas to accomplish limited change would certainly start with some type of analysis of those desirable and different optimum structures and/or cultures. Then we could define the existing organizational structures and sanctions that either support behaviors we want or discourage various kinds of behaviors that are no longer useful.

For example, if the limited intent is to develop new or innovative products, an organizational structure's insistence that time sheets be completed every week by a certain time of the day would probably be a dysfunctional cultural sanction and be among the first things to be changed. Achieving even this relatively trivial change in the limited cultural circumstances in and around the technical group could be difficult, but it is approached most effectively by changing the systems within which people work, not by changing the people in them. In other words, "any approach that singles out people for change when it is just the institution that is determining their behavior is patently unjust" (O'Toole 1979, p. 27).

The change process is best accomplished as an open process, because cooperation of the knowledge workers is essential after you have achieved the easier task of changing the work systems. This process cannot be manipulative, as in Skinnerian conditioning. Even if one assumes that it is possible to manipulate people, it's very difficult to do, because the manipulators don't have complete control over the experimental subjects.

Another point against complete acceptance of Skinnerian conditioning is that any change in the miniculture is almost always pluralistic. You can't change one thing and expect everything else to remain constant. Moreover, the proposed change cannot be monolithic, because it is intended to alter the way things are done now to doing them some other way in the future. Many things happen during the day, and they can all be affected by (and affect) the proposed change. Change cannot be intended primarily to modify the participants themselves; they will still generally be the same loving, nasty, kind, hostile, selfish, or generous individuals before and after the limited change that you have implemented in the miniculture of the technical group.

However, there is always some interaction between the culture in which we work and our personalities, and therefore a change in that culture always has some effect on us. But that effect is neither entirely predictable by management. When the change occurs, the reaction may range from complete acceptance to complete rejection. (The people could decide that, "We're changing the limits within which we'll allow the organization to control our behavior.") When they open those limits, people allow more of themselves to be controlled by the culture. When the limits are closed down, people may leave the organization (either physically, "I quit," or mentally, "I just do my job and not one bit more"). Culture and personality are interactive, regardless of the management intent behind the change, and the interaction always has some unpredictable parts. Therefore, to decrease this human unpredictability, the general rule that "participation, in the long run, minimizes conflict" should be followed. Of course, it may increase conflict in the short run but it's better to get it over with in the beginning rather than let it come up later at a crucial moment.

As technically trained managers, we are not generally an emotionally adaptive group. In most cases, the things that we know and the ways in which we do them are familiar and comfortable. The old saying, "We know the problems we have as opposed to those we might get if we change" applies. Opposition can come from an entirely human response for self-protection and preservation of one's way of life. On the other hand, it's possible that participation and commitment may coexist with some unpredictability during a change in the miniculture. The problem that you face as the leader is to determine if that commitment and/ or unpredictability is *for* or *against* you, and that is generally decided by the self-interest of the people concerned. And that in turn depends to some extent on how you present the proposed change, as well as what the proposed change is.

3.0 HOW TO GET THERE: EQUIFINALITY

When you have completed the process of defining the optimum situation (i.e., what would I like to have happen?) and similarly defining where you are at present (i.e., what is happening now?), the next task is to develop some method for getting there from here. Just as there is no one optimum or one culture, there is no one way to get there. This idea of the existence of multiple ways to reach some organizational goal is called *equifinality*. It represents the notion of many possible processes of growth and development; not any *specific* one. It suggests that organizations are not wholly constrained by their initial structure, can adapt even without losing their basic form, can continually redesign themselves in *more than one way,* and that *change in one part affects the whole.* (Bertanlanffy et al., 1951).

The growth pattern from the small, single proprietorship into the multinational conglomerate can occur through the merger and acquisition path or through internal growth and diversification. These growth or change alternatives or combinations of these and other ways are equally likely according to the concept of equifinality. That seems reasonable when one considers the different ways in which many different kinds of organizations grow or are modified. Usually, the tendency is toward greater organizational complexity and size as a response to a larger range of environmental conditions. In recent times however, there has been an opposing move of larger organizations into smaller divisions with flattened structures. The middle manager is being squeezed out by "re-engineering" and consequent drastic restructuring. But in many cases, desired improvements have not resulted (Floyd and Wooldridge 1994).

We have assumed that the organization is an open system, accepting inputs, modifying those inputs through the technology chosen, and producing outputs. The concept of equifinality indicates that there are many ways for organizational change to take place, and the one you choose should be related to your estimation of the ease of implementation (i.e., the cost to get it done). This estimation must include an evaluation of the potential paths to the goal selected and the possible advantages or disadvantages to the people who have to implement the path selected. Remember that estimates and forecasts do not predict the future. They are only present evaluations of the relative costs and benefits that result from selecting an optimum path from those presently perceived by the estimator or forecaster. Forecasts sometimes don't come true. There are always obstacles to change that seem to be based on the individual's (i.e., the leader's and peers' or subordinates') behaviors and attitudes.

Assuming that equifinality is a reasonable concept, the selection of the path to effect change often seems to be some sort of wish fulfillment on the part of the leaders who desire the change. One leader may try to modify the structure ("What we need is to decentralize."); another, to change the technology ("We really need to computerize."). And then there is (in my opinion) the least productive option, to change people ("We need to have everyone go through sensitivity training."). Unless the change is related positively to the person's self-interest, that person won't change willingly. Nobody sacrifices himself for the glory of the company.

3.1 Consultants, Misconceptions and Other Flaws

There is a common misconception that the leader's way to make the change is the best way. It's almost a religious belief. I have noticed that it is particularly prevalent among leaders who have had no record of any previous disappointment in the work environment (the brilliant wonder-worker who turned that losing division around in six months) and among inexperienced consultants who attempt to repeat past successes without recognizing that the situation is not the same. ("Well, it worked with Amalgamated Iron Works; why shouldn't it be just as effective with this company, the Universal Toy Factory?") This kind of consulting demonstrates a naive approach to solving problems that refutes the undeniable complexity of human beings. I've found some organizations that ignore the solutions that lie within them. The consultant can then be useful by bringing those answers to the attention of senior management. However, when doing so, ethics demand that the source of those answers be revealed by the consultant as coming from the company, though not necessarily disclosing the individual who developed the answer unless that has been agreed to beforehand. What goes on is sometimes amazing.

As an outside consultant, I have discovered situations in some companies where just being able to cross normal structural boundaries has produced answers that were evaluated by senior management as "extraordinarily perceptive." Those answers were really obvious if senior management had looked for them. Like the 1000 pound gorilla who gets anything he wants, the $1000 consultant gets attention when he speaks.

But the potential problems of using inexperienced and overpaid consultants to cause change are minimal compared to the possible disasters resulting from suggestions of the "brilliant wonder-manager." At least the consultant eventually leaves and the organization might recover from his advice. Unfortunately, the "wonder-manager" often remains to continue destructive attempts at manipulation based solely on his naive, self centered viewpoint since failure can be attributed to many other causes. Attempts to implement change will always fail unless those affected participate and accept the new situation. Because much of the work of the knowledge-worker is mental, they always have the ability to sabotage any change with which they do not agree. There may be no change in observable behavior.

Manipulative personal relationships designed to change people's feelings about security, morale, and personal sensitivity usually result in less than optimum change because of two implicit assumptions that are *not* true:

1. Leader controls: The assumption that the leader alone can design and develop an optimum plan (I know what's best for everybody.)

2. Uniform treatment: The assumption that all those who interact with the change can be treated uniformly, because they will all regard the proposal the same way. (Everybody thinks the same way here and we all want what's best for the company.)

Of course, neither assumption is valid. The attempt to change people using only one path for the achievement of some organizational goal will invariably be less than successful because everybody is not the same. Using the "indirect methods" of sounding out potential opposition, gaining converts, changing the plan as new ideas are brought forward, and generally attempting to connect the revised (and undoubtedly improved) plan with the particular individual's self-interest is about as optimal a process as any that the leader can expect. These indirect methods may seem to take longer but they result in lower implementation costs. There are many roads to a goal, and those that are chosen in an open, participative management environment usually are more successful because of the implicit commitment of those involved. The human costs and the consequent financial costs decrease with commitment. The following models can be general prescriptions to use in optimizing change at minimal cost.

4.0 BEGINNING IMPLEMENTATION: MORE MODELS

Congruence model: Obtaining the involvement of the group in the change process can perhaps be started by using a modification of our by-now familiar scientific method. Some researchers (Nadler and Tushman 1980) have developed a step-by-step approach, using a well-named congruence model.[2]

1. Identify symptoms. List data indicating possible existence of problems.
2. Specify inputs. Identify the systems. Determine nature of environment, resources and history. Identify critical aspects of strategy.
3. Identify outputs. Identify data that define the nature of outputs at various levels (individual, group-unit, organizational). This should include desired outputs (from strategy), and actual outputs being obtained.
4. Identify problems. Identify areas where there are significant and meaningful differences between desired and actual outputs. To the extent possible, identify penalties; that is, specific costs (actual and opportunity costs) associated with each problem.
5. Describe components of the organization. Describe basic nature of each . . . component with emphasis on their critical features.
6. Assess congruence (fits). Conduct analysis to determine relative congruence among components. . .
7. Generate and identify causes. Analyze to associate fit with specific problems.
8. Identify action steps. Indicate the possible actions to deal with problem causes. (Nadler and Tushman 1980)

This may seem to be a very structured approach, but it is really not much different from a seemingly undirected and less structured approach (Quinn 1980) because the method suggested here doesn't state that this task falls only to the leader. It may (and I strongly urge it should) include the views and design modifications of the knowledge workers affected. This step-by-step approach simply provides a firmer thinking and estimating framework. The final change is always the result of interaction between the directions resulting from using this framework and the actual implementation procedures.

Over the years, there have been many other management models intended to optimize various situations by changing the behaviors of people, the design

---

[2] Reprinted by permission of the publisher, from "A Model for Diagnosing Organizational Behavior," in *Organizational Dynamics* by D. Nadler and A. Tushman, Autumn 1980, pp. 48–49, © by AMACOM, a division of American Management Associations, NY. All rights reserved.

of the structure and the methods of technology. These include, for example, organizational development (French and Bell 1978), total quality measurement (Fox 1994), and re-engineering (Hammer and Champy 1992). They all have a common aim, to improve the operations of the company. However, they all have common faults: they do not provide for different or unique situations and they assume that all workers have aims that coincide with those of management. On the other hand, these management models can help when modified to fit personal gains by the workers. Organizational development is a typical example of a change model that you might consider.

Organizational development (OD): This is defined as "a long range effort to improve an organization's problem solving and renewal process . . . with special emphasis on the . . . work teams with the assistance of a change agent, or catalyst, and the use of theory and technology of applied behavioral science, including action research. (French and Bell 1978, p. 14).

There are two assumptions about people in this type of change methodology.

1. If provided with a supportive and challenging work environment, most people have drives toward personal growth that are positively valued.

2. Most people can make a better contribution to the organization than the organizational environment will allow.

In my opinion, these assumptions are both the major assets and the major liabilities of organizational development. They depend on the implicit idea that there is some uniformity in the way that people regard their work. In some cases that we have discussed (Lawrence and Lorsch 1967), we have a great similarity within technical groups but great differences among groups in a direct relationship with the way that they perceive change in their organization's environment. It appears that this research might conflict with these assumptions. Therefore, questioning these assumptions can be a major consideration if you intend to use organizational development techniques.

There are other conflicts when using these assumptions. In the mechanistic organization that is optimum in mass production organizations (Woodward 1970), we find that human growth and improved contributions could be dysfunctional. They would probably disturb the machine-like operation of this organizational model. Of course, it might be said that this kind of organization doesn't provide supportive and challenging work either, but we know that the definition of supportive and challenging work is in the mind of the worker, not that of the researcher.

Organizational development, however, has been reported to be used successfully as a change technique in many situations in which it has been attempted. This still doesn't mean that it will be effective in all situations. Those successful situations included positive elements such as open organizational systems, negative elements such as automatic resistance from those affected by any prospective change, and neither positive nor negative, but descriptive elements such as solving nonrepetitive problems that were not of central concern to the organization. Those elements defined as of central concern typically are the responsibility of top management: the survival of the organization, acquisition of new companies, long-range strategic planning, product line design, etc.

Action research: The *action research* process of effecting change (using the terminology of organizational development) is the continuing process of collecting research information systematically about some organizational, goal-oriented system; feeding the findings back to the people in the organization; altering selected variables from the system based upon the hypothesis chosen and the data collected from the organizational people; taking some appropriate action; and then evaluating the results of that action through the acquisition of more data (based on French and Bell 1978, p. 88).

Action research is obviously a participative process and seems to work very well when the situation requires destructuring or loosening up of tight organizations. Our standard assumption that the knowledge worker of the technical organization usually prefers less directive, relations-oriented leadership supports the use of this action research technique in implementing major kinds of change in technical groups, especially when that change is expected to apply immediately. It's almost irreversible once started. Movement in an opposite direction from open, participative groups and leaders toward closed, nonparticipative leadership situations usually decreases the "mental limits" of the knowledge worker that she or he permits the organization (or the boss) to use in controlling his or her behavior. The movement toward openness raises expectations and tends to increase the mental limits. It's difficult to put the escaped genie back in the bottle.

Therefore, action research as a tool in organizational development requires active participative leadership, continuing support of top management, and the need for on-the-job application. Because the result will usually be a move toward more organic structures, there may be a decrease in the speed with which management decisions were made in the past. These kinds of organizations seem to make progress on an overall basis less quickly than mechanistic organizational structures because the top management may be making fewer quick decisions for the whole organization. It's no longer necessary—others are making decisions for their own areas without being told what to do.

5.0 SUMMARY: CHANGE METHODS

We have discussed some of the change methods that may be used to modify the miniculture of the technical organization. Which method is selected depends a great deal upon the leader's forecasts of the "optimum" fit and the problems that are perceived now. The concept of equifinality becomes very applicable here. If the beginnings of the organization do not set it irrevocably upon only one possible path to the future and we understand that there are many potential paths to follow in achieving a goal, the leader is free to suggest any method of change that he or she feels will work. The answer to the question: How do I get there from here? is straightforward enough (and probably quite difficult to use because of that). Just determine where you want to be, where you are now, and use any of the available techniques that you can learn about to get you from here to there. If one doesn't work, maybe another will.

We are concerned initially here with changing the way that groups function, as we have more or less concluded that it is both undesirable and almost impossible to change the individual to fit our goals. We change the way people function by changing systems first. However, when these systems change the group's methods of working, interaction might then make something of a consequent change in the individuals' behaviors. But this takes too much time, according to many managers. However, according to the following author, the time taken will be the same.

> I have encountered some managers who wonder if an organization can take this much time for changing organizational structure. In my experience, although time is a critical factor, it is a false issue. Time will be taken whether management is aware of it or not, by people to ask all the questions, make all the politically necessary moves, develop all the protective devices, and create all the organizational escape hatches that they feel are necessary. The real issue is whether the time will be used constructively and effectively so that the organization can learn from its experiences thereby increasing its competence in becoming a problem solving system. (Argyris 1979, p. 31)

So one might say that the individual is also changed, but that cannot be the primary aim. The individual is much more complex than any group. This would seem obvious, but I have seen a great deal of time and effort expended in organizations on a point of view that is contrary to everything that is reasonable and intuitively obvious (to me, of course). That point of view seems to be concerned specifically with attempts to change people themselves.

A more specific and useful change process, which we have discussed to some extent before, is called "training." And because the intent of this book is to move from general to specific recommendations, we will now consider the specific training methods that organizations can use.

6.0 LEADERSHIP TRAINING: WHAT IT IS SUPPOSED TO DO AND WHAT IT REALLY DOES

In many cases, the purpose of training a person is to cause some change in that trainee that will result in a greater return to the organization providing the training.

The intent is no different from those noted above when working with groups. But critical reviews of research in the training fields show that there is no scientific evidence to support claims that training in various individually oriented techniques is generally effective for the organization (Perrow 1972). Attempts to produce long-lasting and positive change in the organization through typical training methods such as managerial grids, sensitivity graining, job enrichment, and even job enlargement did not succeed.

Although leadership training and leadership experience often result in some positive behaviors, they also do not seem to contribute *uniformly* to group or organizational effectiveness. But even though training does not always produce uniformly positive results for the organization, with the probability that there is minimal overall gain to the organization providing the training, the results seem to have some limited gains because the individual learns something about improvement. Organizational effectiveness in general may not be improved because the training doesn't directly connect with that, but the skills, behaviors, and technical-administrative know-how of the person is probably improved if and when these behaviors can be observed. Rarely will training directly affect overall end results such as organizational profits.

If training is considered within our contingency framework, the person being trained will probably have considerably greater control and influence over the job than someone who is untrained. Leadership training and experience seem primarily to improve the ability of the leader to increase the favorableness of the leadership situation as *perceived by that leader*. The conclusion to be drawn from this is that, although training is intended to benefit the organization, it may not do so directly, but *may* do so indirectly, through benefits for the person. If the person then contributes more effectively, and only that person or the supervisor who closely observes before and after training to note any changes will know that, the organization will gain. Training is just as much a capital investment, in this respect, as the purchase of a machine tool or the improvement of a manufacturing building, but the end benefits are rarely as predictable for the organization as a whole.

It is clear that the complexity of the situation; that is, the interaction of the existing information that people carry into the training situation with the new information they might learn, does not make for easy predictions of positive results. The organization as a whole must be willing to accept a lack of predictability about the training end result; the primary benefit will accrue to the person being trained.

Any improvement in the overall organization will then depend partially upon factors outside the organization's control. That could mean that the optimum result of training would occur when there is a complete coincidence among the organization's goals, the individual's goals and new learning, and an implicit agreement about the best way to achieve these mutual goals. Because this rarely happens, we have a situation in which the individual benefits the most from training and the organization as a whole benefits only when there is goal and methodology congruence. This assumes, of course, that the training has all the good things any training program is supposed to have: a direction, a method to assist the learner in moving in that direction, and techniques to support positive on-the-job applications and feedback at the conclusion of the training.

6.1 Informal Training: Counseling

Managers and leaders use other change methods in technical organizations, methods that are intended to deal primarily with the individual. These methods can all be categorized under the one-on-one interpersonal or communications patterns. A part of the management by objective (MBO) technique includes the leader-subordinate interactions in setting specific goals for the subordinate to achieve in some future time period. It also is intended to include personal counseling.

Although the intent is probably positive, the result often is not. The problems in attempting change through one-on-one discussions are major, because the attributes of an honest, supportive discussion between leader and subordinate rarely exist. Even professionally trained counselors must concentrate to provide an environment that will allow for discussion of the problems of the subordinate, for positive inputs to support personal development for the subordinate and for an unbiased or subordinate-oriented viewpoint. The problems for the manager who attempts this are almost insurmountable—it requires extensive clinical training and an unusually open organizational culture. However, counseling (or supportive criticism) may occur occasionally when the subordinate recognizes its value, when self-interest is aroused and when the subordinate feels that the situation is not punitive. Under those conditions, it probably will change an individual's behavior. But we're never sure.

The initial task would be to differentiate between task-related and interpersonal counseling or evaluations. Task-related information should be handled in an open, problem-solving process. Objective decisions may eventually become subjective as the data about product designs, cost, and delivery schedules are sifted and resolved, but the process of open discussions (such as in the design review meeting—see Jacobs 1979) should always be chosen first. Because problem definition is a first step in problem solution, task-related information can be defined initially through open discussion of the relevant facts.

Interpersonal and subjective information-handling methods, however, are different. They are *always* used on a private, one-on-one basis, because they implicitly contain potential conflict situations that could permanently damage relationships. They involve listening to the other person and understanding what emotions as well as data are being transmitted; e.g., "Let me tell you what you told me. If I can do that and you agree that what I said is what you meant, we will at least be able to understand each other's position. After I have done that, you try it for the things I said and I'll tell you if that's what I really meant."

The training that is done in individual counseling is always interactive. The counselor as well as the counseled is affected. In many organizations, the boss is supposed to be more of a mentor (Schein 1981), as there is a long-term, relationship-oriented leadership situation and the boss trains through cues rather than direct feedback. Direct feedback usually is used in task-related evaluation. Therefore, task-related information is seldom the subject of personal counseling, and usually involves more than the boss-subordinate dyad.

Conversely, interpersonal counseling, unique in that it only concerns two people, the boss and the subordinate, is intended mainly to benefit only those two participants. If the organization benefits, that is a secondary gain.

6.2 Concluding: Training for Leadership and Change

Training is usually the process of choice in organizations when the intent is to change the behavior of individuals by providing new information. On a formal basis, the results have not been uniformly effective for the organization because the information that person brings to the training situation interacts with the new information being transmitted in ways that are not always predictable. However, assuming that it is possible that the training provides new and helpful information to the person, the training results in some type of change intended to help that person be better able to control and improve his or her own leadership or working situation. Therefore, training may lead to individual improvement, and then the desired organizational improvement may or may not occur. The organizational improvement depends on a coincidence of organizational and personal goals, acceptance of the newly learned information and positive reinforcement of any new personal behaviors that seem to match organizational needs.

On an informal basis, individual counseling can be divided into task-related information, which should always be handled in an open forum as it is based primarily on objective data, and interpersonal information, which should always be handled in a private, one-on-one discussion. The results here are similar to those of formal training as they also depend upon goal coincidence, acceptance, and positive reinforcement. The major differences between formal and informal training are the methods used. The end result is always the same: some change for the individual that is hoped to be of benefit to the organization.

7.0 GETTING YOUR IDEAS ACROSS

After considering the general methods to be used in changing the miniculture, groups, and individuals, we come to more specific prescriptions for presenting these change ideas so that they generate minimum opposition. One idea is to gain small, consistent changes or "wins," as the following author calls them.

> The author's research has focused on what many audiences have called "the theory of the small win." Patterns of consistent, moderate size, clear-cut outcomes—patterns of small wins—are a special subclass of managerial activity patterns influencing future change. (p. 17)
>
> Frequency and consistency are two other primary attributes of effective pattern shaping. A pattern of frequent and consistent small successes is such a powerful shaper of expectations that its creation may be worth the deferral of ambitious short-term goals. . . . Support of completed actions typically generates further actions consistent with the rewarded behavior. . . . by varying his patterns of reinforcement he can substantially influence people's behavior over time, often several levels down in the organization. (p. 22) (Peters 1983)

I have found that an adequate plan and preparation before any proposals are presented win many of the battles. Some of the basic planning is fairly obvious. For example, know your audience. Who is going to receive this plan of yours and who has to approve it?

The effectiveness of any proposal depends initially upon the ability of the proposal receiver to understand it. If you are presenting information to someone who is technically qualified to understand the jargon, by all means, use it. Jargon is a form of oral shorthand that can cut down on supplementary explanation time. But use less complex forms of communication if the receiver is not up to it. Also, you might emphasize those parts of the change proposal that would appeal to the receiver. For example, someone in charge of plant operations would certainly be influenced more positively by a discussion of the potential for a decrease in maintenance costs than of an increase in production.

Timing is a vital consideration, so make the presentation as soon as you think you can get a positive reaction. There will be delays enough that are not of your making. Rehearse your presentation; then try a dry run with some cooperative associates to become more familiar with potential problem areas. One manager always had a dry-run staff meeting before presenting his yearly budget to the Board of Directors. It accomplished two things for him: he was able to eliminate some problems before any board member could bring them up and his own team (which served as the "dry-run board") came to appreciate the complexities of their boss's dealing with the Board. The best impromptu speeches are those that have been rehearsed thoroughly. Try listening to yourself reading one of your reports into a dictating machine if you wish to test this idea. What you hear will not be the polished, logical input you thought you were providing.

Use visual aids, if necessary. When slides are projected on a screen in front of the room, you can point to appropriate areas and have everyone's attention on the same thing. Three-dimensional models are also effective, because everyone can then visualize what the final result should be like. A word of caution is necessary, though. These visual aids should be simple and easy to grasp. Colors and perspective drawings should be used to make things apparent, not to dazzle the viewer. The best visual aids quickly become almost invisible as the viewer concentrates on the idea expressed by them, not on their intrinsic value.

No one knows your subject better than you do. Therefore, these techniques are expected to make you more familiar with the act of transmission, rather than with the content of that transmission. Emphasize the receiver's point of view, rehearse your presentation, use visual aids and jargon where appropriate, complete dry runs, do it now, simplify your presentation media, or use any other technique that seems reasonable, but prepare your presentation thoroughly before you make it. Up front planning is the best kind; it costs less and prevents problems later on. Decrease the amount of uncertainty as much as you can at the beginning of any proposal and you'll be absorbing less uncertainty later on.

8.0 REVIEW AND PRACTICAL TIPS

Effecting intended change in an organization is a complex process that doesn't become easier as we move from large organizations through groups to individuals. In my opinion, it becomes more difficult because we deal with greater variability in the "individual" as opposed to the "group." In groups, the extremes can balance out. In any event, the actual change doesn't match the original intent; it may be better or worse, but it is always different. The process itself modifies the original intent into something else. The successful and experienced leader quickly learns that it is the process that is of major concern, and the goal is often achieved by allowing that process to move along, with other people's contributions. But this is not easy. Consider this (even though it is intended to relate to information processing technology, it seems to apply generally here): "Anyone who tells you that it is easy to change the way groups of people do things is either a liar, a management consultant or both." (Economist 1990, p. 11)

We have explored some of the reasons that this happens and also some of the methods intended to keep that intended change as much on the planned track as we can. One of the major supports for a change process in technical organizations is the need for open, participative discussions and commitment in order to respond to both environmental and internal changes. Because the work is done by knowledge workers (both leaders and subordinates are in this classification) and no one can completely predetermine the amount of effort knowledge workers exert, the need for their support is apparent.

Commitment and support come more easily when there is occasional deliberate ambiguity in the behavior of the leader. It's easier to obtain cooperation when one says, "Perhaps you can think about this a bit more," rather than, "You're completely wrong. Redo it!" Laying all the cards on the table may be personally satisfying (especially if you are the one who lays them down), but it may not always be required or even desired if you want others to cooperate. In functional technical operations, directive leadership is not the primary mode; relations orientation seems to be better. Fortunately, the needs of the technical organization and those of knowledge workers often coincide, and in these situations effective techniques for organizational and systems change interact positively between the organization and the people. These suggestions are therefore offered as pragmatic methods for accomplishing change. They have little to do with a claim to making organizations more human-like, but rather with using human-like techniques to gain some organizational goal. In other words,

> Whether deservedly or not, humanism adapted to the management process has the taint of manipulation. It is difficult to imagine management using techniques like organization development, sensitivity training, or job enrichment out of the pure milk of human kindness. (Scott 1974)

But in our examples, the "manipulation" is planned and controlled primarily by the organizational participants for their own self-interest. If that "manipulation" is highly differentiated to account either for the different kinds of people in the organization or for the different tasks that each group within the organization has to do, the organization gains the flexibility and adaptability needed to modify its components and support its own further development. The organization (if we can speak of it as a total thing) is not being altruistic. Its aims just happen to coincide with ours in most technical functions. Those aims are not always completely coincident, however. Each of us, as an individual, has unique aspirations that no organization can satisfy completely. But there is a higher degree of coincidence in our kinds of work now than ever before, and we have the ability to continue to change the organization to suit more of our needs or the freedom to choose another organization. Which would you like to do?

9.0 "ONE-SENTENCE" SUGGESTIONS

1. Define: Why are you making a change? Have you established "ordinal" numbers for "what is" and "what should be? Do you know which difference between those two numbers is greatest? Is it the most important one?

2. If you're absolutely sure of what you want to change, perhaps a task-oriented leadership behavior is best. "This is where we have to go and this is how the road is. Let's get going."

3. Usually, however, we're not sure of everything so, in that case, a relations-oriented leadership behavior might be best. "This is where we want to go

and this is the reason why. Now you can see that it's to our best interests to get there so what suggestions do you have as to the route to be followed?"

4. One of the most straightforward ways to change an individual's behavior is to give him or her a different job with a different set of objectives.

5. Change is not easy. Sometimes we end up at a different place from where we wanted to go. Therefore, if possible, make change slowly in small pieces. Small wins work better in the long run.

And finally: *If there is no wind, row*. (Anon.) and remember " Present fears are less than horrible imaginings" (Shakespeare)

SUGGESTED ANSWERS TO CASE QUESTIONS

1. Lief has four alternatives: (1) He can join the committee meeting and help in the problem diagnosis and solution. (2) He can accept the committee's work and support it. (3) He can have all the committee's suggestions sent to him for approval prior to implementation. (4) He can dissolve the committee.
   (a) I would guess that he will choose alternative (3) first, then find out that the quality of the committee's work immediately drops and go to alternative (4). Of course, he will be firmly convinced that it was all a waste of time.
   (b) The best solution for the technical group and then, of course, for the organization would be for him to choose alternative (2). That choice is rarely made, but when it is, everyone gains.
2. Hank was probably very aware of the reasons for the project cost overruns, but he is a survivor. I don't believe that he was uncertain about his choices, merely those of others.
   (a) His request for the consultant's written report is a typical delaying ploy that allows him to find out how Lief feels about it and act accordingly. With no written report, Hank had to rely on his own thinking processes.
   (b) He has the ability to match the cost of each element of the work breakdown schedule against the contract and find out why costs were so high. All he has to do is compare actual to budget and define the largest variances first.
3. Most organizational participants are familiar with the reasons for and the potential solutions to repetitive problems. Often, it is top management who cannot or refuses to listen to them. If management would follow the decision-making flow diagram in Figure 2-4, the solutions to repetitive problems could quickly be placed in a solution matrix for all to see.

4. The original meeting in Lief's office didn't provide any answers. Has this ever happened in your organization?

REFERENCES

Argyris, C. "Today's Problems with Tomorrow's Organizations." **Matrix Organization and Project Management,** Hill, R. E., and White, B. J. (eds.). Michigan Business Papers, University of Michigan, No. 64, Ann Arbor, MI. 1979. pp. 5–31.

Barnard, C. I. **The Functions of the Executive.** Harvard Univ. Press, Cambridge, MA. 1938.

von Bertalanffy, L.; Hampel, C. G.; Bass, R. E.; and Jones, H. "General Systems Theory: A New Approach to Unity of Science," I–VI. **Human Biology.** December 1951. 23(4):302–361.

Floyd, S. W., and Wooldridge, B. "Dinosaurs Or Dynamos? Recognizing Middle Management's Strategic Role." **Academy of Management Executive.** 1994. 8(4):47–57.

Fox, M. J. **Quality Assurance Management.** Chapman & Hall, London. 1994.

French, W. L., and Bell, C. H., Jr. **Organization Development.** Prentice-Hall, Englewood Cliffs, NJ. 1978.

Hammer, M., and Champy, J. **Reengineering the Corporation: A Manifesto for Business Revolution.** Harper Business, New York. 1992.

House, R. I., and Mitchell, T. R. "Path Goal Theory of Leadership." **Journal of Contemporary Business.** (Autumn 1974):81–97.

Jacobs, Richard M. "The Technique of Design Review." **Proceedings of the Product Liability Prevention Conference.** American Society for Quality Control, 1979. PLP79 Proceedings.

Lawrence, P. R., and Lorsch, J. W. **Organization and Environment.** Harvard Univ. Press, Cambridge, MA. 1967.

Lorsch, J. W. "Organization Design: A Situational Perspective." **Organizational Dynamics,** American Management Association, N.Y. (Autumn, 1977).

————. "Managing Culture: The Invisible Barrier to Strategic Change." **California Management Review.** Winter 1986. 28(2):95–109.

Miller, T. E. "Managing Change Effectively: By Building Teams." **Project Management Handbook** (2d ed.) by Cleland, D. I., and King, W. R. (Eds.). Van Nostrand Reinhold, New York. 1988. pp. 802–822.

Mosier, K. C. II. "Holding Up Technology." **Ohio Business** (November 1988).

Nadler, D. A., and Tushman, M. L. "A Model for Diagnosing Organizational Behavior." **Organizational Dynamics.** (Autumn 1980).

O'Toole, J. J. "Corporate and Managerial Cultures." **Behavioral Problems in Organizations,** Cooper, C. L. (ed.). Prentice-Hall, Englewood Cliffs, N.J. 1979. pp. 7–28.

————. "Employee Practices at the Best-Managed Companies." **California Management Review.** Fall 1985. 28(1):35–66.

Perrow, C. **Complex Organizations: A Critical Essay.** Scott, Foresman, Glenville, IL. 1972.

Peters, T. J. "Symbols, Patterns, and Settings: An Optimistic Case for Getting Things Done." **Perspectives on Behavior in Organizations** (2d ed.), Hackman, R., Lawler E. III, and Porter, L. W. (eds.). McGraw-Hill, New York. 1983. pp. 16–29.

Quinn, J. B. "Managing Strategic Change." **Sloan Management Review.** Summer 1980. 21(4):3–20.

Schein, E. H. "Does Japanese Management Style Have a Message for American Managers?" **Sloan Management Review** (Fall 1981):77–90.

Scott, W. G. "Organizational Theory: A Reassessment. **Academy of Management Journal.** June 1974. 17(2):242–254.

Sproull, L. S., and Hofmeister, K. R. "Thinking About Implementation." **Journal of Management.** 1986. 12(1):43–60.

Woodward, J. **Industrial Organization: Behavior and Control.**, Oxford Univ. Press, London. 1970.

Zilbergeld, B. **The Shrinking of America.** Little, Brown & Co., Boston, MA. 1983.

———. "Information Technology Survey." **The Economist** (June 16, 1990):58, Sect. 1–20.

CHAPTER 15

Speculation on Future Uses
and Personal Thoughts

1.0 OVERVIEW AND INTRODUCTION

We have used the three-part organizational model as outlined first in Chapter 5 and others as a faithful tool to explore, categorize, and select some modern ideas as a basis for your own theories. The by-now familiar components of people, structure, and technology, all held together by leadership and surrounded with information systems, have served us well. But this is a very limited model. In fact, its main use is to supply a framework upon which to hang those ideas and theories that seem to apply to technical management and to illustrate how some of this material can be used by you. However, each time that we have attempted to define a better way to manage, there seemed to be many more ways that could be of equal value. This was noted in Chapter 14 in the explanation of *equifinality* (you can get there using different paths). But even though there are many ways, the one central concept is that you, as the manager, have to decide which path to follow. And, unfortunately, we never know if we have chosen the "right path." It would be nice if we could somehow replicate management by doing it all over again *better than last time* (assuming that the next experience would be pleasant), but we can't do that. We deal with an *art*, not a *science*. Art is usually a one-time activity because the artist can move on to create something entirely new and wonderful. Science, however, is based on replication or re-checking of experiments by anyone who has the technical knowledge. Art is always new. Science is always based on something in the past before it can move on to something else. These two, very human, activities are interrelated in management. While we cannot replicate management findings (an art), we can learn new and better techniques and based on our learning, become better managers (a partial scientific approach). Management then combines past research (science) with the individual (art). It's a typical bilateral model.

2.0 THE BILATERAL MODELS OF MANAGEMENT

In technical operations, under the *science* division of the model, we have the logic and consistency of nature. Under the *art* division of the model, there are the emotions and creativity of human beings. The organizational structures can also be divided into bilateral related models: the division of mechanistic vs. organic structures into the functional product vs. project prototypes. Technology also can be divided into unit (one of a kind) and process (nonstop) production. I believe that mass production, the manufacturing of large lots of products, will soon be obsolete, so I don't consider it as important now. Even in that original prototype of mass production, the automobile assembly line, there are really two different kinds of production. If you consider that each automobile is built differently now to suit each customer's needs, you have defined unit production. If you consider that automobile assembly lines rarely stop (except for yearly model changeovers), you have process production. Leadership also seems to follow "bilateralism." It is either task or relations oriented or conversely can be classified as initiating vs. consideration behaviors. And even information systems can be somewhat arbitrarily divided into financial and management accounting systems. But even though bilateral models seem to be dominant, to make it a bit confusing, there are many other models.

2.1 Equifinality: Not Limited to Bilateralism

I am *not* suggesting that equifinality is limited to either of any two paths as suggested by the bilateral models. It is not. Our descriptions of two different models of ideas in this book were presented merely as contrasting theory and to assist in presenting new ideas to you. There are many paths and that's what *equifinality* really means. It just seems that at this stage of the development of management theory, it is easier to have two general paths to consider. But you have many because your situation may not match either of any of the paths *exactly*. The use of two general categories simplifies our ability to define, understand, and then begin to build our personal theory. It's a defining and clarifying process similar to a general description/diagnosis. The two-sided, bilateral theories are not precise, merely guides. For example, we will probably never see a pure mechanistic organizational structure or an equally pure relations-oriented behavior pattern in a leader. But these definitions give us two general, easy-to-use categories in which to classify all the research that we have reviewed. This semiartificial kind of dividing process is selected as only one way to start your understanding of the management ideas we have discussed.

2.2 Bilateral Model of Technical Operations

If we can begin with a bilateral division of ideas, it would seem possible to theorize that another idea would be that the technical operation itself might also be bilateral, with one section intended to handle problems primarily in the natural sciences, using logic and consistency, and another in the social sciences, being primarily concerned with creativity and innovation. As noted before, this could be "functions" vs. "projects." These sections or parts are not, of course, entirely separated in any organizational structure. They are related because their different kinds of assignments require both general kinds of skills within the same technical function.

We already know that science is not absolutely objective, because it is practiced by human beings called "scientists." These scientists are subjective to some extent in what they perceive and in what they allow themselves to perceive even though they have been trained to follow socially accepted "objective" methods.

Therefore, the personal theory of the technical manager must always include both the rational tasks of the *relatively* well-defined natural sciences and the less well-defined, emotional tasks to produce required innovations. The model of Chapter 5 could therefore really be a section view through an organizational "ball." Figure 15-1 shows the ball cut apart to demonstrate both the continuity of the total technical organization, since it contains both aspects, and the vital differences that should deliberately be included in technical management theories.

Speculatively (I have no objective data about this idea), the "ball" model can be a starting point for integrating many of the diverse concepts discussed in the book. It provides for many different kinds of ideas about people, structure, technology, etc., but it also indicates that these diverse (differentiated?) components are tied together. When there is unequal weight that is unintentionally given to any component, the model should be able to show this and simultaneously indicate the components that have to be corrected. As our model, at this point, is bilateral, the above example considers only two very general areas, functions and projects.

2.3 Measuring the Amount

Assume that you have been able to reduce the major descriptions of each component to an ordinal measurement of some kind.

In effect, you measure *what you have* against *what you would like to have* for each component. The difference in these scores would tell you which of the components has the highest score and is therefore the primary problem. If you made the model area size for a component equal to the score number for that component, the component with the biggest imbalance would be the largest one in your drawing of the model (e.g., if the sum of the difference in points for technology is larger than the point difference for either structure or people, technology has to be brought into line first). The bilateral split of this organizational ball model between the somewhat arbitrary definitions of functions and projects refines the subjective measurements that you place on the measurement of each component.

To reiterate, the model of the organization as a bilateral design is not really being suggested here, as it is only a speculation. Other designs might be even more applicable, and that point is vitally important. Whatever you choose, it should be understandable to others (because they will want to know how you came to your conclusions) and it should be measurable (for the same reason). No one design can ever have all the answers. Conversely, there are no useless designs. Some are just better than others. They are all incomplete, and that makes you, as the technical manager, invaluable. You can handle all that complexity, build and adapt your personal theory, and do your management job by making decisions.

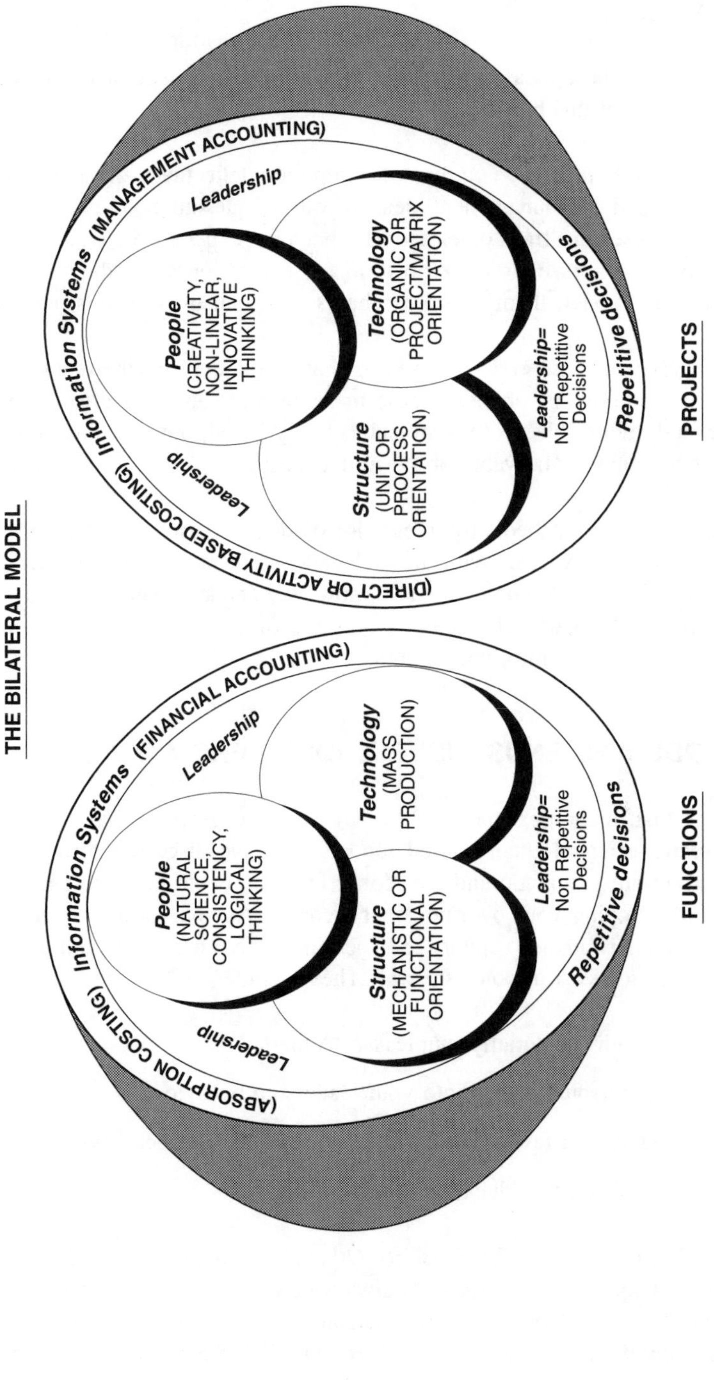
THE BILATERAL MODEL
Information Systems (FINANCIAL ACCOUNTING)
(ABSORPTION COSTING)
Leadership
Leadership
People (NATURAL SCIENCE, CONSISTENCY, LOGICAL THINKING)
Technology (MASS PRODUCTION)
Structure (MECHANISTIC OR FUNCTIONAL ORIENTATION)
Leadership= Non Repetitive Decisions
Repetitive decisions
FUNCTIONS
Information Systems (MANAGEMENT ACCOUNTING)
(DIRECT OR ACTIVITY BASED COSTING)
Leadership
Leadership
People (CREATIVITY, NON-LINEAR, INNOVATIVE THINKING)
Technology (ORGANIC OR PROJECT/MATRIX ORIENTATION)
Structure (UNIT OR PROCESS ORIENTATION)
Leadership= Non Repetitive Decisions
Repetitive decisions
PROJECTS

FIGURE 15-1

3.0 SPECULATIONS

Speculatively, these ideas incorporate two major themes that have been emphasized throughout this book:

1. Integration of ideas: You must develop the intellectual tools that allow you to integrate and use management ideas in your unique situation. These ideas can come from research, from other people's personal experiences, which you have observed, or from trying out your own ideas. The major task is the development of a personal model, theory, or a mechanism of your own for this purpose.

2. Improving your theories: The ideas that you adopt must be measurable by whatever methods you choose. These measurements don't have to be exact or even exactly replicable, as they are subjectively based, but you must have them in order to compare the value of dissimilar ideas.

Without a personal theory for integration of ideas and some method of measurement, we are as savages, living in accordance with magic rites that seemed to have worked once, although we don't know why and are unable to predict when they will happen again. That is not a proper posture for a manager. I hope that I have helped you in improving yours.

4.0 ODDS AND ENDS: USEFUL IDEAS FROM OTHERS

In a less than theoretical mode, these are some suggestions that don't fit into the research-oriented–theory-based models we have discussed throughout the book. They are pragmatic and anecdotal, but they could be quite useful when you are developing your own personal theories. As you know, sometimes the world is neither fair nor supportive. Therefore, these are "odds and ends" that are intended to assist in your survival. They include

(a) Dealing with (potentially) unpleasant "others,"

(b) Injecting personal values into your leadership behaviors,

(c) Understanding a few of your own qualities as a manager, and

(d) Realizing that your education process never stops.

Dealing With Potentially Unpleasant "Others": Obviously, it is possible that the organizational situation may not always be supportive of the goals that you, as the manager, want to accomplish. Management meetings can become difficult. One problem is that of "backstabbing," using underhanded means such as accusation, innuendo, refusal to acknowledge prior commitments, etc., in an attempt to discredit.

> Backstabbing . . . seems to be prevalent in all kinds of organizations, including families, churches, businesses, government bureaucracies, academic institutions, and voluntary associations. People in managerial and leadership roles, in particular, seem to complain a lot about it. (p. 271)
>
> . . . As I look back on my experiences, the pattern for getting stabbed was absolutely predictable. It was only at the behest of my bureaucratic buddy, however, that I took to hear B. F. Skinner's dictum that the major difference between rats and people is that rats learn from experience. (p. 272)
>
> . . . However, given the collusive, antisocial nature of the backstabbing process, I do know that breaking the pattern of destructive complicity by taking some sort of action . . . is an absolute necessity.
>
> 1. Confronting Potential Perpetrators with your beliefs regarding their intentions. . . . Whatever happens, they can't stab a Potential Victim in the back.
>
> 2. Confronting Messengers about their role in the impending crime. Such confrontation, I think, requires several actions, including:
>
> > a. Refusing to abide by any request from the Messengers to keep the content of their messages secret.
> >
> > b. Explaining the reason for your refusal. For example, we might say something like, "If you give me the information under a restriction that makes it difficult or impossible for me to solve the problem, there is no reason to give it to me in the first place. However, if you want to give it to me and then let me use my own good judgment as to how to use it, I will accept it with thanks."
> >
> > c. Pointing out the way in which the combination of Messenger's failure to express their avowed disagreement with the Potential Perpetrator's actions and their attempt to prevent you from using the information contained in their messages makes them parties to the potential crime.
>
> 3. Confronting oneself about one's own role in the process. (p. 276) (Harvey 1989)

This works very well.

Injecting personal values into your leadership behaviors: We have reviewed many management theories and suggested some leadership behaviors that seem to flow from them. But, as in the idea of equifinality, there are many ways to lead. Because motivation, as the inferred (not observed) mental force that is behind behavior is an internal force, perhaps you can consider your own motivations and thoughts when acting as a manager.

> Surely, whatever managerial leadership is, it is more than a coldly analytical procedure for deciding when to play tough guy and when to act like one of the gang. The right answer is not just waiting out there in the external situation. Some

of the leadership must reside inside the manager. The fact that those inside beliefs and values are hard to observe or measure shouldn't trap even dedicated empiricists into denying their functional relevance. The external situation is unquestionably important, and much of managerial behavior deserves to be contingent upon it. But it is important mostly because a thorough situational analysis can help the manager to plot course and circumvent obstacles, enroute toward a pre-envisioned and desired future state. . . . (Leavitt 1989, p. 45)

Not every situation and every individual is completely predictable. Our inside beliefs are powerful drivers and should be considered as best we can as we manage others. Self-reflection is also a prescription.

A Few Qualities of Management: Continuing the idea of personal beliefs and consequent drives, we might contemplate the following:

1. Each individual is fundamentally in charge of his or her own life.
2. All models are transient. Scientific theories change; e.g., Newtonian physics was replaced by Einstein. Management and organizational theories change rapidly. As a result you cannot reasonably hope for a reliable, stable catalogue of organizational facts and prescriptions from which you can draw your answers. You must plunge ahead. . . .
3. Honesty is the best policy. To thine own self be true.
4. Personal development is a matter of filling in holes (correcting your own individual faults). (Adapted from Herman 1994)

Realizing that your education process never stops: Besides the obvious requirement to maintain our technological competencies, there is the never-ending problem (opportunity?) that the education process does not end for us. In one way or another, we will be in a learning situation our whole lives.

. . . this new pattern of education would be distinguished by . . . unique departures from past attitudes, i.e. . . . uninterrupted commitment to formal education on the part of engineers . . . (p. 2) Engineering is a participatory rather than a spectator game (p. 13)

. . . The work community (should) view formal study as an integral and expected part of productive work and (should) allocate the necessary time to it. This practice would be equivalent to budgeting preventive maintenance and scheduling downtime for equipment (p. 22)

. . . In this report we are focused exclusively on engineering and scientific education. Obviously there is a need for similar instruction in various aspects of management, just as there is a need for a variety of cultural subjects to enrich the lives of engineers. (p. 40) (Bruce et al. 1982)

5.0 SHORT COMMENTS ABOUT LIFE IN GENERAL

Have you ever thought of this? "I present an engineer's solution. Lay out the problem, lay out possible solutions and then present a proposed answer. The trouble is that by the time that I get to the answer, most people have quit listening." (Carter 1982)

And—as a Spanish proverb says, "The stands are filled with critics, but only one person fights the bull."

And—"In the physical world, there are neither rewards nor penalties, merely consequences." (Albert Einstein)

And—". . . in this world we have only ourselves and each other. It may not be much, but that's all there is. . ." (Kopp 1971, p. 119).

And finally:

The following case study might be interesting as a sum-up.

Case Study:
Señor Payroll
by William E. Barrett[1]

Larry and I were Junior Engineers in the gas plant, which means that we were clerks. Anything that could be classified as paperwork came to the flat double desk across which we faced each other. The Main Office downtown sent us a bewildering array of orders and rules that were to be put into effect.

Junior Engineers were beneath the notice of everyone except the Mexican laborers at the plant. To them we were the visible form of a distant, unknowable paymaster. We were Señor Payroll.

Those Mexicans were great workmen; the aristocrats among them were the stokers, big men who worked herculean eight-hour shifts in the fierce heat of the retorts. They scooped coal with huge shovels and hurled it with uncanny aim at tiny doors. The coal streamed out from the shovels like black water from a high-pressure nozzle, and never missed the narrow opening. The stokers worked stripped to the waist, and there was pride and dignity in them. Few men could do such work, and they were the few.

The Company paid its men only twice a month, on the fifth and the twentieth.

---

[1] Reprinted by permission of Harold Ober Associates Incorporated by W. E. Barrett from **Southwest Review,** Autumn 1943, © 1943 by University Press.

To a Mexican, this was absurd. What man with money will make it last fifteen days? If he hoarded money beyond the spending of three days, he was a miser and when, Señor, did the blood of Spain flow in the veins of misers? Hence it was the custom for our stokers to appear every third or fourth day to draw the money due to them.

There was a certain elasticity in the Company rules, and Larry and I sent the necessary forms to the main office and received an "advance" against a man's paycheck. Then one day, Downtown favored us with a memorandum:

> There have been too many abuses of the advance against wages privilege. Hereafter, no advance against wages will be made to any employee except in a case of genuine emergency.

We had no sooner posted the notice when in came stoker Juan Garcia. He asked for an advance. I pointed to the notice. He spelled it through slowly, then said, "What does this mean, this 'genuine emergency'?"

I explained to him patiently that the Company was kind and sympathetic, but that it was a great nuisance to have to pay wages every few days. If someone was ill or if money was urgently needed for some other good reason, then the Company would make an exception to the rule.

Juan Garcia turned his hat over and over slowly in his big hands, "I do not get my money?"

"Next payday, Juan. On the twentieth."

He went out silently and I felt a little ashamed of myself. I looked across the desk at Larry. He avoided my eyes.

In the next hour two other stokers came in, looked at the notice, had it explained and walked solemnly out, then no more came. What we did not know was that Juan Garcia, Pete Mendoza, and Francisco Gonzalez had spread the word and that every Mexican in the plant was explaining the order to every other Mexican. "To get the money now, the wife must be sick. There must be medicine for the baby."

The next morning Juan Garcia's wife was practically dying, Pete Mendoza's mother would hardly last the day, there was a veritable epidemic among children and, just for variety, there was one sick father. We always suspected that the old man was really sick; no Mexican would otherwise have thought of him. At any rate, nobody paid Larry and me to examine private lives; we made out our forms with an added line describing the "genuine emergency." Our people got paid.

That went on for a week. Then came a new order, curt and to the point:

> Hereafter, employees will be paid ONLY on the fifth and the twentieth of the month. No exceptions will be made except in cases of employees leaving the company.

The notice went up on the board and we explained its significance gravely. "No, Juan Garcia, we cannot advance your wages. It is too bad about your wife and your cousins and your aunts, but there is a new rule."

Juan Garcia went out and thought it over. He thought out loud with Mendoza and Gonzalez and Ayala; then in the morning, he was back. "I am quitting this company for a different job. You pay me now?"

We argued that it was a good company and that it loved its employees like children, but in the end we paid off, because Juan Garcia quit. And so did Gonzalez, Mendoza, Obregon, Alaya, and Ortez, the best stokers, men who could not be replaced.

Larry and I looked at each other; we knew what was coming in about three days. One of our duties was to sit on the hiring line early each morning, engaging transient workers for the handy gangs. Any man was accepted who could walk up and ask for a job without falling down. Never before had we been called upon to hire such skilled virtuosos as stokers for handy gang work, but we were called upon to hire them now.

The day foreman was wringing his hands and asking the Almighty if he was personally supposed to shovel this condemned coal, while there in a stolid patient line were skilled men—Garcia, Mendoza and others waiting to be hired. We hired them, of course. There was nothing else to do.

Every day we had a line of resigning stokers, and another line of stokers seeking work. Our paperwork became very complicated. At the Main Office they were jumping up and down. The procession of forms showing Juan Garcia's resigning and being hired over and over again was too much for them. Sometimes Downtown had Garcia on the same payroll twice at the same time when someone down there was slow in entering a resignation. Our phone rang early and often.

Tolerantly and patiently we explained: "There's nothing we can do if a man wants to quit, and if there are stokers available when the plant needs stokers, we hire them."

Out of chaos, Downtown issued another order. I read it and whistled. Larry looked at it and said, "It's going to be very quiet here." The order read:

Hereafter, no employee who resigns may be rehired within a period of thirty days.

Juan Garcia was due for another resignation, and when he came in we showed him the order and explained that standing in line the next day would do him no good if he resigned today. "Thirty days is a long time, Juan."

It was a grave matter and he took time to reflect on it. So did Gonzalez, Mendoza, Ayala, and Ortez. Ultimately, however, they were all back and all resigned.

We did our best to dissuade them and we were sad about the parting. This time it was for keeps and they shook hands with us solemnly. It was very nice knowing us. Larry and I looked at each other when they were gone and we both knew that neither of us had been pulling for Downtown to win this duel. It was a blue day.

In the morning, however, they were all back in line. With the utmost gravity, Juan Garcia informed me that he was a stoker looking for a job.

"No dice, Juan," I said. "Come back in thirty days. I warned you."

His eyes looked straight into mine without a flicker. "There is some mistake, Señor," he said. "I am Manuel Hernandez. I work as the stoker in Pueblo, in Santa Fe, in many places."

I stared back at him, remembering the sick wife and the babies without medicine, the mother-in-law in the hospital, the many resignations and rehirings. I knew that there was a gas plant in Pueblo, and that there wasn't any in Santa Fe; but who was I to argue with a man about his own name? A stoker is a stoker.

So I hired him. I hired Gonzalez too, who swore that his name was Carrera, and Ayala, who had shamelessly become Smith.

Three days later, the resigning started.

Within a week our payroll read like a history of Latin America. Everyone was on it: Lopez and Obregon, Villa, Diaz, Batista, Comez, and even San Martin and Bolivar. Finally, Larry and I, growing weary of staring at familiar faces and writing unfamiliar names, went to the Superintendent and told him the whole story. He tried not to grin, and said "Damned nonsense!"

The next day the orders were taken down. We called our most prominent stokers into the office and pointed to the board. No rules any more.

"The next time we hire you hombres," Larry said grimly, "come in under the names you like best, because that's the way you are going to stay on the books."

They looked at us and they looked at the board; then for the first time in the long duel, their teeth flashed white, "Si, Señores," they said.

And so it was.

REFERENCES:

Bruce, J. D.; Siebert, W. M.; Smullin, L. D.; and Fano, R. M. **Report of the Centennial Study Committee.** Dept. of Electrical Engineering and Computer Science. Massachusetts Institute of Technology, Cambridge, MA. October 2, 1982.

Carter, J. Washington Whispers. **U.S. News & World Report** (November 28, 1982).

Harvey, J. B. "Some Thoughts About Organizational Backstabbing: Or How Come Every Time I Get Stabbed in the Back My Fingerprints Are on the Knife?" **The Academy of Management, Executive.** 1989. 3(4):271–277.

Herman, S. M. **A Force of One: Reclaiming Individual Power in a Time of Teams** in **Work Groups, and Other Crowds.** Jossey Bass Publishers, San Francisco, CA. 1994.

Kopp, S. B. **Guru, Metaphors from a Psychotherapist.** Science & Behavior Books, Palo Alto, CA. 1971.

Leavitt, H. J. "Educating Our MBAs: On Teaching What We Haven't Taught." **California Management Review** (Spring 1989):38–50.

Index